Safety Management System

and

Documentation Training Programme

HANDBOOK

Safety Management System
and
Documentation Training
Programme
H A N D B O O K

SV Paul
Safety Advisor

CBSPD

CBS Publishers & Distributors Pvt Ltd

New Delhi • Bengaluru • Chennai • Kochi • Kolkata • Lucknow • Mumbai
Hyderabad • Jharkhand • Nagpur • Patna • Pune • Uttarakhand

Safety Management System and Documentation Training Programme
HANDBOOK

ISBN: 978-81-239-2344-4

Copyright © Author and Publisher

First Edition: 2013
Reprint: 2017, 2019, 2023

Published by **Satish Kumar Jain** and produced by **Varun Jain** for
CBS Publishers & Distributors Pvt Ltd
4819/XI Prahlad Street, 24 Ansari Road, Daryaganj, New Delhi 110 002, India
Ph: 011-23289259, 23266861
Website: www.cbspd.com
e-mail: delhi@cbspd.com

Corporate Office: 204 FIE, Industrial Area, Patparganj, Delhi 110 092, India
Ph: 011-4934 4934   Fax: 011-4934 4935   e-mail: publishing@cbspd.com;
publicity@cbspd.com

Branches

- **Bengaluru:** Seema House 2975, 17th Cross, KR Road, Banasankari 2nd Stage, Bengaluru 560 070, Karnataka, India
  Ph: +91-80-26771678/79   Fax: +91-80-26771680   e-mail: bangalore@cbspd.com
- **Chennai:** 7, Subbaraya Street, Shenoy Nagar, Chennai 600 030, Tamil Nadu, India
  Ph: +91-44-26680620, 26681266   Fax: +91-44-42032115   e-mail: chennai@cbspd.com
- **Kochi:** 42/1325, 1326, Power House Road, Opp KSEB, Power House, Ernakulum Kochi 682 018, Kerala, India
  Ph: +91-484-4059061-65,67   Fax: +91-484-4059065   e-mail: kochi@cbspd.com
- **Kolkata:** 147, Hind Ceramics Compound, 1st Floor, Nilgunj Road, Belghoria, Kolkata-700056, West Bengal, India
  Ph: +91-033-25633055, 033-25633056   e-mail: kolkata@cbspd.com
- **Lucknow:** Basement, Khushnuma Complex, 7 Meerabai Marg (Behind Jawahar Bhawan),Lucknow-226001, UP, India
  Ph: +91-522-4000032   e-mail: tiwari.lucknow@cbspd.com
- **Mumbai:** PWD Shed, Gala no 25/26, Ramchandra Bhatt Marg, Next to JJ Hospital Gate no. 2, Opp. Union Bank of India Noorbaug, Mumbai-400009, Maharashtra, India
  Ph: 022-66661880/89   e-mail: mumbai@cbspd.com

Representatives

- Hyderabad   0-9885175004
- Patna   0-9334159340
- Jharkhand   0-9811541605
- Pune   0-9923910676
- Nagpur   0-9421945513
- Uttarakhand   0-9716462459

Printed at: Glorious Printers, Dilshad Garden, Delhi, India

Preface

This book is the result of my long-term dedicated effort. My long years of experience in renowned multinational companies in Gulf countries was strictly of OSHA 18001: 2007 standards in the safety management field which helped me in presenting the subject in a quite authentic and orderly manner to the deep understanding of the safety personnel in service and those entering in the field in future.

The book consists of two parts. The first part covers in detail the scientific aspects of the safety management systems. The safety operations of the different kinds of companies have been separately treated to suit the safety personnel working in different companies. In part two the method and procedure of safety documentation are elaborated under separate heads to suit safety personnel working in different environments. As safety documentation is the essential part of the safety management in any company, a deep knowledge in this programme is inevitable.

As per OSHA 18001 Standards, ISO Standard and 1948 Factories Act specification, each company has obligation to maintain well organized safety management system and to document their programmes in a legally appropriate pattern. The aim of the book is to provide them enough information and procedural practice to maintain the safety operations in conformity with the international standards and OSHA regulations. This book also serves as a study material for the students who are engaged in safety engineering or safety management system courses.

SV Paul

Contents

PART II

SAFETY DOCUMENTATION TRAINING PROGRAMME

Part I

SAFETY MANAGEMENT SYSTEM

Basic Knowledge about Safety: Definitions and Prevention Strategies

1

1.1 DEFINITION OF SAFETY AND BRIEFINGS

As per OSHA 18001 standards, safety is defined as "a tool or device designed to prevent accidents, incidents and injuries".

This can be achieved only by eliminating or reducing the hazards and risk involved in every working activities that are conducted on industrial or work premises. Hazard identification and risk assessment and safe work procedure are essential for every activity to prevent accidents and incidents. The frequency exposure limit of potential hazards cannot be changed but their harm and consequence limit can be controlled by taking precautionary measures. This means, to bring down high hazard level of exposure of each and every activity to a workable level control and additional control measures are to be undertaken.

1.2 DEFINITION OF ACCIDENT, INCIDENT, HAZARD AND RISK

Accident can be defined as "an unexpected or unforeseen event which causes or likely to cause an injury, fatality or any other loss".

Incident can be defined as "an occurrence or event that interrupts normal procedure or precipitates a crisis".

Hazard can be defined as "a potential to cause harm or injury".

Risk is defined as "likelihood of consequence":

- Likelihood is probability or frequency of an injury occurring.
- Probability is the actual outcome of an event happened.

Near-miss can be defined as "an incident which does not result any injury, death or any other loss".

1.3 CAUSES OF ACCIDENTS

The major causes of accidents are mainly:
a. Unsafe working condition
b. Unsafe act
c. Environmental and personal factors.

Accident statistics reveals that 60% of accidents occur due to unsafe acts and 38% due to unsafe working condition and the balance 2% due to environmental and personal factors.

(a) Unsafe working condition

Unsafe working condition can be avoided by the contractor or management by taking responsibility of maintaining safe working condition on work premises as detailed in safe work procedure. Proper planning from the part of management is necessary to constantly maintain safe working environment.

Factors leading to unsafe conditions are

a. Lack of safety awareness from the part of the management
b. Negative attitude of management personnel towards safety
c. Improper selection of key personnel
d. Lack of proper safety inspection and audits.

Some of the unsafe working conditions are

a. Defective equipment, machinery and tools
b. Improper design or construction
c. Improper plant layout
d. Insufficient illumination and ventilation

3

e. Inadequate PPEs (Personnel Protective Equipments)
f. Lack of supervision
g. Bad housekeeping
h. Inadequate guarding
i. Lack of scaffoldings
j. Lack of training according to safe work procedures
k. Excessive noise, etc.

(b) Unsafe acts

The major part of the accidents are mainly caused due to unsafe acts. Unsafe acts are due to the unsafe practices committed by the workers because of lack of knowledge or skills. This can be avoided by giving proper education and training to line and low management according to the concerned safe work procedures.

Factors leading to unsafe acts

a. Lack of safety instructions
b. Lack of proper communication
c. Lack of effective supervision
d. Lack of interest in job
e. Lack of knowledge
f. Inadequate job instructions
g. Ignorance on safety rules and practices
h. Improper motivation
i. Negligence towards safety
j. Overconfidence
k. Arson, etc.

Some of the unsafe acts are

a. Operating without authority
b. Improperly using tools and equipment
c. Working with defective tools and equipment
d. Not wearing proper PPEs
e. Smoking in prohibited areas
f. Taking up unsafe position
g. Unsafe manual lifting practices
h. Violating safety rules and regulations
i. Discarding safety of others
j. Negligence in use of safety appliances
k. Working with defective PPEs
l. Operating machineries and equipment with extra speed

m. Horseplay
n. Making safety devices inoperative
o. Negative approach and temptation in conducting safe work practice.

(c) Environmental and personnel factors

Some of the environmental and physiological conditions can also lead to accidents. Environmental factor is not in the hands of human beings. But the warnings of atmospheric condition from the legal authorities should be considered seriously.

The personnel factors can be avoided by proper selection of personnel, regular and periodical medical check-up. Actions like pest control, clear drainages and waste clearance systems are required to maintain good hygienic practices.

Environmental factors

a. Atmospheric conditions
b. Noise and vibration
c. Thermal radiation
d. Movement of air
e. Humidity
f. Dust and fumes, etc.

Personnel factors

a. Age
b. Health
c. Body structure
d. Tiredness
e. Experience, etc.

1.4 ACCIDENT PREVENTION

The costs that may be generated from accidents are lot. The employer loses production and productivity, the employees loses their wages and may face fatality, permanent disability lost time-injuries and severe injuries. If accident arises employer also needs to spend a lot of time for legal consequences and to pay compensation to injured workers as per statutory requirements. So accident prevention is an essential element in work premises.

Basically, steps taken to eliminate unsafe condition and unsafe act are the motto of accident prevention method.

Methods of Accident Prevention based on Five E's

1. Engineering

Planning and designing on safe work practice, prior to starting of any activity is essential to avoid unsafe act and condition.

2. Education

Effective training program according to safe work procedures should be conducted among the low line management to ensure safe work practice.

3. Evaluation

Every company wants its own policy, and to make HSE (Health Safety Environment) plan according to that policy. As per HSE plan, it needs to conduct hazard identification, risk assessments and safe work procedures for every activity as per standards. These HIRA and SWP should be approved by client and legal authorities (if necessary). Committing of safe work procedure to practice may cause some difficulties, then the safe work procedures should be revised and reapproved officially.

Evaluation program is exercised aiming improvements in committing safe work procedure for better performance. The first line management should set an example to others in implementing safety rules and regulations, then only the workers too practice them.

4. Enforcement

Visibly felt leadership is to be enforced on work premises to discipline workers if they fail to perform their duties as per safe work procedure and safety trainings conducted. So organization should employ some safety specialists for enforcing discipline on work premises.

5. Enthusiasm

It is a program that can be developed and maintained among work force by proper competition, contest, publicity and providing suitable incentive schemes to individual worker as well as group of workers. This program is planned to provoke safe attitude toward safety of employees for better efficiency. It will also provide encouragement and confidence to the entire workforce.

1.5 SAFETY MANAGEMENT SYSTEMS

As per OSHA standards, every industry or company shall have their own 'Policy' detailing aims, objectives and responsibilities. According to the policy, a detailed 'HSE plan' shall be made after conducting 'Hazard identification, Risk assessment and Safe work procedure' of each activity in the work premises. These processes shall be made while on designing stage of a project and shall be subjected to the approval the client, consultant and legal department (if necessary).

If any difficulty or deficiency is experienced in the work face while conducting the procedure as detailed in 'Safe work procedure', then the procedure shall be revised with additional points or a new safe work procedure shall be activated in the premises, after taking the reapproval from client, consultant and legal department (if necessary).

Most of the management systems are based upon the following model

$$\text{PLAN} \rightarrow \text{DO} \rightarrow \text{CHECK} \rightarrow \text{ACT}$$

First plan the activities—then activate work activities—check for any deviations while on conducting activities—and if any deviations found, correct or revise the procedure before re-activating the plan.

Five steps leading to a successful health and safety management systems are:

1. HSE policy

Generally organization's top management needs to prepare HSE policy to state their aim, objectives and commitment to improve health and safety performance. This piece of approved document should compromise with safety legislations, local rules and concerned standards. The contents in the document should be implemented and maintained and communicated to all employees working under them. Policies are to be reviewed periodically to ensure that they remain relevant and appropriate to the organization.

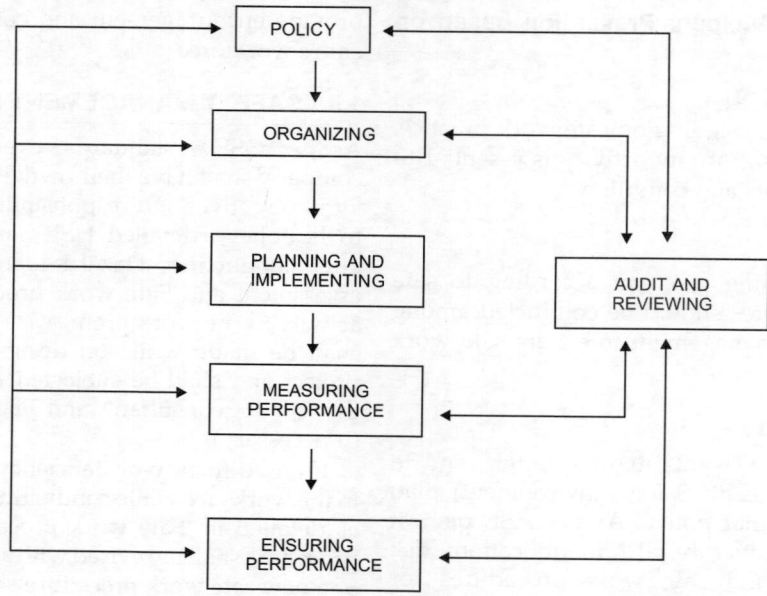

Fig. 1.1: Health and safety management system

2. Organizing

The effectiveness of management systems largely depends upon the organizing process. The elements in the organizing process are:

a. Competence: Right person is recruited for right job. Training is to be conducted periodically to the selected employees and subsequent advisory support is given to them.

b. Control: Allotting responsibilities to each, securing commitments by instruction and supervision

c. Co-operation: Needs to build up effective co-operation between management, staff and work groups

d. Communication: All decisions are to be communicated to line and low management in their understanding languages, in writing or in other modes.

3. Planning and implementing

Planning and implementing are essential for maintaining an effective safety culture and to control risk. All work activities are to be well planned according to the approved safe work procedure and implemented through right professionals with positive safety approach. They should also compromise with the changing demands.

4. Measuring performance

Measurement is an essential element to maintain and improve health and safety performance. There are two types of performance measuring methods:

a. Proactive monitoring: This monitors the achievement of plans which is in compliance with standards before arising any accident, incident or ill-health.

Example: Training assessment, Site inspections, System audits and health checks, etc.

b. Reactive monitoring: This monitors accident rates, ill-health rates, property damage rates and incident rates to find out the system failures occurred by calculating frequency and severity rates for realistic improvements.

5. Audits and review

This determines whether the HSE policy ensures that the Organization and Systems are actually achieving the required reliability and effectiveness and it also aims at improving their ability to manage hazards and risks.

1.6 ASSIGNMENT OF DUTIES

When a person is recruited for job position according to his professional talent, the company has to define to the recruited person about, what company needs from him and what he must do exactly for the company. For this course, assignment of duties is held. This document should be defined, discussed and signed by both the senior management person of the company and by the recruited person. According to the company's point of view, it is a part of company policy and if the recruited person fails to perform his duties as signed and agreed the company can discipline him or terminate him from his position. If a serious accident occurs and it occurs as a result of incorrectly performing his duties, he is to be considered responsible to the accident and legal department can take action against him.

1.7 TRAINING PROGRAMS

Training Programs are intented for making every worker of a company to perform his job activities as per Safe Work Procedure and for developing their expertness, professionalism and safe work practice according to OSHA standards.

According to the company's point of view, it is a part of company policy to train their employees prior to committing job activities, for performance of safe work procedure according to the job activity, hazard identification and risk assessment practice.

From the part of employees, after training they will get clear information about the job activity they are to perform, risk and hazards they need to avoid and safe work practice they have to follow.

Types of training programs that are conducted in the Organizations are:

a. Induction training: Training given to the newcomers and visitors on work premises.
b. Competence training: Training given to build up the expertness and professionalism among work force in their area of work.
c. Special training: Training given on occasions like the occurrence of accident, stoppage of site, near miss, installment of new plant or machinery or tools, etc. to the site and introduction of new rules and regulation to industry.
d. Refresher training: Same as special and competency training but given after a period of time for reminding and refreshing duties, rules and practices of work force.
e. Emergency evacuation drill: Drill program is conducted to each and every employee of the Company, in the appropriate premises for knowledge and practice of the Emergency and Evacuation procedure to be followed prior to emergency situation for their own survival. This can be achieved only by periodical drill and training programs for their remembrance as well as practice.
f. Fire extinguisher training: To show which types of fire extinguisher is available for use and to train how to use extinguisher and what to do if fire arises.
g. Toolbox talk: A talk is given to the work force by their concerned supervisor in the presence of safety personnel in the beginning of the shift or work activity.

1.8 HAZARD IDENTIFICATION, RISK ASSESSMENT AND SAFE WORK PROCEDURE

Every activity in any company premises co-link with high risk and hazard nature and so planning and designing is crucial to avoid accidents and incidents due to the use of highly sophisticated equipment, machineries, tools and also of different atmospheric conditions. The hazard identification, risk assessment and safe work procedure document is a tool to bring down the actual exposure limit to low exposure limit (workable level) by following the precautions of the risks and hazards which are identified in each and every activity.

According to the control and additional control remedies of each activity, the top management needs to conduct safety training and safety talks prior to starting work, to the line and low management for good work practice.

1.9 ACCIDENT INVESTIGATION

Each and every accident, incident and near miss in the Company premises should be investigated to find out the direct, indirect and root causes of the accident/incident happened, to find out the failures in operating safe working procedure and remedial measures to be taken against the failures happened according to OSHA standard. The bottom line of accident/incident investigatio.n is to stop the recurrence of the accident/incident. Accident/Incident investigation is to be done by a group of members assigned by the top management.

This group of members need to visit the accident/incident place and find out the actual reason of accident by clarifying all witnesses and activities and need to note all points regarding accident and to fill investigation report, by considering all relevant details found and collected by communicating with each members of the group.

1.10 JOB SAFETY ANALYSES CHECKLIST

It is an operational control document and a part of Company policy, HSE Plan and Safe Work Procedure to visibly check and record daily/weekly/monthly the Plant/Machinery/Equipment/work and welfare activities for checking their accuracy and wear and tear for better efficiency.

In this procedure Job safety analyses checklists help to find out any deviations from its Safe work procedure. Every company needs to prepare standard and approved Job safety analyses checklists as per OSHA standard according to their activity requirements, plant, machinery, equipment and tools used, and after a detailed hazard identification, risk assessment and safe work procedure analyses.

The occupational and epidemic disease is common in work premises which can cause fatalities, disability and lost-time incidents. The importance of checklist is to visibly find and eliminate any deficiency in wear and tear originated in machines and lack of provisions in work and welfare activities. So the logic behind Job Safety Analyses is "Prevention is better than cure". A competent, trained and expert person has to deal with these checklists.

1.11 PERMIT-TO-WORK SYSTEMS

Some work activities and area of work have high accidental and incidental nature because during the course of carrying out work in such areas and activities can produce high disaster such as fatalities, disabilities, multiple loss times and heavy damages regularly. For eliminating hazards and controlling risk on hazardous and danger zone, the procedure of Permit-To-Work specific-job systems come into play.

The Permit-to-Work system procedure activity is mainly needed in the circumstances of carrying out jobs in confined space entry/hot works/working in the area where gas and vapor are present in the atmosphere/working with or near chemical and compressed gas cylinders/deep excavation, working on charged electric system, working near or overhead live services, working with pressurized pneumatic and hydraulic systems, working in access to restricted areas/road works/cold works, etc.

1.12 SAFETY SIGNAGES AND POSTERS

Signages are considered as the mirror of the risks and hazards shown to the employees as well as visitors involved in work place activities. Every person's responsibility is to obey the signages placed in different locations of work activities, if not, the harm and consequence will be very high because the unknown activities and procedures may take their life.

Safety posters are designed near the work place to build up the awareness of the work force. If the safety posters are exhibited on main places of work face, the workers in their absconding mind should be always aware that if they do this activity in negative way, this tragedy will happen and hurt them. Much more unsafe acts and unsafe condition can be avoided to a great extend.

1.13 MATERIAL SAFETY DATA SHEET

While receiving each and every hazardous chemical and hazardous substance, its material safety data sheet is to be obtained from the concerned chemical, petroleum and compressed gas manufactures, to know the risk and hazard involved in handling, stacking, in fire prevention, first aid treatment to be given in case of accident and the usage of that particular product.

Copies of MSDS sheets are to be filed in the Safety Office, First aid room and one copy should be displayed on the storage area as per OSHA standards to give information to the employees and visitors about the risks and hazards involved in handling such materials, and to make them prepared to meet any emergency that may arise.

1.14 SAFETY COMMITTEE MEETINGS

It is the conference of assigned and selected internal site members of a company to discuss and clear past defects happened in their work premises, and to discuss, plan and make decision for the future site work, safety and welfare activities. The decision made should comply with company policy, HSE plan, and safe work procedure and legal requirements.

The safety committee members should instruct and discuss within their sections, the new work, safety and welfare procedures which are decided in the committee meetings for their further adherence

So the importance of the safety committee meeting is that, every individual employee of the company site work premises will get a clear picture of what work procedure he has to follow, what safety precaution he needs to take and what is his responsibility in achieving safety.

1.15 SAFETY PUNISHMENT NOTICES

It is the prime duty of the safety personnel to correct unsafe working condition and unsafe working acts to avoid accidents. While on inspection any activity is a found deviating from the actual safe work procedure, Safety personnel and top management need to discipline the workers as well as supervisors to prevent accident occurrence.

Training, safety talks and safety awareness will help to prevent accident to a great extent. But the human mind can forget everything which had been inducted, in a moment due to various reasons like tension, negligence, horseplay, production pressure, lack of knowledge, amnesia, absence of mind and habit, etc. The fault of even one person can also cause accidents to others.

These human errors may lead to major accidents and damages and need to be disciplined. It is the basic nature of the work force that, if they are warned and they lose a part of wages as penalty for committing their own unsafe work or mistakes, they will not forget that in their life time and will not commit that activity again. For this, the concept the safety punishment notice is implemented.

Fire Safety

2

2.1 WHAT IS FIRE

Fire is the rapid oxidation of a fuel evolving heat particulates, gases and non-ionizing radiation. Fire is also known as a chemical reaction called oxidation or combustion.

Fire results from a chemical reaction between a fuel and oxygen and it is usually accompanied by heat and smoke generation and by emission of light or flames.

2.2 FIRE PYRAMID OR TRIANGLE

There are three elements involved in fire. They are Oxygen, Fuel and Heat. A fire can start only when the following elements are combined:
- **A fuel,** being a gas, a liquid or a solid
- **An oxidant,** usually (air/oxygen) (21 % of air)
- **A heat source,** such as a flame, a spark or the like.

Together they form the so-called **Triangle of Fire,** where each leg represents a parameter. Every fire involves a chain reaction, in which a multitude of chemical reactions generate chemical compounds or elements that sustain the process. These are called "free radicals".

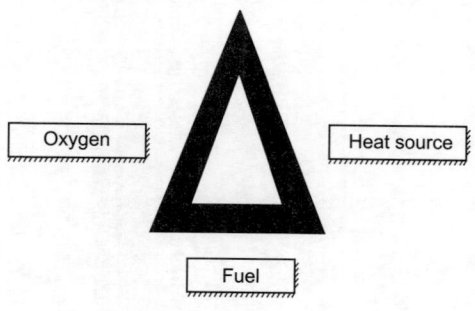

Fig. 2.1

Removal or elimination of one of the parameters prevents or ceases fire. Insight in these parameters allows for fire prevention and for efficient extinguishing of fire.

2.3 SOURCES OF IGNITION

Main sources of ignition on work premises are:
a. Smoking
b. Electrical equipment
c. Tools and equipment
d. Heaters
e. Arson, etc.

Main fuels that can accumulate on work premises are:
a. Paper and boxes
b. Plastics
c. Solvents
d. Waste materials
e. Furniture and carpets, etc.

2.4 FIRE PREVENTION

Theory of Fire Extinguishing

To understand fire prevention it is important to recognize that every fire consists of a chemical reaction between a fuel and oxygen, activated by some form of energy.

With reference to the fire triangle, the principles of fire extinguishing is the elimination of one of the three parameters fuel, oxygen, heat source which will halt the reaction. Every means of fire extinguishing destroys the triangle of fire.

1. Elimination of the fuel (starvation)

Lack of fuel will extinguish any fire. However, it is often almost impossible to remove fuel from an ongoing fire.

Examples where this means can be used effectively:

1. Cutting the supply by closing a gas or oil valve
2. Pumping liquid fuel from the bottom of the fuel tank to another reservoir
3. Spreading out the fuel: timber fires can be extinguished easier when the wood is spread out.

2. Elimination of oxygen (smothering)

Most fires extinguish spontaneously when the oxygen level drops below 14%.

Preventing air or oxygen from reaching the fire will extinguish it. One way this can be accomplished is through partial replacement of the air by an inert gas, such as carbon dioxide (CO_2), steam or nitrogen. This method is effective only for smaller liquid fuel fires using portable fire extinguishers. Class A fires cannot be treated this way.

Another way of using this "suffocation method" is by cutting the supply of oxygen to the fire. Covering a burning frying pan with a lid or with a moist towel will effectively stop the fire.

Examples of preventing oxygen from feeding the fire:

1. Putting a lid on a burning recipient
2. Covering a fire with foam or powder, thus preventing generation of flammable fumes.

3. Elimination of the heat source (cooling)

This method is used most often. Cooling the fuel below its ignition temperature will cease the fire. Water is used most frequently to cool the fuel, as it boils at 100 C with a large heat of evaporation. This is the most efficient method for Class A fires. Spraying sufficient amounts of water over a fire is more efficient than using a steady water jet.

4. Neutralizing the chain reaction

Some means of extinguishing, such as powders or halon, can scavenge these very reactive radicals, thus preventing propagation of the fire. These means have a negative catalytic effect on the fire.

2.5 CLASSIFICATION OF FIRES AND APPROPRIATE METHODS FOR EXTINGUISHING

The type of the fire and the properties of the burning materials determine the choice of appropriate method to extinguish the fire. Fires are classified into different categories, that feature each appropriate methods of extinguishing.

1. Class A (carbon fire)

Class A comprises fires of solid fuels, such as wood, paper, cardboard, coal, fabrics, etc. (burning dry matter).

Appropriate means of extinguishing
Water, foam or some special powders.

2. Class B (liquid fire)

Class B comprises fires of liquid fuels more precisely of the vopors in equilibrium with the liquids such as oil, gasoline, alcohol, paints, varnishes, tar, grease, etc. Class B fires proceed at the surface of the fuel, where to a more or lesser degree the vapor phase of the liquid acts as the actual fuel.

Appropriate means of extinguishing
DCP (Dry Chemical Powder), CO_2, foam.

3. Class C (gas fire)

Class C comprises fires of flammable gases, such as natural gas, methane, propane, butane, etc.

Appropriate means of extinguishing
First of all, the gas supply should be cut off.

The fire will extinguish automatically. If this is impossible, the fires can be kept under control by powders, CO_2, halon. It is dangerous to extinguish the fire as long as the gas supply is not cut, for accumulation of gas may pose an explosion hazard.

4. Class D (metal fire)

Class D comprises fires of chemicals, such as flammable metals like magnesium, sodium, titanium, etc.

Appropriate means of extinguishing
Commonly used methods are inappropriate and may worsen things. Powders and dry sand are excellent materials.

5. Class E (electric fire)

Class E fires are very commonly due to short circuit on any combustible materials, but also comprise fires in electrical apparatuses, such as motors, transformers, switch boards, cables, etc.

Appropriate means of extinguishing

Never use conductive materials such as water or foams.

Preferred methods use DCP, CO_2 or halon. Occasionally non conductive high pressure methods can be used.

2.6 FIRE EXTINGUISHING

Knowledge of the procedures to operate fire extinguishers is important to enable swift reaction to a developing fire, before it becomes big to handle.

Apparatuses containing powder or CO_2 are ideal to deal with beginning fires. These apparatuses are abundant in the laboratories.

Note: **They are effective only when used to extinguish small scale fires or developing fires.**

1. Powder extinguishers

Performance

Powder is forced out under pressure and it forms a dense cloud around the burning mass. The powder contains approximately 80% of $NaHCO_3$. When in contact with fire, it dissociates into CO_2 and H_2O, thus depleting the surrounding atmosphere from oxygen necessary to sustain the fire and at the same time lowering the temperature by absorbing energy to dissociate. In addition, a layer of powder will be deposited on the mass. Direct contact with the surrounding air becomes impossible, thus preventing re-ignition of the fuel.

Powder fire extinguishers have a range of 6 to 8 meters, but the nozzle should not be more than 3 to 4 meters away from the fire to be effective.

Procedure

- Move the apparatus to the fire and pull or push down the safety pin to pressurize the extinguisher
- Approach the fire as close as possible (depending on the type and intensity of the fire)
- Ascertain the availability of an open escape route in case of need
- Point the funnel at the base of the flames and activate the lever
- Under the protection of the powder cloud in front approach makes the effectiveness of the method.

Wrong

Wrong

Right

Right

Fig. 2.2

2. CO₂ extinguishers

CO_2-apparatuses do not leave traces as do powder types. Hence they are more suitable to deal with fires in rooms containing sensitive measuring and control equipment, communication devices and the like. Powder spread over the room requires thorough and expensive cleaning. CO_2 is an inert gas and disappears in the atmosphere.

CO_2 extinguishers have a limited effective range of approximately 1.5 meters, forcing the operator to approach the fire closely. In the case of electrical fires where heavy smoke prevents the operator from clearly distinguishing where exactly the fire source is located, the risk of electrocution by inadvertent contact of the metal funnel with live leads turns this method inappropriate.

Performance

CO_2 extinguishers contain CO_2 gas under a pressure of approximately 60 kg/cm². When released to atmospheric pressure, the temperature of the gas drops to – 79 C.

This temperature drop has a twofold effect. CO_2 extinguishers are powerful weapons. Never engage in horseplay or point the funnels to people or people's faces. CO_2 snow can freeze human tissue and cause wounds comparable to third degree burns.

On the one hand, the temperature of the burning mass can be lowered beneath the ignition temperature of the fuel, thus extinguishing the fire. More importantly however, CO_2 gas is inert and heavier than air, so it displaces the source of oxygen to the fire and builds a protective blanket over the fuel.

Procedure

- Move the apparatus to the fire and remove the safety pin from the handle.
- Approach the fire as close as possible (dependent on the type and intensity of the fire).
- Ascertain an open escape route is available in case of need.

Point the funnel at the base of the flames from a distance of approximately 1.5 meters and activate the lever. Do not fire the gas jet from too close a distance at especially Class B liquid fires, as this may result in fuel splattering and fire propagation to adjacent areas. Always apply an uninterrupted jet of gas.

3. Water extinguisher and hoses

Water acts as a coolant. It has the most pronounced cooling effect of all extinguishing means through its high specific heat and low boiling point of 100 C. It succeeds in dropping the fuel temperature to below the ignition temperature, thus preventing further propagation of fire. Evaporation of water requires lots of energy to be delivered by the burning mass. Spraying water is more efficient than jetting, because a larger amount of water can come into contact with the fire and extract heat from it. Steam formation has a secondary effect of displacing oxygen, thus suffocating the fire. One liter of water can produce 1,650 liters of steam.

Wrong

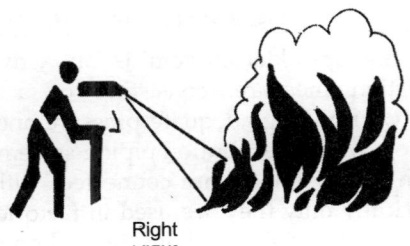

Right

Fig. 2.3

Advantages of water

- Lowest cost
- Abundance
- Not toxic
- Allows extinguishing from long distances

Disadvantages of water

- Can cause severe damage to properties
- Freezing point of 0 C
- Many fuels (paper, foamed plastics, fabrics) absorb water readily thereby increasing their weight
- Ineffective and even dangerous with flammable liquids, some metals, electrical fires

4. Automated fire systems

When a fire develops, automated systems are most effective because they are sooner activated upon, the onset of a fire.

Systems based on fire sprinklers

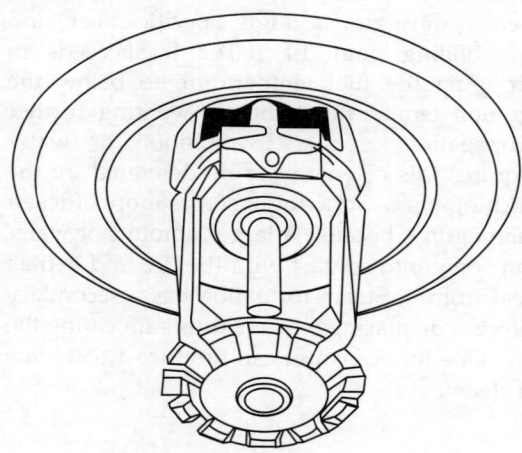

Fig. 2.4: Sprinkler

A fire sprinkler system is an active fire protection measure, consisting of a water supply, providing adequate pressure and flow rate to a water distribution piping system, onto which fire sprinklers are connected. Although historically only they are used in factories and large commercial buildings, also home and small building-systems are now available at a relatively cost-effective price.

Sprinkler systems based on powder

At K.U. Leuven this type is in use primarily in fireplaces and in engine rooms of elevators. Detection happens through fuses Programed at 70 C, 90 C or 120 C. When the fuses melt, powder is pressurized and spread over the surfaces to be protected. Simultaneously all power supply is cut.

Sprinkler systems based on CO_2

These systems are suited for rooms with high degree of occupation, such as electrical cabinets, engine rooms, computer rooms, etc. As they displace oxygen from the rooms, they must be inactivated prior to entering the room.

Sprinkler systems based on halon (already this product has been banned)

Due to their environmental hazards these systems are gradually replaced at K.U. Leuven. As for CO_2 systems the working principle is based on suffocation. Halon is a liquefied gas, freon 1301. A concentration of less than 5% is adequate as fire extinguisher, enabling personnel to enter the room without incurring health risks.

2.7 ACTIVE FIRE PROTECTION

Active fire protection (AFP) is an integral part of fire protection. AFP is characterized by items and/or systems, which require a certain amount of motion and response in order to work., contrary to passive fire protection. AFP falls in to two categories. They are:

1. Fire suppression

The fire is extinguished by manual or automatic means, such as a fire extinguisher or a fire sprinkler system, which automatically releases water, or a gaseous or foam-based fire suppression system, to suppress a fire which is activated by pressure.

2. Fire detection

The fire is detected either by locating the smoke, flame or heat, and an alarm is sounded to enable emergency evacuation as well as to dispatch the local fire department. Example: Fire alarm, False alarm, etc.

2.8 BURNS

Fire causes injury in forms of first-, second-, and third-degree burns. A first-degree burn damages the epidermis only, while a second-degree burn goes through the epidermis and dermis. A third-degree burn destroys both the epidermis and dermis, and kills all nerve receptors underneath the skin. A common result of second- and third-degree burns is large amounts of granulation tissue, or scar tissue, in place of the burnt skin.

2.9 FIRE EXTINGUISHERS

A fire extinguisher is an active fire protection device used to extinguish or control small fires, often in emergency situations. It should never be used on an out-of-control fire, such as

S - Squeeze the handle

S - Sweep the extinguisher from side to side while aiming at the base of the fire

2.10 TYPES OF EXTINGUISHING AGENTS

(a) Water extinguisher

Effect: Cooling Effect—cools burning material APW (Air pressurized water) cools burning material by absorbing heat from burning material. Effective on only Class A fires, but has the advantage of being cheap, harmless, and relatively easy to clean up.

(b) Foam extinguisher

Effect: Cooling and Smothering—cools by means of water content in it and smothering

Table 2.1: Color coding of extinguishers							
As per British Standards							
Type of Extinguisher	Old Code	Color Code	Fire Class				
Water	Signal Red	Signal Red	A				
Foam	Cream	Red with a Cream panel above the operating instructions	A	B			sometimes D
Dry Powder	French Blue	Red with a Blue panel above the operating instructions	A	B	sometimes C		E, D
Carbon Dioxide CO_2	Black	Red with a Black panel above the operating instructions	B				E
Wet Chemical	Yellow	Red with a Canary Yellow panel above the operating instructions	A	C			

one which reaches the ceiling, endangers the user, or otherwise requires the expertise of a fire department. Typically, a fire extinguisher consists of a handheld cylindrical pressure vessel containing an agent which can be discharged to extinguish a fire.

How to operate a fire extinguisher

The typical steps for operating a stored pressure fire extinguisher (described by "PASS") are the following:

P - Pull the safety pin

A - Aim the nozzle at the base of the fire, from a safe distance

by means of cutting down the oxygen to the burning source.

Applied to fuel fires as either an aspirated (mixed and expanded with air in a branch pipe) or non aspirated form to make a frothy blanket or seal over the fuel, preventing oxygen reaching it. **Type of Foam agent in use:**

1. **AFFF**—(aqueous film forming foam), used on A and B fires and for vapor suppression, the most common type in portable extinguisher

2. **AR-AFFF**—(Alcohol-resistant aqueous film forming foams), used on fuel fires containing alcohol. Forms a membrane between the fuel

and the foam preventing the alcohol from breaking down the foam blanket.

3. **FFFP**—(film forming fluoroprotein) contains naturally occurring proteins from animal fats to create a foam blanket that is more heat resistant than the synthetic AFFF foams.

4. **CAFS**—(compressed air foam system) Any APW style extinguisher that is charged with a foam solution and pressurized with compressed air. Used on class A fires, and with very dry foam, on class B for vapor suppression.

5. **Arctic-Fire**—is a liquid fire extinguishing agent that emulsifies and cools heated materials

6. quicker than water or ordinary foam. Used extensively in the steel industry. Effective on classes A, B, and D.

7. **Fire-Ade**—a foaming agent that emulsifies burning liquids and renders them nonflammable.

8. Able to cool heated material and surfaces similar to CAFS. Used on A and B (said to be effective on some class D hazards), although not recommended to metal fires because it contains amounts of water which will react with some metal fires.

(c) Dry chemical powder extinguisher

Effect: Smothering—by cutting the oxygen supply to the source.

The thin powder float on the top of fire fuel and thus cut the oxygen entering the fire and eliminate the fire down.

Powder based agent also extinguishes fires by separating the four parts of the fire triangle. It prevents the chemical reaction between heat, fuel and oxygen, thus extinguishing the fire.

(d) CO$_2$ extinguisher

Effect: Smothering—by cutting the oxygen from the surrounding.

Displaces oxygen, or inhibits chemical chain reaction. It is labeled as clean agents because they do not leave any residue after discharge which is ideal for sensitive electronics and documents because after extinguishing it escapes in air. Not intended for class A fires.

2.11 MAINTENANCE OF FIRE EXTINGUISHERS

Most countries in the world require regular fire extinguisher maintenance by a competent person daily or weekly to operate it safely and effectively, as part of fire safety legislation. Lack of maintenance can lead to an extinguisher not discharging when required, or rupturing when pressurized. Deaths have occurred, even in recent times, from corroded extinguisher explosion.

2.12 FIRE HYDRANT

A fire hydrant is an active fire protection measure, and a source of water provided in most urban, suburban and rural areas with municipal water service to enable firefighters to tap into the municipal water supply to assist in extinguishing a fire.

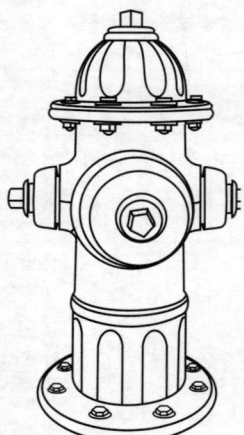

Fig. 2.5: A hydrant

2.13 SMOKE DETECTORS

Optical smoke detector

(1) optical chamber, (2) cover, (3) case moulding, (4) photodiode (detector), (5) infrared LED

A smoke detector or smoke alarm is a device that detects smoke and issues an alarm to alert nearby people that there is a potential fire. Most smoke detectors work either by optical detection (photoelectric) or by physical process (ionization), but some of them use both detection methods to increase sensitivity to smoke.

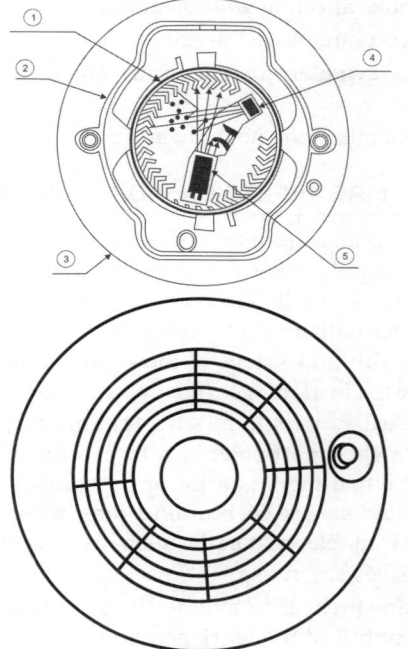

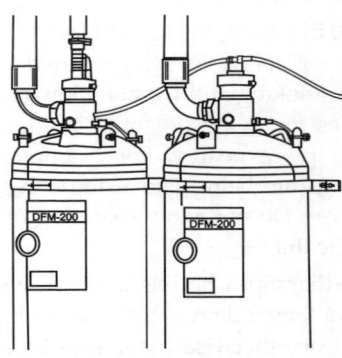

Fig. 2.6: Smoke detectors and components

2.14 GASEOUS FIRE SUPPRESSION

It contains Argon gas for use in extinguishing fire in a server room without damaging equipment. Gaseous fire suppression is a term to describe the use of inert gases and chemical agents to extinguish a fire. The system typically consists of the agent, agent storage containers, agent release valves, fire detectors, fire detection system, agent delivery piping, and agent dispersion nozzles.

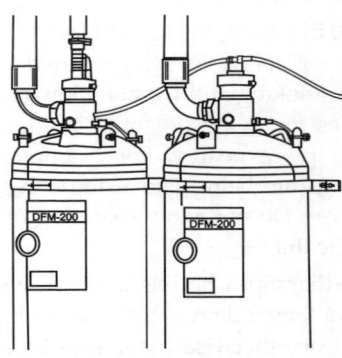

Fig. 2.7: Gaseous fire suppression system

They act on fire by

a. Reduction of heat—representative agents—Novec 1230.
b. Reduction or isolation of oxygen—representative agents—argonite IG-55, Carbon Dioxide and NN100.
c. Inhibiting the chain reaction of the above components—representative agents—FE-13, FE-227, FE-241, FE-25, FE-36, FM-200, Halons, Halon 1301, Freon 13T1, NAF P-IV, NAF S-III, and Triodide.

2.15 PASSIVE FIRE PROTECTION

The aim for passive fire protection systems is structural fire protection and fire safety in a building. PFP attempts to contain fires or slow spreading. PFP systems must comply with the effectiveness expected by building codes.

Fig. 2.8: Fire proofing

2.16 FLAME DETECTORS

A flame detector is a device that uses optical sensors to detect a flame.

Types

a. UV flame detection: Ultraviolet detectors work with low wavelengths. Detect flames at speeds of 3–4 milliseconds due to the UV high-energy radiation emitted by fires and explosions at the instant of their ignition. False alarms from random UV sources such as lightning, arc welding, radiation, and solar radiation may occur.
b. IR flame detection: IR flame detectors work within the infrared spectral band. The mass of hot gases emits a specific spectral pattern in the

infrared spectral region which are sensed with a thermographic camera. False alarms by any other "hot" surface in the area may occur.

c. Video flame detection: Web camera video detection. Like humans the camera can be blinded by smoke or fog.

2.17 DRY AND WET RISERS

A riser is a main vertical pipe intended to distribute water to multiple levels of a building or structure as a component of the fire suppression systems. A standpipe when kept with full of water for automatic operation will support the system and refer to vertical main pipe connected to fire sprinkler system or to fire horse hydrant on each floor.

Wet risers will always carry water at all times and in some cold countries the risers are kept dry because the water carried inside riser will convert to ice.

2.18 FIRE PROCEDURE FOR EMPLOYEES

a. Leave the building immediately
b. Use the nearest exit
c. Walk quickly but do not run closing doors behind you
d. Do not delay to collect your personnel belongings
e. Attend the fire assembly point and report to Fire Marshal standing behind your supervisor
f. Do not return to work premises until told to do so.

2.19 ACTION FROM THE PART OF EMPLOYEES ON DISCOVERING A FIRE

a. Shout fire and raise the alarm by breaking the glass of FIRE CALL POINT
b. Call the Fire service
c. Fight the fire if you are competent and if you consider it is safe to do so
d. Follow the evacuation procedure.

2.20 DO NOT FIGHT THE FIRE IF

a. It is bigger than a waste paper bin
b. One extinguisher is not enough for fire fighting

c. Smoke affecting your breathing
d. If you cannot see the exit
e. Gas cylinders or chemicals are involved in fire
f. Your efforts are not reducing the size of the fire.

2.21 FIRE SAFETY ON WORK PREMISES

Many solids, liquids and gases can catch fire and burn. It only takes a source of ignition, which may be a small flame or an electrical spark, together with air. Any outbreak of fire threatens the health and safety of those on site and will be costly in damage and delay. It can also be a hazard to people in surrounding properties. Fire can be a particular hazard in refurbishment work when there is a lot of dry timber and at the later stages of building jobs where a lot of flammable materials such as carpets and adhesives are in use.

Many fires can be avoided by careful planning and control of the work activities. Good house-keeping and site tidiness are important not only to prevent fire, but also to ensure that emergency routes do not become obstructed. Making site rules can help.

To prevent fires

a. Use less easily ignited and fewer flammable materials, for example, use water-based or low solvent adhesives and paint
b. Keep the quantity of flammables at the workplace to a minimum
c. Always keep and carry flammable liquids in suitable closed containers
d. If work involving the use of highly flammable liquids or solids is being carried out, stop people smoking and Do not allow other work activities involving potential ignition sources to take place nearby. For example, if floor coverings are being laid using solvent-based adhesives, Do not allow soldering of pipes at the same time
e. Ensure that pipes, barrels, tanks etc which may have contained flammable gases or liquids are purged or otherwise made safe before using hot cutting equipment

f. Minimize the risk of gas leaks and fires involving gas fired plant:
   - dose valves on gas cylinders when not in use
   - Regularly check hoses for wear and leaks
   - Prevent oil or grease coming into contact with oxygen cylinder valves
   - Do not leave bitumen boilers unattended when alight.

g. Store flammable solids, liquids and gases safely. Separate them from each other and from oxygen cylinders or oxidizing materials. Keep them in ventilated secure stores or an outdoor storage area

h. Do not store them in or under occupied work areas or where they could obstruct or endanger escape routes

i. Have an extinguisher to hand when doing hot work such as welding or using a disc cutter which produces sparks

j. Check the site at lunch-time and at the end of the day to see that all plant and equipment that could cause a fire is turned off. Stop hot working an hour before people go home, as this will allow more time for smoldering fires to be identified

k. Remove rubbish from the site regularly. Collect highly flammable waste such as solvent soaked rags separately in closed fire-resisting containers.

2.22 SAFE PLACE OF WORK

Any place where work is carried out should be safe and free of risks to health at all times. This includes access to and egress from the workplace.

Can the hazards be avoided?

This is the first question. For example, the need to paint at height can be reduced if materials are brought to site ready-finished.

If not, can the risks be controlled?

In many cases it will not be possible to avoid the hazard altogether, so the risks need to be controlled by making sure the work can be done from a safe place. Consider the size and position of the place where work is to be carried-out. People need adequate space to do their work, and space for the plant and equipment that will be used for the tasks.

When planning for a safe workplace, also consider access by people other than those doing the work. Where it is possible that members of the public or other contractors may be at risk, take precautions to exclude or protect them.

Construction Safety

3

3.1 ORGANOGRAM OR ORGANIZATIONAL CHART

Before starting activities on projects, the organizational charts of the project shall be formulated and displayed on relevant locations as per OSHA standards for recognizing the position of personnel for proper channel of communication, responsibility and authority of persons working under the concerned organization.

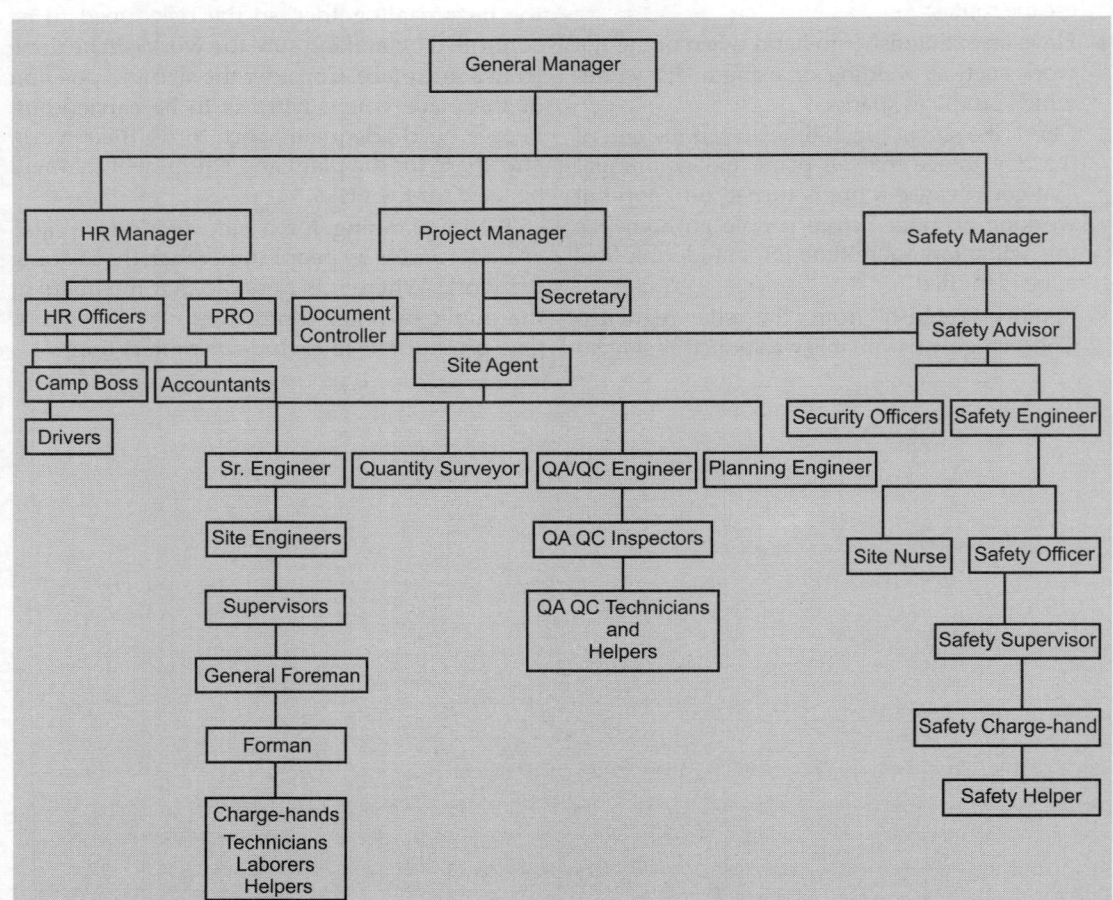

Fig. 3.1

Legally all activities on organization are the overall responsibility of General Manager. Besides General Manager is also a human and he cannot control all activities on organization by himself and so he needs to transfer his responsibility to different competent persons by assigning their duties under different managerial heads. They are called as Top management.

These assigned managerial heads have to select and assign the employees under their section, to design, organize, implement, supervise and revise the activities, proceed with approved safe work procedure. They constitute line management.

The persons who are recruited for committing work activities on premises as per their qualification, talent and experience are called Low Management. Top Management will take the decisions and implemented them through line management by the means of low management.

3.2 HSE RESPONSIBILITIES IN WORK PREMISES

Generally safety is everybody's responsibility on work premises. Top Management and Line Management personnel's should set them as an example to Low Managements for promoting safety culture on work premises. The Line management shall exercise continuous efforts to educate the work force on safe method of work to improve the safe working condition and to maintain positive attitude to avoid unsafe acts.

Project Manager Responsibility

Inputting safety on the work premises and to liaise with client, consultant and local authorities on HSE matters is the overall responsibility of Project Manager. Placing 'right persons for right job' and facilitate 'right tools and equipment for right job' for the smooth and safe working environment is his prior responsibility. He is considered as the chairman while on the scenario of Safety Committee Meeting as well as Accident Investigation Procedure. He has to ensure that the Project HSE Plan, Site Safety rules and regulations is implemented at the site premises and to take initiative in changing of existing HSE procedure if necessary.

He has the responsibility to carry-out hazard identification and risk analysis (HIRA) for the tasks at hand and to ensure that appropriate and correct controls are in place and are communicated to all relevant personnel and to take the necessary corrective action where required. Wherever it is not possible to remove any remaining hazards he should inform employees thereof and what precautionary action is to be taken. He wants to ensure that all HSE Operational Controls, Method Statements and Work instructions are communicated to all employees and adhered to at all times.

HSE Manager Responsibility

Overall responsibility for ensuring HSE Management System is implemented and maintained in accordance with the standards, and for reporting directly to the General Manager for the management of the Systems. He has to verify whether activities on site premises comply with planned arrangements and Safe Work Procedure. He need to Collect and present to the Management Review team data on Customer complaints, corrective actions, supplier problems and any other measurements of the effectiveness of the system.

Provide schedule, agendas and Minutes for Management Review Meetings and to develop and control documentation relating to the System. He has to ensure that the policies are reviewed and communicated to all management team members and staff. He need to provide training in Safety, Health, Quality and Environmental matters to staff, including auditor training for internal auditors. He wants to deal with suggestions for improvements proposed by staff and subcontractors and to make appropriate recommendations to the management.

HSE Advisor/ Engineer/Officer Responsibility

Their prior responsibility is to stop unsafe acts being undertaken by anyone on the site or work to commence and continue in safe working conditions. They needs to conduct Safety Inductions, training programs and to make their presence while conducting Toolbox Talks for building up the awareness, competency and safety culture among the work force on site premises. They want to ensure that workers use correct PPE for the type of work being carried out and site welfare facilities (rest areas, water, toilets, etc.), sufficient for the number of personnel, are being provided by the site supervisory staff.

Overall responsibility of inspecting on-site safety and promote the safe conduct of site work and to ensure that management (project supervisory and management staff) are advised with written compliance with the law, local regulations and client's requirements. Responsibility for overseeing the implementation of the project, site or facility safety plan and keeping of up to date records for the safety file to enable auditing to be carried out. They need to promote a positive safety attitude to personnel, subcontractors, consultants and clients and to facilitate communication and coordination between the company and subcontractors.

They have the authority to investigate incidents, accidents and near miss events to determine the root, basic and immediate causes and ensure that corrective, preventative actions are implemented and to report any incidents, accidents or near miss immediately, as per the Company procedure, this may include liaison with police and legal authorities. They wants to inform the Construction, Contracts, Snr. Projects, Plant, HR Manager and Safety Manager of any areas where realistic improvements can be made for site working conditions.

Site Agents/Engineers Responsibility

The prior responsibility of continuously thinking of and reviewing the consequences and risks associated with the execution of respective activities, whilst ensuring that the appropriate actions and controls are in place before any tasks are carried out. Overall responsibility of safely implementing the duties what Project Manager transfers to them in the matter of planning, designing and scheduling work activities. They have to ensure that no unsafe acts to be undertaken by yourself or any of your

subordinates, or any unsafe conditions exist that could cause harm, damage or endanger any persons.

Ensure all plant and equipment is inspected daily and the findings recorded on the register provided. They need to assist with the compiling of Method statement and the development of Safe Working Procedures. They also want to ensure that all Permits to work systems are obtained prior to proceeding with any hazardous job carrying out for taking additional precaution and inspections and its results recorded daily. They are responsible for receiving, signing and closing-out procedure for any safety inspection findings, NCR's raised or any safety improvement notices issued.

Supervisor/Foreman Responsibilities

The prior responsibility to carry-out all decisions, Safe Work Practices, rules and regulations that made by Top Management through Line Management by means of work force on work premises and to transits what ever recommendations and complaints from the part of work force to Top Management via Line Management. So they are known as a bridge connection between Low Management and Top and Line Management.

They are responsible for conducting Pre-Start Review Meeting with teams, gangs prior to commencing daily activities whereby orientation of the tasks at hand is being carried out. They need to ensure that no unsafe acts or conditions are undertaken or allowed either by yourself, your subordinates or any other persons performing tasks on the premises under your control. They want to provide appropriate PPE for the relevant job being executed and specialized or additional PPE are provided to execute hazardous job being performed.

Overall responsibility to notify or inform the Safety Department of any areas of the operation where achievable and realistic improvements can be implemented to improve the working conditions for eliminating or minimizing employee risk and exposure because they always present on work premises. They also want to ensure that the right gang is utilized for the right job while ensuring that the tasks are executed in a controlled and safe manner. They have to accept, sign and perform immediate corrective or preventative actions with regards to any safety notices issued to your gangs or area of responsibility.

Technician/Work Force Responsibility

The prior responsibility is to observe and obey HSE rules, regulations, instructions and Safe Work Procedures while executing site work activities. They want to attend all training programs that conducted by company for building up their awareness and competency. They should not put themselves and others at risk by committing risk work, shortcuts and unsafe acts on work premises. They have to wear all PPE as per regulation and wants to discard defective PPE, tools and equipment on work premises.

Overall responsibility of cleaning up the work area immediately after finishing work rather than go to the end of the shift to avoid unsafe condition that can led to accident or incident. They want to be alert at all times on work premises in the case of emergency alarm and act as per the Emergency Response Procedure. They have to report all incidents and accidents immediately to the concerned supervisor or safety department.

3.3 CONSTRUCTION SITE MOBILIZATION

Mobilization is an essential matter on construction site. Every country have their on rules and regulations in placing different temporary compartments on site premises especially in the matter of hazardous materials like chemical, compress gas cylinder and hydrocarbon stacking procedures, to be followed. Mobilization have to comply also with client requirements.

Welfare activities like eating area, rest rooms, site toilets and drinking water cooler area wants to provide according to the number of employees working on the premises and as per local regulations. Proper drainage facilities, waste management systems and pest control systems have to provided for good hygienic practice for avoiding occupational diseases.

On workshop area and hazardous areas, appropriate fire prevention systems to be installed. Material stacking area to be provided on every construction site for avoiding congestion on work premises.

3.4 SITE SECURITY

Every site premises should provided with round the clock security systems, such as fencing and gates, competent security personnel, fire watchers, good illumination facilities and emergency assembly points. Unauthorized entry on premises to be prohibited because the unfamiliar activities on site can turn the visitor to danger. For this purpose signages such as "Construction Area Keep Away", "Unauthorized Entry Strictly Prohibited" and "Safety First" to be installed in front of entrance gate.

Emergency assembly point has to be provided probably near exit gate for the easy clearance of employees from the site in case of any emergency situations. Theft on the site can be avoided to a great extend by providing tight security.

Separate registers to be provide on each gate for marking pedestrian and vehicle entered and exit for further clarification. Construction vehicles which have valid permits, certificates and obtained entry permit pass from client, shall be allowed on site premises.

3.5 SAFE ACCESS

In all areas of work to be provided with safe access on the premises and kept free from obstructions, slip and trip hazards. If there have more than one floor temporary stairway, ladder, ramp and hoist access to be provided according to the number of persons working in that floors, because in case of emergency situations they want to reach to the assembly point without any trouble. These accesses to be built with strong materials holding four times its intended load capacity and securely braced to prevent swaying or displacement. Means of access constructed by metal parts shall not used for electric works for avoiding electrocution.

There are mainly two types of access to be maintained on premises. They are:

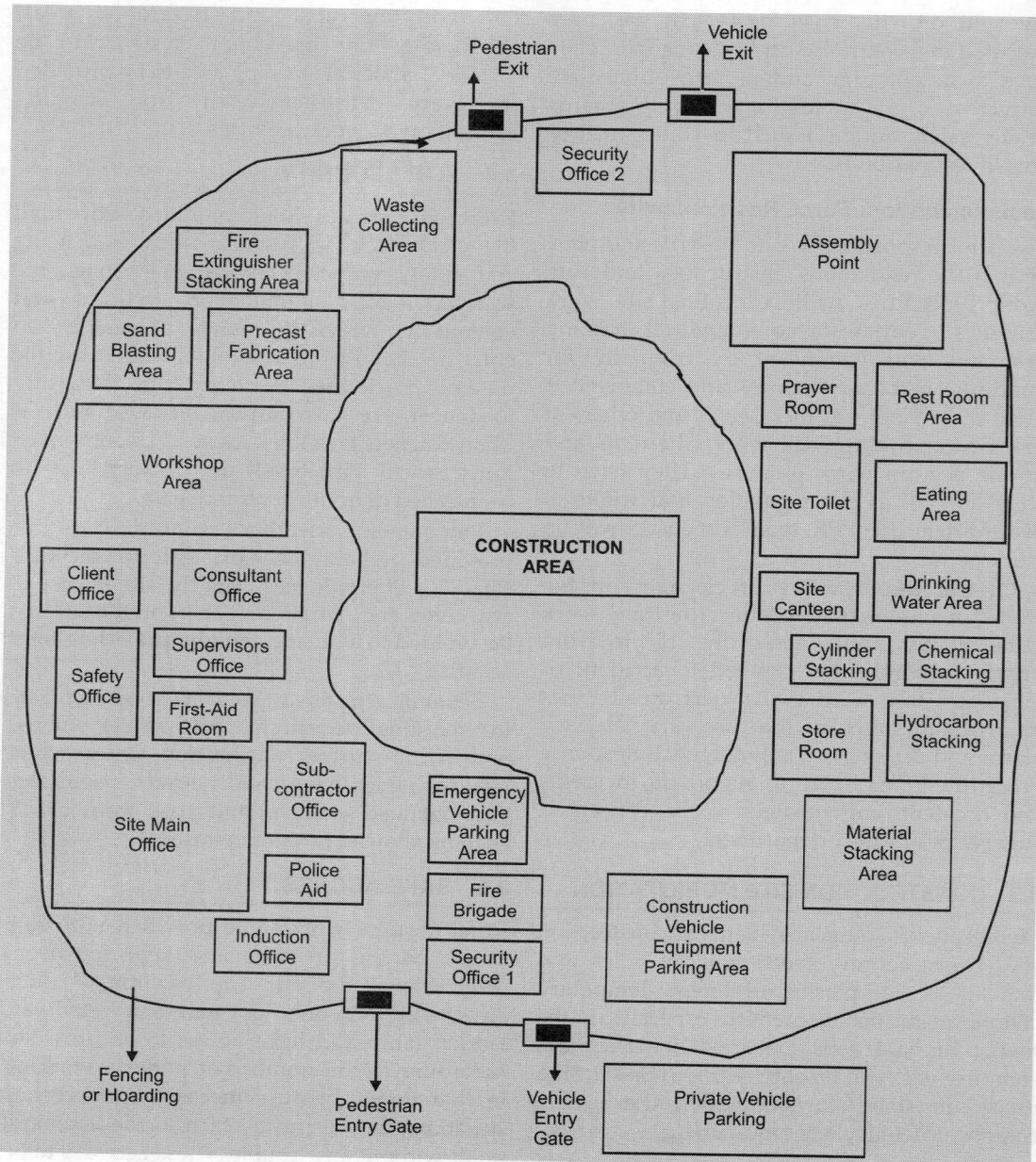

Fig. 3.2: Construction site mobilization

3.5.1 Pedestrian Routes

Only allow pedestrians to safely access to work areas and should maintained a safe distance away from the areas in which vehicles operate. It should be built wide enough to accommodate the number of persons likely to use them at peak times and to be kept barricade on two sides.

If pedestrians route crossing across any vehicle routes, provide with clear view of traffic movements at crossing and to place a banksmen to control the traffic.

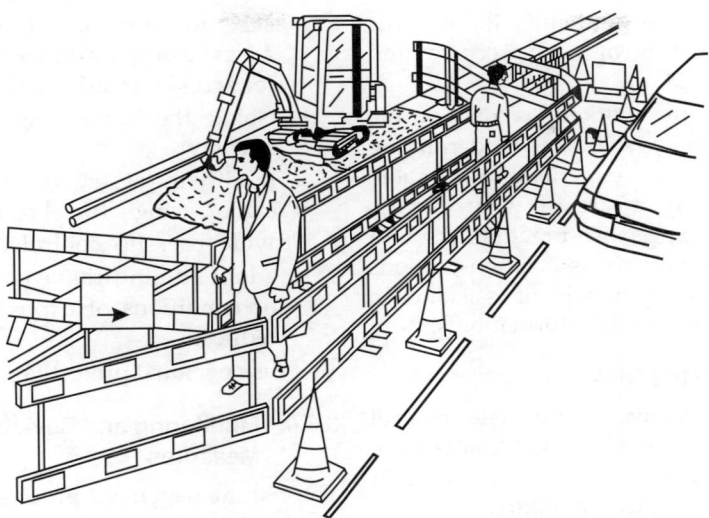

Fig. 3.3: Pedestrians walkway is separated from plant working area

3.5.2 Vehicle Routes

The risk associated with vehicle movements on work areas can be controlled by building up safe vehicle routes around the work premises and should always separate from pedestrian routes. Private vehicles are to be prohibited on site and the maximum speed limit of the construction vehicles that can ride on work premises should be set as 25 km/hour. While reversing ensure that the vehicle should accompanied with a banks-man.

Vehicle routes should be built avoiding steep gradients and tight bends where possible for avoiding incidents and accidents. Needs to ensure that operators of vehicles are licensed, competent and prohibit unauthorized operators on premises. When the vehicles are parked, operator has to ensure that the power and reverse break are own, remove key and place locked.

3.6 RISK MANAGEMENT

Every industry or company shall have their own 'Policy' detailing aims, objectives and responsibilities. According to the policy a detailed 'HSE plan' shall be made after conducting 'Hazard identification, Risk assess-ment and Safe work procedure' of each activity under the work premises.

If any difficulty or deficiency is facing on work face while conducting the procedure as detailed in 'Safe work procedure', then the procedure shall be revised with additional points and new safe work procedure shall be activated on premises after taking the reapproval from client, consultant and legal department (if necessary).

This 'Safe work procedure' of each activity is considered as the parameter or control measures of safety concern in risk management. Every activity on premises shall be designed, planned, conducted and inspected according to the approved Safe work procedure. All training programs shall be designed and conducted for train the employees to work as per approved Safe work procedure.

The steps of risk management while in the process of 'Hazard identification, risk assessment and Safe work procedure' are:

- Identify the hazards
- Assess the risks
- Control the risks monitor and review the effectiveness of the control measures.

3.6.1 Identifying the Hazards

Hazards must be identified prior to commencing work. There are a number of ways to identify potential sources of injury. Selection of the most appropriate procedure to identify the hazards will depend on the type of work processes and the hazards involved.

Consultation with workers is one of the easiest and most effective means of identifying hazards. Based on their experience with a job, workers are usually aware of what can go wrong and why. Specialist practitioners and representatives of industry associations, unions and government bodies may be of assistance in gathering health and safety information's.

3.6.2 Assessing the Risk

Risk assessment allows appropriate control measures to be developed. Once hazards have been identified, they should be assessed in terms of their risks to incur an injury.

To assess risk, consideration should be given to the:

- Likelihood that injury will occur
- Severity of the injury should it occur.

Factors to consider when assessing the likelihood and severity of risk are:

- Condition of the work surface
- Bad weather conditions, e.g. heavy rain or wind
- Number of people who may be at risk
- Location of the work area
- Location of access routes
- Type of work to be carried out
- Work practices in use
- Scheduling of work
- Type of plant, machinery and equipment to be used
- Training and experience of the people carrying out the work.

3.6.3 Controlling the Risk

Upon completion of hazard identification and risk assessment, ensure that risks are eliminated or minimized. To achieve this, implement the following controls in the order set out below:

- Eliminate the hazard, e.g. work on the ground where possible
- Minimize the risk

- substitute a work method or process with a less hazardous one, e.g. using walkways for access instead of using ladders
- isolate the hazard, e.g. using a physical barrier
- modify the system of work or equipment, e.g. by using a travel restraint
- Provide 'back-up' controls
  - adopt administrative controls so the time or conditions of exposure to the risk is reduced
  - use personal protective equipment.

3.6.4 Monitoring and Reviewing Control Measures

The risk management process requires regular monitoring to ensure the control measures that have been implemented have performed as intended. Regular reviewing also ensures that the risk management process continues to prevent or adequately control the risk.

3.7 SAFE MATERIAL STACKING

Materials shall not be stored in a manner that could create a hazard. All materials should be stacked on material stacking area or store rooms and only permitted to transfer materials that needed for a shift or day work to work premises for avoiding congestion. Storage areas shall be kept free from tripping, fire, explosion and pest harborage.

Access to materials shall be kept clear and free from risks to workers or property. All material shall be stacked separately on racks, secured to prevent sliding, falling, or collapse during storage and on transit. Employees shall not allow climbing on shelves, bins and pallets as an access for the reach of materials stacked and Precautions shall be taken against unauthorized entry to stacking area for avoiding injury, fire, and property damage.

Where personnel accessing to heavy, large on shelves more than 5 feet off the floor is required, walkways, safety ladders and mechanical equipment shall be provided. Material stored inside buildings under construction shall not be placed within 6 feet of any hoist way and near floor openings. Brick stacks and cement sacks shall be stacked not more than 7 feet high and it shall be tapered at 2 inches for every foot of height above the 4th foot level. Stacking shall comply with sufficient clearance from fire doors, extinguishers, manual fire alarms and sprinklers.

3.8 PERSONNEL PROTECTIVE EQUIPMENT (PPE)

Must be worn to provide protection when necessary by reason of hazards encountered that is capable of causing injury or impairment. Use of PPE does not eliminate the hazards or not a substitute of engineering or work practice, but considered as the last line of defense against hazards.

Personnel protective equipment must be worn at all times, fit correctly and kept clean by you. Do not use damage PPE. Common PPE's to be worn on work premises are:

a. **Head protection**—safety helmet—to protect head from striking and falling objects
b. **Eye protectors**—safety glass, goggles, Face shield,—to protect eyes from dust, ultraviolet rays, Welding shield, Hood smoke, splinters and fumes
c. **Respiratory protectors**—dust mask, Breathing apparatus—to protect respiratory system from dust, metal dust, toxic gasses and fumes
d. **Ear protectors**—earplug, Ear muff—to protect ears from noisy atmosphere
e. **Body protectors**—coverall—for the protection of body from cuts, burns and injuries
f. **Hand protectors**—gloves—for the protection of hand from cut, burns and injuries
g. **Fall Protectors**—safety harness—personnel fall arrester systems (PFAS) are designed to protect falling from heights
h. **Foot protectors**—safety shoes with steel toecap—for the protection of foot from striking hazards and mid sole inserted the body from electrocution.

The following guidelines apply to personal protective equipment

1. Certified hard hats with properly adjustable suspensions must be worn on worksites

2. Workers must wear non-conductive hard hats
3. Workers must wear appropriate eye protection all the times on the site
4. Workers must not wear contact lenses where gases, vapors, flying objects, dust or other materials are present that may harm the eyes and be absorbed by the lenses
5. Hearing protection must be selected, maintained and worn in accordance with Safe Work Procedure
6. Workers must use additional protective equipment such as fall-arresting equipment, respiratory protection, aprons, etc. as the work requires it
7. Workers must wear appropriate footwear and ensure that is in a condition to provide the required protection.

3.9 HOUSEKEEPING

Housekeeping means to maintain a clean and orderly work area. The definition of housekeeping is general care, cleanliness, orderliness, and maintenance of business or property. In work premises more than 20% of accidents like fall, slip and trip hazards are caused due to bad housekeeping. Bad housekeeping can cause big fires on site due to the accumulation of oily rags and combustible materials on work premises. Good housekeeping can prevent incidents, accidents and fires on work premises by keeping tidy.

It can be achieved by removing all non-hazardous solid waste and recyclables at regular intervals or at the end of the shift to waste collecting area or out of work premises. For these Purpose Company needs to provide waste skips or bins on work areas. Access to electrical panels, fire extinguishers, safety showers and eyewash stations, fire hydrants and points of egress is to be kept always free and clear of obstructions.

The following rules on housekeeping will be strictly enforced

a. Floors, steps and stairs shall be kept clean and free from oil, grease and other slippery substances.

b. Steps, walkways, passages and doorways shall be kept clear of obstructions

c. Soiled cleaning materials, scrap and waste oil shall be placed in the appropriate containers for proper disposal

d. All offices and workshops shall be kept clean and clear of scrap

e. Escape routes and access to safety equipment shall not be restricted in any way

f. Clear lightweight rubbish regularly to reduce Foreign Object Debris (FOD) because it always create danger to eyes and respiratory systems of the persons working on these areas

g. All bags, sacks, boxes and other goods that are required to temporally positioned on floors shall be placed so as not to cause a harm to persons

h. All drawers and other openings should keep closed except when access is necessary

i. All edges should protect with guardrails

j. Materials, tools and equipment must not impede access or egress routes

k. All materials must be staked and secured to prevent sliding, falling or collapse. Pipe, conduit, and bar stock will be stored in racks or stacked and blocked to prevent movement.

3.10 EMERGENCY RESPONSE SYSTEM

The Emergency Response requirements procedure establishes the minimum requirements for the implementation of an emergency management plan including evacuation. Emergencies and disasters can occur any time without warning. However the more you are prepared for them, the better you will be able to act, minimizing panic and confusion when an emergency occurs. All staff members need to be familiar with these procedures, and trained and drilled in any particular procedure that they may need to undertake in case of emergency.

The importance of Emergency and Evacuation drill program is, each and every employee of the company premises should have the knowledge and practice of the Emergency and Evacuation procedure to follow prior in case of emergency

situation for their own survival. This can be achieved only by periodical drill and training programs for their remembrance as well as practice.

Assign a competent and expert person to act as the leader or Marshall, control all site activities and correspond with all legal departments including media in case of Emergency situation until the site returns to the original stage. He has the authority and responsibility to select and train a group of persons as a team to perform under him in case of Emergency. As a practice or drill program matter also he will be the chief and command the whole site.

3.10.1 Essential Points for Successful Emergency Response System

(a) Communicating the emergency plan

All employees must be aware of procedures required to when heard an alarm. In addition, exit routes, fire extinguisher locations, and outside designated assembly areas must be discussed. Employees also should be instructed not to use elevators, and provisions should be made for "search and rescue" while the evacuation is in progress. Disabled employees, visitors, and employees who are in transit between work areas must be included in the evacuation plan. For this purpose Evacuation instructions by concern supervisor with briefing to the employees is essential.

A typical briefing might include the following key points

1. **Preliminary evacuation instructions**
   - Review exit locations and emergency travel routes
   - Identify fire extinguisher and fire alarm locations
   - Provide information about emergency lighting
   - Establish an outside assembly area
   - Designate Emergency Assistance Teams and Search and Rescue Personnel.

2. **Evacuation instructions**
   - Review evacuation instructions posted in hallways
   - Outline fire emergency response procedures

- Advise employees not to use elevators as exit routes
- Instruct employees to assemble in a designated area following evacuation
- Report any missing employees.

3. **Reentry instructions**
- Instruct employees not to re-enter the building until the "all clear" signal is given by a designated individual.

(b) Providing building 'safety information bulletin boards'

A "building safety information board" is one way to have a centralized source of updated safety information. It can contain emergency evacuation instructions, floor plans and exit routes, and fire emergency response procedures.

The "safety information boards" should be conspicuously located in places where they can be seen by all employees who are assigned in the immediate work area. It may also be desirable to frame the posters and to cover them with a red or green transparent acetate covering. This will draw additional attention to the safety information.

(c) Train your employees

The quick and orderly execution of your evacuation plan is the primary objective. Employees who practice emergency evacuation procedures, who are trained to operate fire extinguishers, and who know how to assist in the evacuation of employees will be of valuable assistance in accomplishing the overall objective. In addition training to be conducted among search and rescue teams for aware them the purpose and scope.

(d) Evaluating evacuation drills and emergency performance

There is no emergency that does not contain a list of things that went wrong or could have been done better. It is absolutely essential that this information be collected, evaluated, and adjustments made to ensure that future emergencies or drills do not contain the same faults. A performance evaluation meeting should be conducted following each practice drill or emergency.

3.11 SUPERVISION

Supervision deal with giving responsibility to competent person by Top Management, for controlling other people, not only responsible for his own safety, but that of the people he is overseeing and supervising.

These assigned supervisors want to provide adequate preventative equipment to protect workers against the dangers of occupational accidents and diseases that may occur during the work, also against the fire hazards and other hazards that may result from the use of machines and other equipment and also adopt all other preventative methods ordered by the Ministry of Labor and Social Affairs.

3.11.1 Selecting Supervisors/Team Leaders

It is unwise to assume that the person with the best technical knowledge necessarily has the right leadership qualities. A prospective supervisor should:
- Honestly want the job
- Have some record of achievement
- Have the ability to develop
- Be acceptable to the work group
- Know what the job involves.

3.11.2 Training Supervisors

Ideally supervisors should be introduced gradually to the job by:
- Standing in for an existing supervisor
- Taking charge of a small section
- Taking charge of a project
- Progressive delegation.

Supervisors should be involved in planning their own induction and training. It is important that their progress is monitored and they are given feedback and counseling as necessary.

3.11.3 Support for Supervisors

Managers should take care to involve supervisors in the management chain. This can be achieved by giving the supervisor responsibility for:

- Communicating management information to the work group
- Dealing with day to day matters such as holidays, sick leave, safety and training
- The first stage in disciplinary proceedings and grievance solving.

3.12 WELFARE ACTIVITIES

The term 'Labor Welfare' refers to the facilities provided to workers inside the premises such as canteens, rest rooms, recreation facilities, toilets and all other services that contribute to the well being of workers. Welfare measures are concerned with general well being and efficiency of workers.

It also deals with

a. Cleanliness of the construction premises
b. Disposal of wastes and debris
c. Providing proper ventilation and noise controls
d. Temperature control systems on work area
e. Dust and fume controlling systems
f. Heat stress controlling procedures
g. Limiting overcrowding on each compartment
h. Ensuring proper illumination facilities on work areas
i. Providing adequate drinking water facilities
j. Providing adequate latrines and urinals
k. Providing first-aid center
l. Providing adequate rest rooms and eating area with tables and chairs
m. Providing child care units for woman wor-kers
n. Providing transportation facility for employees
o. Providing adequate training for concern workers
p. Providing medical, insurance, holiday leaves and pension to the employees, etc.

3.12.1 Toilet Facilities

The numbers of toilets required will depend on the number of people working on the site. Wherever possible toilets should be flushed by water, but if this is not possible, use chemical toilets. Rooms containing sanitary conveniences should be adequately ventilated and lit. Separate toilets shall be provided for men and women. A wash hand basin with water, soap and towels or dryers should be close to the toilets if the toilets are not near the other washing facilities provided on the site.

3.12.2 Washing Facilities

On all sites, provide basins large enough to allow people to wash their faces, hands and forearms. All basins should have a supply of clean hot and cold, or warm, water. If mains water is not available, water supplied from a tank may be used. Soap and towels (either cloth or paper) or dryers should also be provided. Where the work is particularly dirty or workers are exposed to toxic or corrosive substances (for example, during work in contaminated ground), showers may be needed. Rooms containing washing facilities should be sufficiently ventilated and lit.

3.12.3 Drinking Water Facilities

Make sure there is a supply of drinking water. It is best if a tap direct from the mains is available. Otherwise bottles or tanks of water may be used for storage. If water is stored, it should be protected from possible contamination and changed often enough to prevent it from becoming stale or contaminated. If it is possible to confuse the drinking water supply with other water supplies or other liquids.

3.12.4 Rest Facilities

Facilities for taking breaks and meal breaks should be available. The facilities should provide shelter from the wind and rain and be heated as necessary.

The rest facilities should have:

- Tables and chairs
- A kettle or urn for boiling water
- A means for preparing food (for example, a gas or electrical heating ring, or microwave oven).

It should be possible for non-smokers to use the facilities without suffering discomfort from tobacco smoke. It may be possible to prevent discomfort by increasing ventilation. If this cannot be done, it may be necessary to provide separate facilities for smokers and non-smokers, or prohibit smoking in the presence of non-smokers.

3.13 ENVIRONMENT PROTECTION PLAN

Environment consist of different living organisms, non living components and their inter relationship. Environmental protection plant is activated for the purpose of protecting and improving the quality of the environment by preventing or limiting emission of environmental pollution and improving the standards for emission or discharge of environmental pollutants from the industrial sectors or operations for protecting earth.

It is now become a global issue considered as a high percentage in the change of climatic conditions due to the emission of pollutants on past decades through sky, sea and land by industrial sectors around the world.

3.13.1 Common Potential Sources of Pollution from Site

a. Dust, smoke and fumes emission
b. Waste material disposable
c. Food waste, water waste and sewage waste discharge
d. Solid waste disposable
e. Chemical waste disposable
f. Noise generation
g. Fires, etc.

Management has to input several measures or precautions to prevent or minimize the discharge of environmental pollutants. For this purpose management has to activate Environment protection plan.

3.13.2 Management Plan Consists of the Following Activities

(a) Specific action plan for implementing measures

Plan covers all aspects of the construction and operation phases related to environment. The plan needs to be implemented right from the beginning of project and should continue till the end. An implementation task list to be formatted concerning the time frame for implementation and also the responsibilities of the concerned authority.

(b) Monitoring of environmental quality

The success of environmental control measure can only be taken care by proper monitoring of the environmental parameters. Environmental Audit systems will add merit for monitoring purpose. Environmental Impact Assessment (EIA) should exercise on site activities to improve the quality of plan. It deals with detailed monitoring for different environmental parameters such as noise control checks, waste disposal systems, air checking's and fire prevention methods will be carried out on premises as per the direction of Top Management.

(c) Training

Training is of much importance in environmental management. Environmental science is a developing subject and the people implementing environmental strategies should remain up to date with the environmental control processes. Training to be conducted among the line and low management for ensuring the importance of environmental protection plan.

(d) Statutory requirements and implementation

All laws under Water (Prevention and Control of Pollution) Act Air (Prevention and Control of Pollution) Act, Environment (Protection) Act, Hazardous Waste (Management and Handling) Rules, Manufacture, Storage and Import of Hazardous Chemicals Rules, legal authorities and client requirements to be complied and implemented in environmental protection plan.

(e) Documentation

Documentation is an important step in implementing Environmental Management Plan. All statutory norms should be kept at one place for quick references. All monitoring results should be kept at selected files which can be easily accessed. The presentation of the results should also be planned. Graphs and diagrams can be used to show the trend in environmental quality or achievement.

Documentation will include

- Major technical information in operation
- Organizational charts
- Environmental monitoring standards
- Environmental and related legislation
- Operational procedure
- Monitoring records
- Quality assurance plan for monitoring (ElA records)
- Emergency plans.

(f) Social responsibility

Corporate social responsibility is now an important factor in company's project operation. Medical checks has to conduct periodically for the employees as well as the persons living around the site for finding any evidence of occupational diseases as per plan.

3.14 ILLUMINATION OR LIGHTING

Every part of the site which is in use should be lighted so that people can see to do there work and move about the site safely. As far as possible the site should be arranged so that natural light is available. Where work will continue outside daylight hours or the building or structure is enclosed, artificial lighting will be required.

Where failure of the primary artificial lighting would be a risk to the health or safety of anyone (for example, someone working on a tower scaffold in a basement may fall while trying to descend in the dark), provide secondary lighting. Where it is not possible to have lighting which comes on automatically when the primary lighting fails, torches or other similar lights may provide suitable lighting.

In addition, emergency routes (the corridors, passageways, etc. which people must follow in an emergency to escape from danger), should be kept well lighted while there are workers on the site. Where daylight provides adequate lighting, no further action is required. Where emergency routes need artificial light, also provide emergency lighting which comes on if the primary lighting fails (for example, battery or emergency generator powered lighting).

3.15 NOISE PROTECTION

Sound is a form of energy made up of invisible vibrations which travel through the air as sound pressure waves and enter the ear and create the sensation of hearing. In ear the organ of hearing mechanism comprises three distinct compartments such as the outer (external), middle and inner (internal) ear.

3.15.1 The Two Properties of Sound Waves

- Frequency (pitch) which is calculates in 'Hertz' (Hz)
- Intensity (loudness) which is calculates in 'Decibels' (db).

If the human ear does not respond equally to all frequencies, sound level meters on ears are designed to respond to noise. So Noise is considered as an unwanted or unpleasant or nuisance sound. Noise is a source of irritation and stress for many people and can even damage temporarily and permanently our hearing if it is loud enough.

Noise can be effect on human body by

a. Physically: Cause pain in ears which can lead to permanent deafness.
b. Mentally: Can disrupt concentration, causing undue fatigue and interference with communication which reduces both performance and safety on job.

3.15.2 Protection of Noise at Work

On planning stage of the project, noise level can be controlled by proper design of project compliance with legal requirements, phasing the operations (dividing contractors), working hours controls and by providing provision (PPE) for controlling noise.

Under the Control of Noise at Work Regulations 2005, employer has a duty to assess noise levels. For the purpose noise detectors or devise to be used for checking continuously the noise level on work premises. In all activities on premises must be supplied hearing protectors such as ear plug and ear muff and instruct to wear it when noise levels reach 85 dB and above.

Instruction and training shall be provided to employees who are likely to result in exposure to noise levels above safer level, remembering the importance of ear protectors.

3.16 VIBRATION PROTECTION

Oscillating motion of a body above its fixed position called vibration. The number of oscillating motion in a second is referred as frequency of vibration and measured in Hertz (Hz). The human body is more sensitive to vibration in the frequency between 1 and 80 Hz.

3.16.1 There are Two Type of Exposure to Vibrations

(a) Effects of exposure on whole body

The effect of exposure is loss of balance, blurred vision, headaches, shakiness and loss of concentration. This type is mainly affected to the operators of plant, machinery and vehicle drivers due to the long usage.

(b) Hand-arm vibration also called

1. Vibration - induced white finger (VWF)
2. Hand arm vibration syndrome (HAVS)

Hand-arm vibration damages blood vessels in the hands and fingers, reducing the flow of blood and harming the skin, nerves and muscles and can leads to amputation to effected part in few years. The cause is due to the continuous use of vibrating electric power hand tools, hydraulic Jack hammers, operating rotary tools, compressed air tools and chain saws, etc.

3.16.2 Reducing Exposure to Hand-arm Vibration

The best way to avoid injury is to work with non-vibrating tools whenever possible. If a vibrating tool must be used, use one that has effective anti-vibration features built in. Some new designs can reduce tool vibration by more than 50 percent.

To reduce exposure to hand-arm vibration

a. Limit the amount of time (hours per day and days per week) vibrating tools are used
b. Take a 10-minute break for every hour spent working with a vibrating tool

c. Alternate work with vibrating and non-vibrating tools
d. Let the tool do the work. Use as light a grip as possible to keep the tool under control. A tight grip restricts blood flow in the hands and fingers and allows more vibration to pass from the tool to the body
e. Maintain tools properly. Tools that are worn, blunt or misaligned vibrate more.

3.17 HEIGHT WORK SAFETY

Working at heights cannot be avoided in any form of industrial and constructional premises and 50% of reported accidents are due to the course of conducting height works. Falling from heights is one of the most common causes of death on construction sites, but in any type of business, working at height is a high risk activity. Any work or movement at height near an exposed edge should be considered a hazard and this includes work above or below ground.

Hazards associated while working at heights are mainly

- Fall of personnel, equipment, materials and tools from height.
- Collapse of the platform by stacking more weight than its designed capacity or striking against any equipment.

In order to safeguard your workforce and minimize the risk to those exposed to height works the following guidelines should be followed:

(a) Carry-out a risk assessment

Factors to consider when assessing the likelihood and severity of risk that may cause a person to fall include:

- Condition of the work surface, e.g. an uneven surface
- Bad weather conditions, e.g. heavy rain or wind
- Number of people who may be at risk
- Location of access routes
- Type of work to be carried out
- Scheduling of work
- Type of plant, machinery and equipment to be used

- Training and experience of the people carrying out the work
- Capability of the platform to support the load
- Type of PPE to use
- Lighting
- Likelihood of being struck by a moving or falling object.

(b) Establish a suitable control method for controlling the risk

Upon completion of hazard identification and risk assessment, ensure that risks are eliminated or minimized. To achieve this, implement the following controls in the order set out below:

- Eliminate the hazard, e.g. work on the ground where possible
- Minimize the risk
- Substitute a work method or process with a less hazardous one, e.g. using stairways for access, instead of using ladders
- Isolate the hazard, e.g. using a physical barrier
- Modify the system of work or equipment
- Provide 'back-up' controls
- Adopt administrative controls so the time or conditions of exposure to the risk is reduced.
- Use personal protective equipment.

3.17.1 Personnel Fall Protection

Systems of work and equipment that secure a person to a building or structure are known as personal fall protection. Work at height should be carried out from a scaffold or mechanical platform with suitable edge protection. Occasionally this may not be possible and a ladder may have to be used. However, ladders are used as a means of getting to a workplace, but they should only be used as a workplace for light work of short duration.

Personal fall protection systems should be used to minimize the risk of:

- A person falling from a height (travel restraint devices)
- Injury to a person after they have fallen from height (fall-arrest systems)
- Catch platforms or Safety nets

Fall-arrester systems

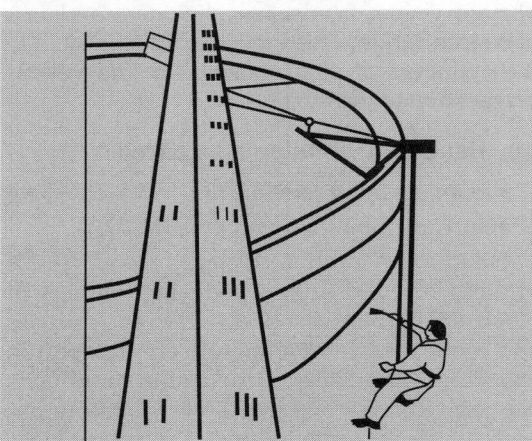
Travel retaining device

Fig. 3.4

(a) Travel restraint device

Prevents a person from reaching an unprotected edge by tethering them to an eye-bolt or other suitable anchorage point. This type of personal fall protection system is preferred over those that arrest a person after they have fallen.

(b) A fall-arrest system

Is designed to arrest the fall of a person. An important factor in the safe use of a fall-arrest system is to reduce the free fall distance as far as possible. Correctly installed fall-arrest equipment will only safely arrest a fall if there are no obstructions in the fall path. The longer the free fatting obstructions. Any obstruction should be removed from the fall path area.

A safety net must be installed as close as possible to the underside of the work area, but not in contact with the surface. The safety net must cover an area extending beyond the work area.

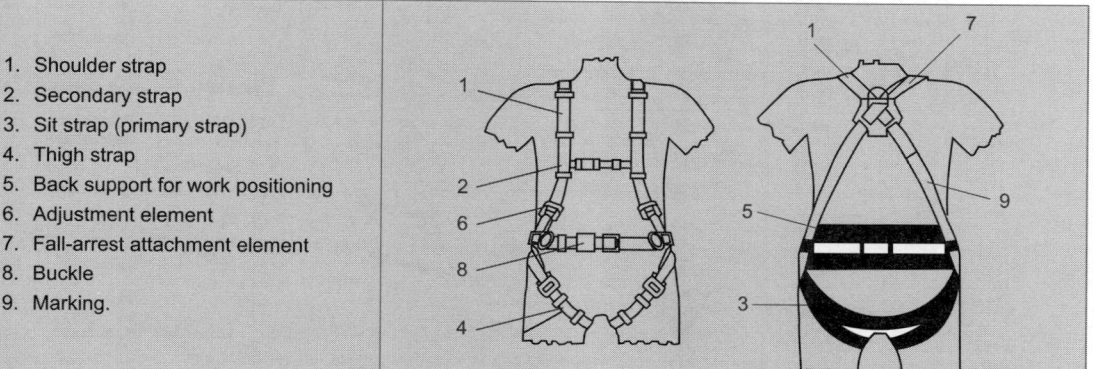

1. Shoulder strap
2. Secondary strap
3. Sit strap (primary strap)
4. Thigh strap
5. Back support for work positioning
6. Adjustment element
7. Fall-arrest attachment element
8. Buckle
9. Marking.

Fig. 3.5: Components of fall-arrester system (PFAS or safety harness)

The following points describe the different components of a fall-arrest system

- A fall-arrest harness is an assembly of interconnected shoulder and leg straps, with or without a body belt designed to spread the load over the body and to prevent the wearer from falling out of the assembly.
- A lanyard is a line used, usually as part of a lanyard assembly, to connect a fall-arrest harness to an anchorage point or static line.
- A lanyard assembly consists of a lanyard and a personal energy absorber.
- The lanyard assembly should be as short as practicable and the working slack length not more than 2 m under a free fall condition.

(c) Catch platforms or Safety nets

Catch platforms or safety nets should only be used where it is not possible to provide any more reliable means of fall protection. For example, the erection of physical barriers or personal protection systems.

A catch platform is a temporary platform located below a work area to catch a person after they have fallen. It should be of robust construction and designed for the potential impact of the load.

3.17.2 Ladder Safety

(a) Portable step-ladders should

- Not be used on working platforms to gain height above the protected edge, e.g. next to floors with penetrations or the edge of the floor
- Only be used in the fully opened position
- Be of a length that ensures a person's feet are not positioned any higher than the second top rung.

(b) Portable single and extension ladders should be

- Pitched at a slope of 1 horizontal to 4 vertical (75 degree angle).
- Extend 900 mm above the last surface where a person can gain access and should not be used:
  - In access areas or within the arc of swinging doors
  - On working platforms to gain height above the protected edge
  - To support a work platform.

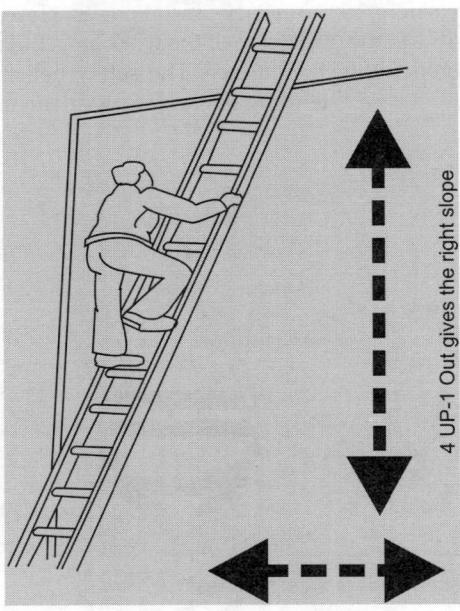

Portable extension ladder

4 UP-1 Out gives the right slope

Portable step ladder (A-type ladder)

Fig. 3.6

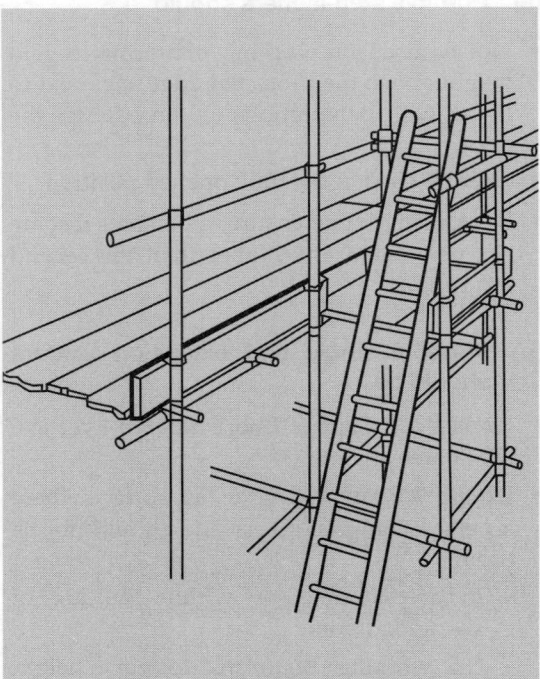

Fig. 3.7: Portable extension ladder fixing on working platforms

When using a ladder you should

a. Always face towards the ladder while climbing up or down it
b. Only move up or down the ladder one rung at a time
c. Keep three point secure at all times
d. Keep your body centered within the ladder
e. Climb down the ladder if you need to reposition it
f. Check always ladder and its rung for any sign of looseness, cracks or bends
g. Ladder with metal parts should not use near electricity
h. While placing ladder beware of any protruding or sharp edges that can harm you
i. Take precautions to prevent vehicles or people knocking against the ladder
j. Clean mud or grease from boots
k. Ensure that tools cannot fall from your pockets and strike the persons standing below.

3.17.3 Scaffold Safety

Scaffold is defined as "a temporary working platform designed to work at heights".

Three basic types

a. Supported scaffolds or Tower scaffolds or Moving scaffold—platforms supported by rigid, load bearing members, such as poles, legs, frames, and outriggers.

b. Suspended scaffolds or Cantilever scaffolds — platforms suspended by ropes or other non-rigid, overhead support.

c. Aerial Lifts or Mechanical Aid—such as "cherry pickers" or "boom trucks'

3.17.3.1 Scaffold Erection, Dismantling, Maintaining or Modification

a. Shall be done in accordance to the client's Safe Work Permit System, the Occupational Health and Safety Act and legal requirements

b. Shall made with ISO standard materials and erect on stable and level ground

c. Scaffold should support by post, supports, brackets, poles, uprights, bracings, runner and bearer. Scaffold posts must be fixed on base plates and mud sills or other firm foundation

d. The height of the scaffold should not be more than four times its minimum base dimension unless ties or braces are to be erected

e. Shall be erected, modified and dismantle by certified scaffolders

f. In the event that the scaffold is modified or repaired in any way, the date of modification shall be entered on the tag

g. Install along all open sides and ends with guardrails unless using PFAS. Ensure that front edge of platforms has the clearance distance not more than 14 inches from the work

h. Platforms must:
   • Use scaffold grade wood and ensure that no paint on wood platforms

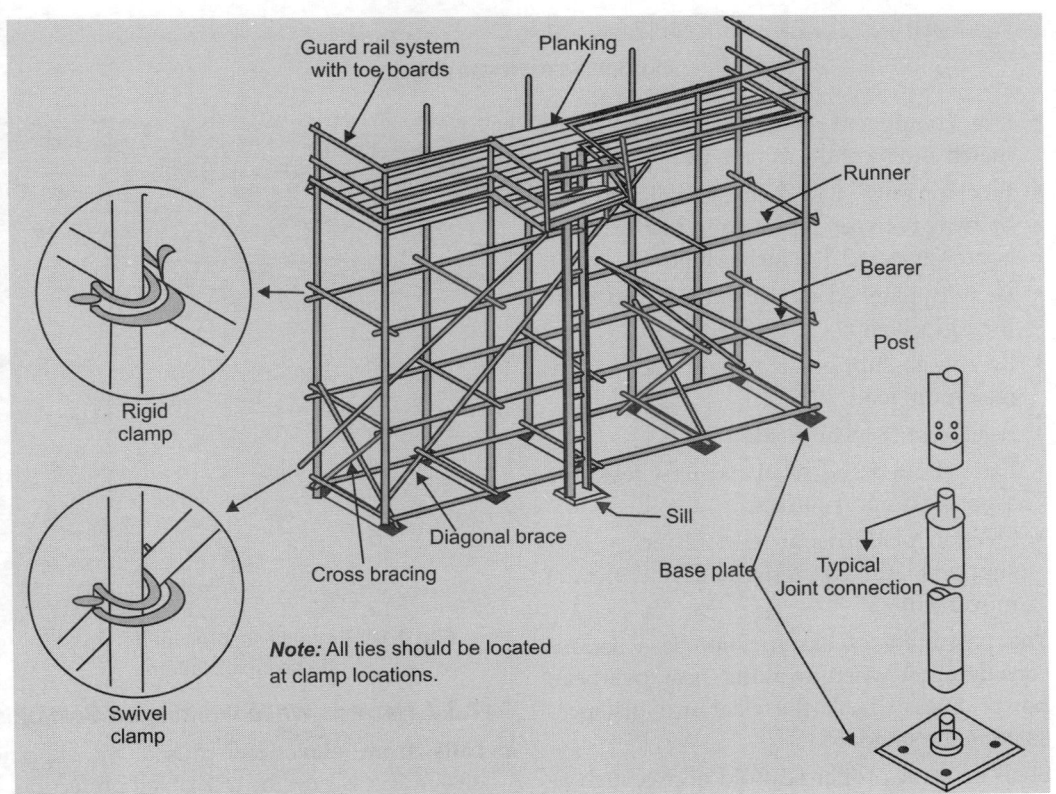

Fig. 3.8: Tower scaffold

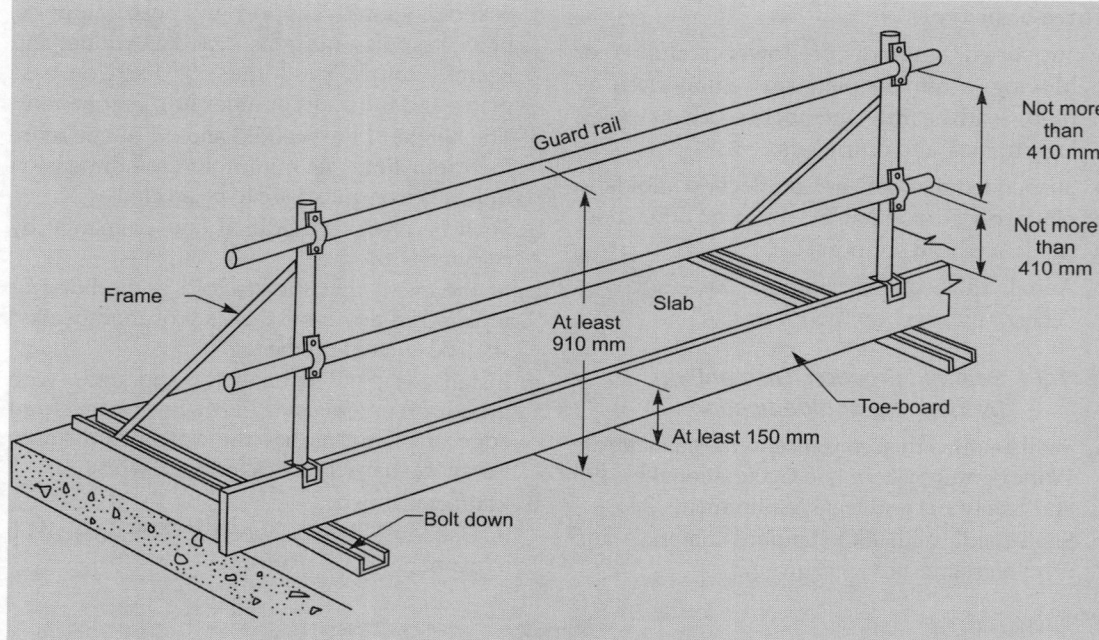

Fig. 3.9: Safe working platform

- Use component plank pieces and must match and be of the same type
- Erect top rails—38 to 47 inches tall, mid-rails halfway between top-rail and platform toe-boards at least 3-1/2 inches high
- Be fully planked or decked with no more than 1 inch gaps
- Be able to support its weight and 4 times maximum load
- Be at least 18 inches wide
- Each abutted end of plank must rest on a separate support surface
- Overlap platforms at least 12 inches over supports, unless restrained to prevent movement.

i. The possibility of electrocution is a serious consideration when working near overhead power lines. Check the clearance distances listed in the standard

j. Shall be erect proper scaffold access such as ladders or stairways as the access to go top.

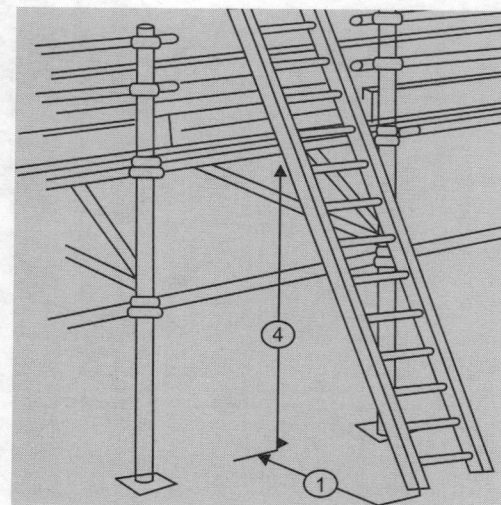

Fig. 3.10: Properly secured ladder access to scaffold platform

3.17.3.2 Hazards While Working on Scaffolds

a. **Falls from elevation**—caused by slipping, unsafe access, and the lack of fall protection.

b. **Struck by**—falling tools/debris

c. **Electrocution**—from overhead power lines

d. **Scaffold collapse**—caused by instability or overloading

e. **Bad planking** giving way for falling from heights.

Fig. 3.11: Scaffold plank collapse

3.17.3.3 Safety while Working on Scaffold

a. Ensure that the scaffold is green tagged by scaffold inspector

b. Train workers to recognize hazards and selects qualified workers to conduct work

c. Use panels or screens if material is stacked higher than the toe board

d. Barricade area below scaffold to forbid entry into that area

e. Do not work on snow or ice covered platforms or during storms or high winds

f. Ensure guardrail is required on all sides except the side where the work is being done

g. Build a canopy or erect a net below the scaffold that will contain or deflect falling objects

h. Appoint a competent supervisor capable of identifying, checking its components and promptly correcting hazards. Defective parts must be immediately repaired or replaced

i. Never allow to monkey-climb on a scaffold, always use the ladder provided

j. If you think the scaffold is unsafe, immediately report it to your supervisor or Safety officer.

3.17.3.4 Moving Scaffolds (Fig. 3.13)

When working on moving scaffolds

a. Ensure that moving scaffolds are using on level and stable ground

b. Ensure that scaffold is certified by scaffold inspector and green tag is on entrance

c. Ensure that the scaffold is not higher than three times of its smallest base, if taller out-riggers shall fitted

d. Ensure that caster brake in the wheels is applied while workers are working on scaffold

e. Ensure that nobody is standing on platforms while scaffolds are being moved from one place to another

f. Ensure that scaffold maintain clearance from charged electric lines as per standards.

3.17.3.5 Scaffold Tagging and Inspection

SCAFFOLDING IDENTIFICATION TAG		
SCAFFOLD ID #................................		
DATE ERECTED D/M/Y:	Expected Removal Date	
W.O. or Project #		
I have inspected and approved the scaffold built and consider it to be safe and adequate for completion of the work specified.		
Inspected By:	D/M/Y:	
REINSPECTED		
Name: Date:	Name: Date:	Name: Date:
Name: Date:	Name: Date:	Name: Date:
MODIFICATION DATE		
Name: Date:		Name: Date:
Name: Date:		Name: Date:

Fig. 3.12: Scaff-tag

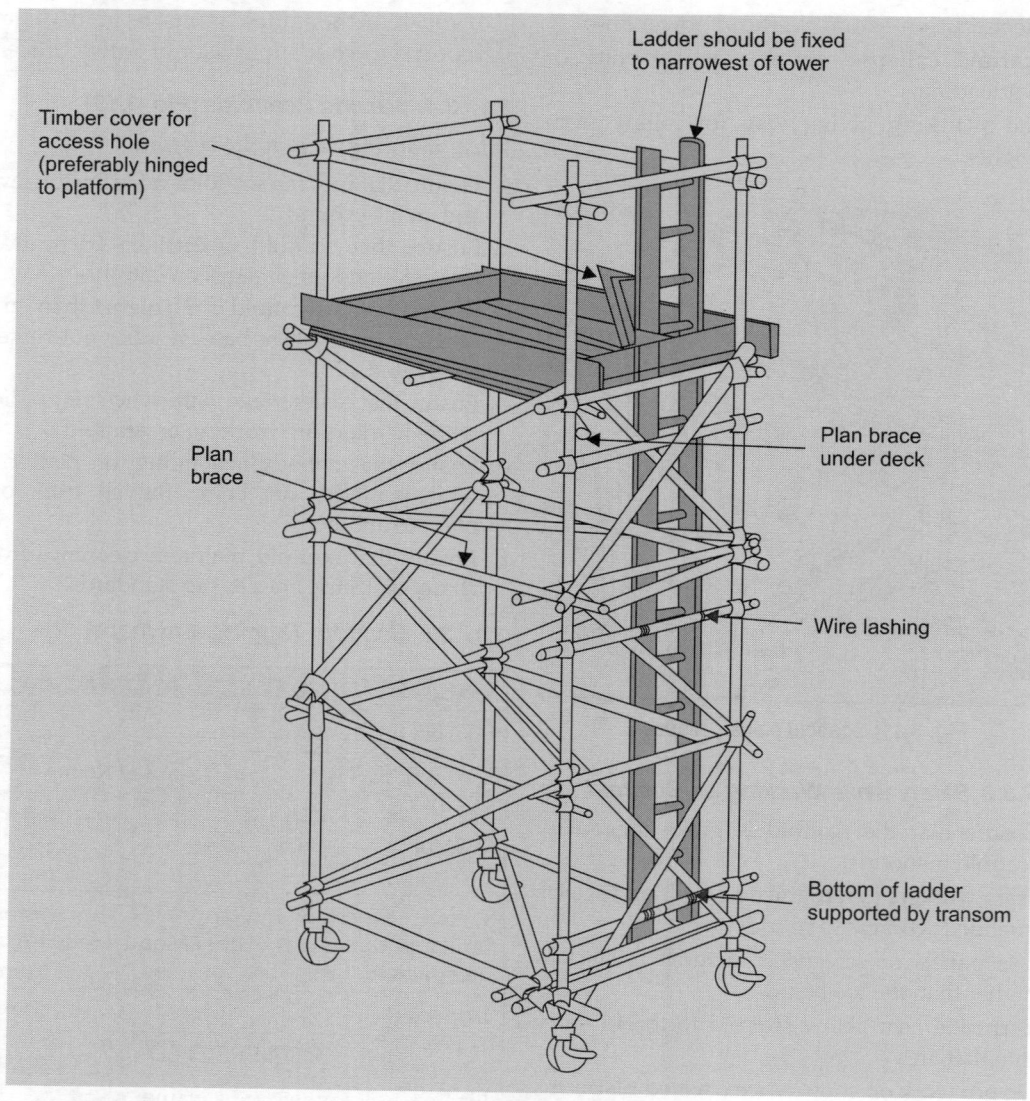

Timber cover for access hole (preferably hinged to platform)

Ladder should be fixed to narrowest of tower

Plan brace

Plan brace under deck

Wire lashing

Bottom of ladder supported by transom

Fig. 3.13: Moving scaffold

a. Inspection and tagging of the scaffold is to be performed by a Scaffold Inspector who is experienced in the erection of scaffold

b. A unique scaffold identification tag number must be clearly identified on all tags for tracking purposes

c. All scaffold identification tags will be of a solid green, yellow and red color with black lettering

d. All scaffold identification tags will have the front information displayed as per and the following must be completed for each tag.

Scaffold tag system

a. **Green Tags:** will be hung on scaffolds that have been inspected and are safe for use. A green **"SAFE FOR USE"** tag will be attached to the scaffold at each access point after the initial inspection is completed.

b. **Yellow Tags: "CAUTION",** will replace all green "Safe Scaffold" tags whenever the scaffold has been modified to meet work requirements, and as a result could present a hazard to the user. Therefore the tag should be considered a "Supervisory Tag".

c. **Red Tag: "DANGER—UNSAFE FOR USE"** tags, will be used during erection or dismantling or in the event of scaffold has been deemed unfit for use.

3.17.4 Suspended Platform Safety

A 'suspended working platform' is a scaffold or a working platform suspended from a building. Suspended working platforms are widely used in the construction and maintenance of buildings. A suspended working platform should be of sound construction, adequately supported and properly maintained. A suspended working platform should be of sufficient length and at least 440 mm wide. To prevent fall or slip, its sides must be provided with suitable guard-rails to a height between 900 mm and 1150 mm and toe boards to a height of not less than 200 mm.

The mechanical-aid suspended platforms is structure by means of lifting gear and capable of being raised or lowered by lifting appliances, wire ropes, counterweights, ballast, outriggers and the whole mechanical and electrical supports are required. There have been serious and fatal accidents caused by failure of suspension ropes or improper maintenance of operating mechanisms.

A suspended working platform should be inspected by a competent person before use. It should also be subject to a load test and thorough examination by a competent examiner before use or after substantial repair, re-erection, adjustment, failure, collapse, or exposure to weather conditions likely to have affected its stability. To prevent overloading, a suspended working platform should be marked with its safe working load and the maximum number of persons that may be carried at any one time.

3.17.4.1 Cantilever Suspended Platform Safety

To widen working levels of single scaffold platforms, cantilever platform is erected. Cantilever platforms are constructed with brackets, ledgers and vertical standards. Install a ledger at the required height of the cantilever bracket connection (the ledger connection). Use only ISO marked materials.

Cantilever bracket shall be at least 0.32 m wide deck platform. A support for the cantilever bracket is not required. The cantilever bracket must be supported by a braces. Alternatively, this support can be made from tubes and couplers.

Periodical tests and examinations thereafter are essential to ensure that it is in safe working order. The erection, dismantling and alteration of a suspended working platform should only be carried out under the supervision of a competent Scaffold Inspector. Tag systems to be followed.

Fig. 3.14: Cantilever scaffold

If cantilever scaffods are erected on another exiting scaffold (mother scaffold), ensure that :

- Text as per rules and regulations shall be conducted whether, mother scaffold can hold a cantilever scaffold.
- Use only ISO certified materials.
- Mother scaffold should be well supported as per legal requirements and method statements.
- Mother scaffold shall be braced well enough to the existing structure.
- A competent engineer shall inspected both scaffold dailywise.

If cantilever scaffolds are erected on another exiting scaffold (Mother Scaffold), ensure that:

- Test as per rules a regulations shall be conducted whether mother scaffold can hold a cantilever scaffod.
- Use only 150 certified materials
- Mother scaffold should be well supported as per legal requirements and method statements.
- Mother scaffold shall be braced will enough to the existing structure.
- A competent Engineering shall inspected both scaffolds dailywise.

3.17.4.2 Cradles Safety

Cradles are special suspended platform which is also called special temporary motorized access mobile equipment designed to meet working on heights. It is made with rigid aluminum or steel and Ideal for construction or maintenance works in the difficult to access areas, such as exterior walls of high-rises buildings, chimneys, boilers and pits, etc. The main hazards are falling from heights and collapse of the equipment. The working load limit (WLL) is to be clearly marked on the cradle for the acknowledgement of load carrying capacity.

Recommended operating speeds, special hazard warnings or instruction shall be conspicuously posted on all equipment. The operator should aware of these instructions as mentioned.

Reciprocating, rotating or other moving parts or equipment shall be gaurded, if such parts are exposed to the fellow workers.

Equipment should be selected, installed and tested to ensure that it is suitable for its intended purpose. Cradles should only be used by suitably trained and competent workers.

When using cradles check that

a. Equipment is load tested and all valid certificates are checked by competent engineer
b. The equipment is capable of fitting closely to the building and buffers or rollers are fitted, they will run against suitable features on the building
c. The building is capable of carrying the loads placed upon it, particularly under the counterweights and under the fulcrum (or pivot point) of the outrigger. The advice of a structural engineer may be needed

Fig. 3.15: Cradle

d. When the cradle is moving up and down, adequate stops are provided to prevent the cradle running off the end of the track or rope

e. A secondary safety rope fitted with a fall arrest device is provided and used

f. The cradle is not overloaded and loads are placed on the platform as uniformly as possible

g. There is safe access into the cradle. Access at ground level is safest. If access is from the roof, the cradle should be secured to prevent it swinging away from the building

h. Safe access must be possible without the need to climb up or down through suspension ropes

i. The cradle has adequate guard rails and toe boards and material cannot fall from or through the cradles base

j. Operatives has to wear safety harness while entering cradles and to hook lanyard to the anchor point fixed in it

k. A competent person has to check daily its efficiency and any sign of wear and tear and its checklist to be updated as per standard.

3.17.5 Mobile Equipment Working Platform (MEWP) Safety

Where it is not possible to work from the existing structure and the use of a scaffold working platform is not appropriate, a range of mobile access equipment including Mobile Equipment Working Platforms can be used. Those using this type of equipment should be trained and competent to operate it.

Before work starts, check that

a. A hand over certificate is provided by the installer. The certificate should cover how to deal with emergencies, operate, check and maintain the equipment, and state its safe working load

b. Equipment is installed, modified and dismantled only by competent specialists

c. Areas of the site where people may be struck by the platform or falling materials have been barricade off or similar

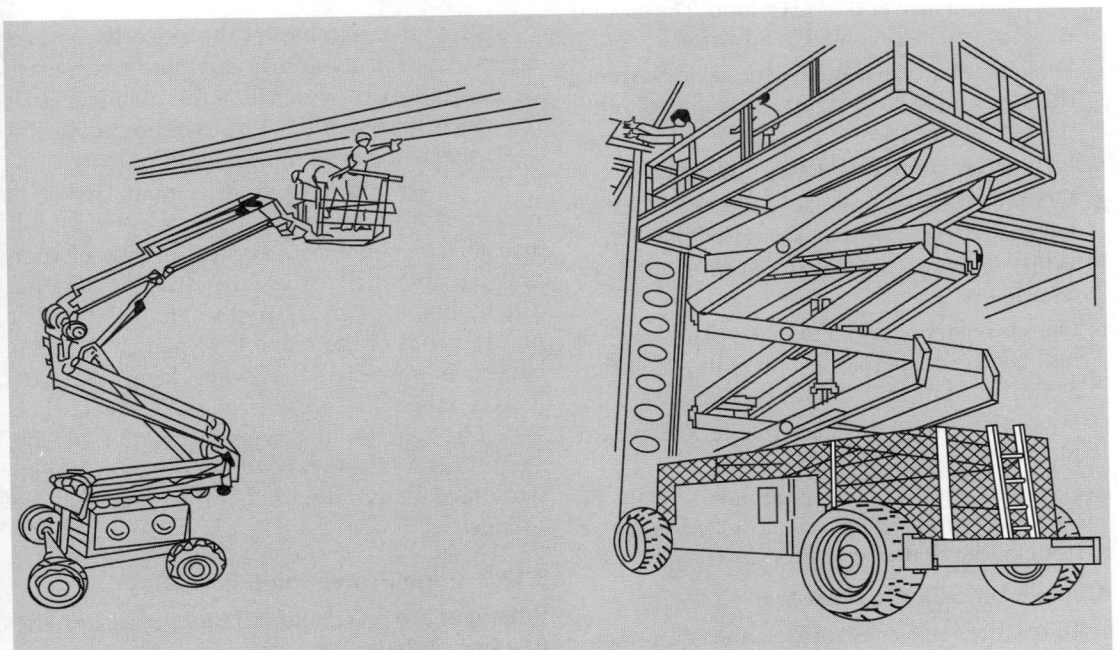

Fig. 3.16: Mobile equipment working platform (MEWP)

d. Systems are in place to prevent people within the building being struck by the platform as it rises or descends and prevent the platform coming into contact with open windows or similar obstructions which could cause the equipment to trip

e. Supports of the equipment shall protected from damage (for example, by being struck by passing vehicles)

f. The equipment shall not operate on adverse weather condition such as storms and snow. High winds can tilt platforms and make them unstable

g. Shall establish a maximum safe speed for operation.

When using a MEWP make sure that

a. Who ever is operating it is fully trained and competent

b. SWL to be marked in platforms as identification for carrying loads

c. The work platform is provided with guard rails and toe boards or other suitable barriers

d. It is used on firm and level ground. The ground may have to be prepared in advance

e. Shall maintain standard electric cable clearance distance and keep away from vehicles or other danger machineries

f. Its tyres are properly inflated and air filled

g. Everyone knows what to do if the machine fails with the platform in the raised position

h. While moving, raised platform should be shutdown

i. Objects bigger or wider than the platform shall not been transported for avoiding collision against existing structure or building

j. Workers to wear all PPE while on equipment platform

k. Ground level of the equipment shall be barricade to avoiding unauthorized entry or from falling or flying objects.

At the end of each day check that

a. The platform is cleared of tools and equip-ment

b. All power has been switched off

c. Equipment was secured properly.

3.17.6 Man Basket Safety

A man basket can also be used where it is not possible to use of a scaffold working platform by the help of crane or forklift (if the forklift designs to carry man basket). The man basket in use shall be ISO certified and has valid load tested certificate from legal authorities. Prohibited home made Man Basket on work premises.

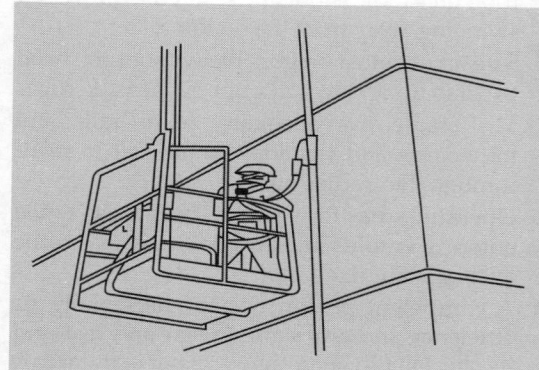

Fig. 3.17: Man basket suspended on crane hook

SWL of the man basket shall clearly marked on basket for avoiding carrying over load. Objects bigger or wider than the platform shall not been transported for avoiding collision against existing structure or building.

A worker working from a man basket is protected from falling by using a personal fall arrest system such as a vertical lifeline or rope grab combination or a self-retracting lifeline. The lifeline is often secured to an anchor point on the boom of the crane from which the man basket is suspended. Always keep the gate locked while the worker entered the basket. When it reaches the work face, the person should hook his lanyard to the solid surface on work face for avoiding fall, if the basket fails or collapse.

3.17.7 Edge Protection (Fig. 3.18)

Wherever anyone could fall more than 2 m, the first line of defense is to provide adequate edge protection. It needs to meet minimum legal standards of, or be equivalent to:

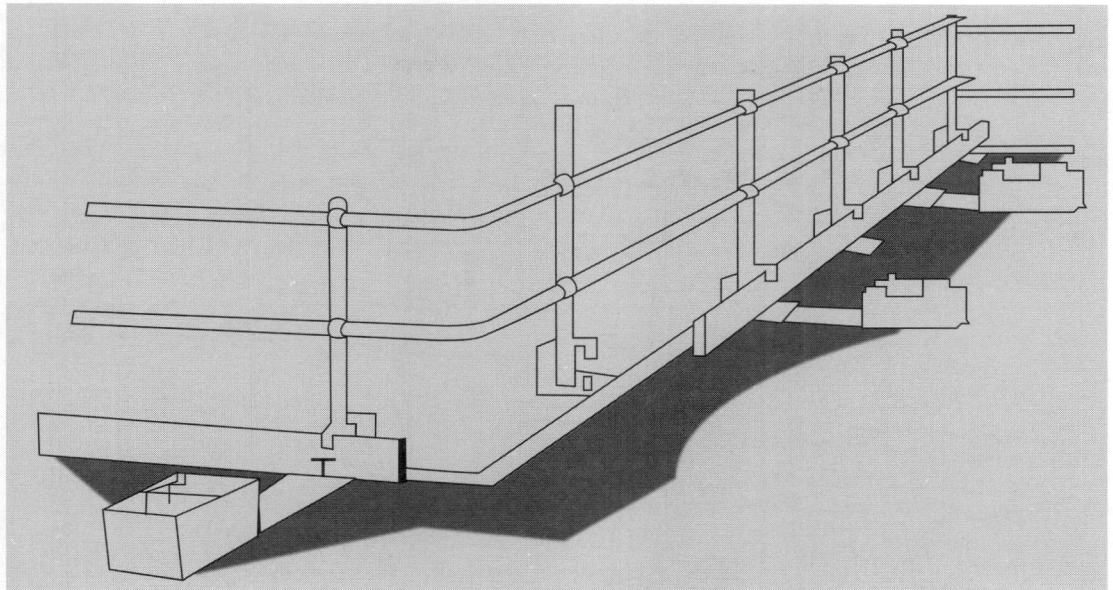

Fig. 3.18: Edge protection

- A main guard rail at least 910 mm above the edge
- A toe board at least 150 mm high
- An intermediate guard rail or other barrier so that there is no gap more than 470 mm.

It will often be more appropriate to securely cover openings rather than put edge protection around them. Any protection should be:

- In place from start to finish of the work
- Strong enough to withstand people and materials falling against it.

Whatever system of work is chosen the presence of dangerous gaps is always a possibility as space is created to place the next leading edge sheet. Options to deal with this include:

- Temporary barriers at the leading edge, such as trolley systems
- Birdcage scaffolds
- Safety nets
- Safety harnesses used with running line systems.

3.17.8 Man/Material Hoists Safety (Fig. 3.19)

Same like lift, a hoist is temporary equipment designed to lift man and heavy materials from the ground level up to the top level of constructing building or structure. Proper Selection of a hoist which is suitable for the site and capable of lifting the loads is required.

Set the controls of hoist from ground level, so that the hoist can be operated from one position and can see all the landing levels from the operating position. The hoist is thoroughly examined and tested after erection, substantial alteration or repair and at six month intervals. Regular checks should be carried out and the results recorded.

When using hoist, ensure that

a. The hoist is erected by trained and experienced people following the manufacturers instructions and properly secured to the supporting structure

b. The hoist operator has been trained and is competent

c. Loads are evenly distributed on the hoist platform

d. SWL shall be marked on the hoist platform for avoiding overloading

e. The hoist way is fenced where people could fall down or might be struck

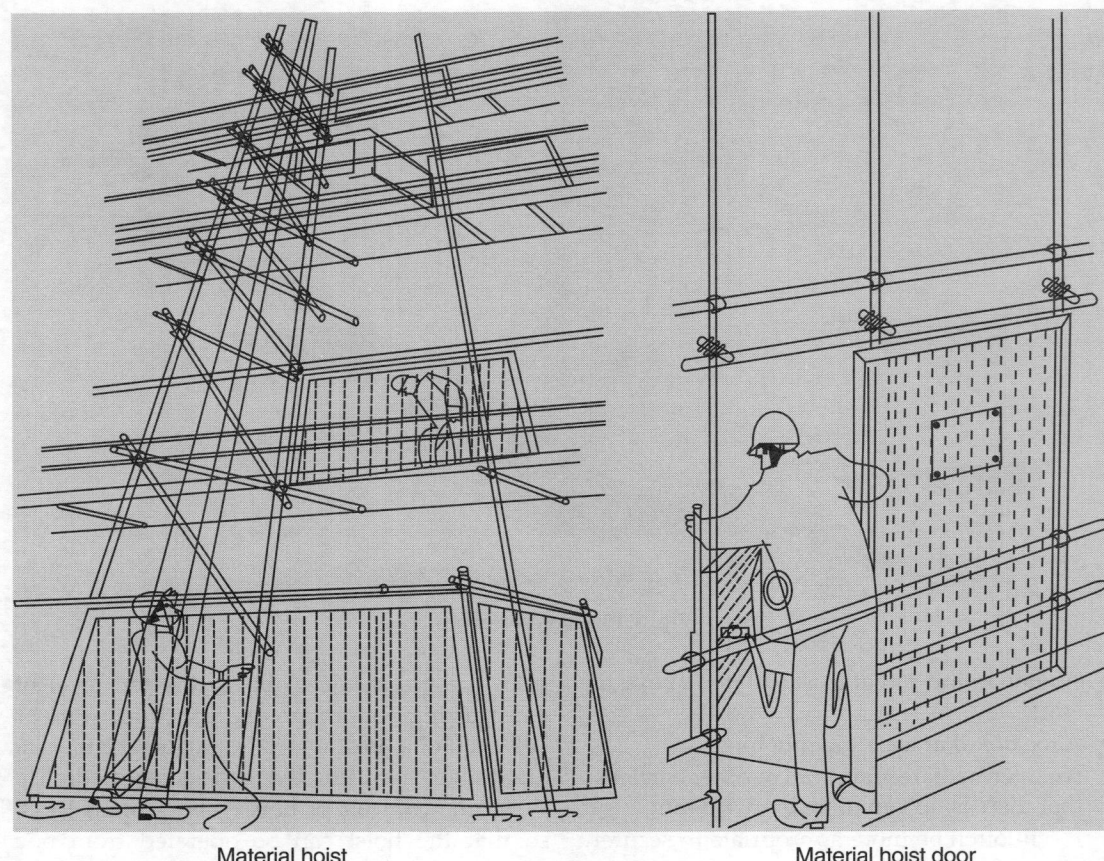

Material hoist   Material hoist door

Fig. 3.19

f. Gates shall provide at all landings from the ground level and should keep closed at all time when not in use

g. The edge of the hoist platform is close to the edge of the landing so that there is no gap to fall through.

h. Hoist platform provide with enough barricading (guards or nets) for loads falling from the platform

i. Hoist is not carrying loads more than its size, capacity or loose loads

j. Hoist door has to be kept closed while lifting.

3.18 EXCAVATION SAFETY

Excavation means digging large open holes or trenches on ground for special purpose. Definition of excavation is "a man-made cut, cavity, trench, or depression formed by earth removal". Definition of trench is "a narrow excavation", the depth is greater than the width, but not wider than 15 feet. Excavating is one of the most hazardous construction operations and most accidents occur in trenches 5–15 feet deep. There is usually no warning before side wall collapse or cave-in.

Every year people are killed or seriously injured while working in excavations. Many are killed or injured by collapses and falling materials, some are killed or injured when they contact buried underground services. Groundwork has to be properly planned and carried out to prevent accidents Excavations

Fig. 3.20: Deep excavation

more than 1.5 m depth shall be considered below points:

3.18.1 Hazards Associated with Excavation Work on Planning Stage and its Remedies

a. Location of underground services—such as water pipes, gas pipes, high voltage electric cables, underground fuel tanks and pipes, sewerage pipes, telephone and communication cables.

Hazards (consequence)—fatality, disability, damage to equipment or tools, loss of services supply to surrounding areas, injury, electric shock, short circuit and pressure hammer.

Possible remedies—obtaining permission and area clearance certificate from munici-pality, client and local authorities prior designing excavation work. Check with under ground service detector prior starting excavating work. Approved drawings of the existing ground services of site premises also be considered before designing excavation. Dig trial holes prior committing excavation to know the underground service routes.

b. Toxic or inflammable contaminants—such as poisonous or hydrocarbon mixtures mixed with soil on ancient time.

Hazards (consequence)—fatality, Asphyxia-tion, Lung disease, unconsciousness, shock, irritating, fire, explosion and disability.

Possible remedies—collect data's of ancient work activities that followed on the site premises to know the hazardous chemicals or materials used. Workers need to wear breathing apparatus compulsory in doubtful areas. Appropriate fire extinguishers to be placed. Always check with toxic gas detector in every consecutive 2 meter depth of digging by competent persons. Unauthorized entry of persons should be prohibited.

c. Under ground water flow—such as natural or environmental condition of under ground water presence in planned area.

Hazards (consequence)—drowning, fatality, Unconsciousness and Shock.

Possible remedies—company has to collect all relevant details about the atmospheric and under ground geographic condition to recognize the possible natural hazards. Dewatering to be done if there is presence of water using dewatering pumps up to the safe level by accessing permission from concern Municipality.

d. **Supporting existing structures**—if not supported, a small excavation can create dander to existing structures or buildings.

Hazards (consequence)—fatality, lose time accidents, injuries and damage to properties, etc.

Possible remedies—all existing structures and buildings shall be well supported for avoiding from collapse or damage or fall down, before digging an excavation. Site Sr. Civil Engineer should check the area and execute plans for supporting existing structures.

Fig. 3.21: Supporting existing structures

3.18.2 Hazards Associated while Excavating and its Remedies

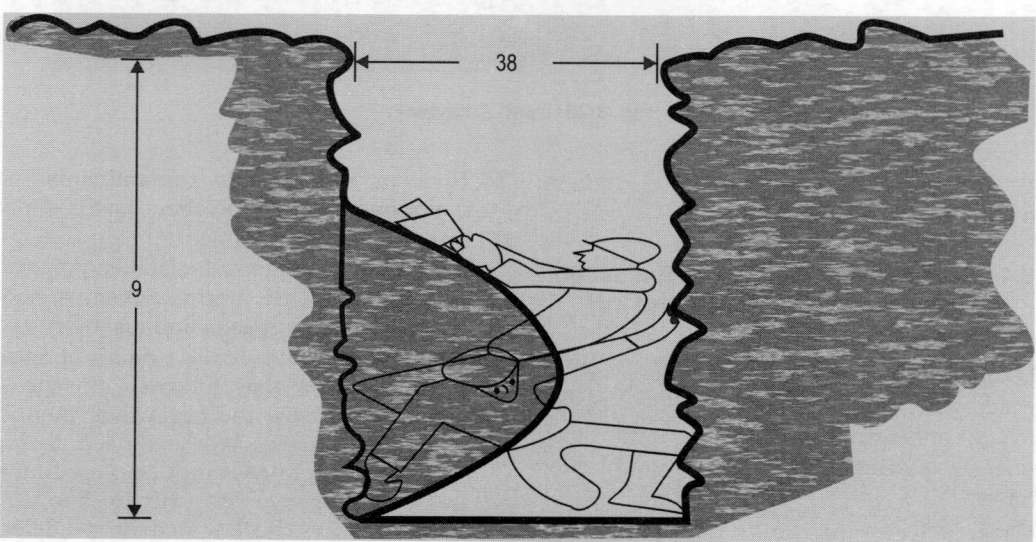

Fig. 3.22: Excavation side wall collapse

The stability of an excavation face is affected by the:

a. **Loose soil Condition**—generally Stable rock, A, B, C type—loose soil, Rock, Clay type simultaneously. Loose soil or clay type soil is more danger than rock soil type because side wall can collapse without warning.

Hazards (consequence)—fatality, Shock, Disability, Injury, Unconsciousness.

Possible remedies

1. **Benching or Stepping**—It is used for excavations that are deeper than 3 m. Benches must be set at 3 m vertical intervals and must not be less than 1.2 m wide. Benching is only allowed in Type B or Type A soils.

- Type B requires a 1 to 1 bench
- Type A requires a % to 1 bench
- Maximum depth of 20 feet, unless approved by a Registered Professional Engineer.

2. **Battering or Slopping**—it is the simplest method of stabilizing excavation faces and involves digging the faces so they are

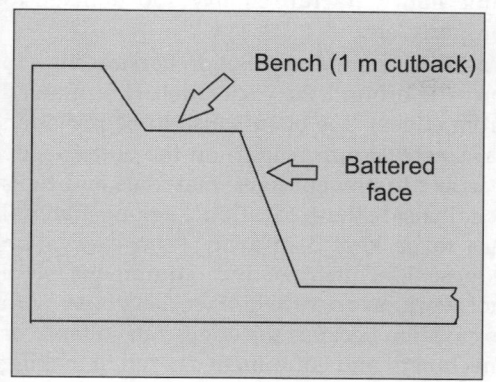

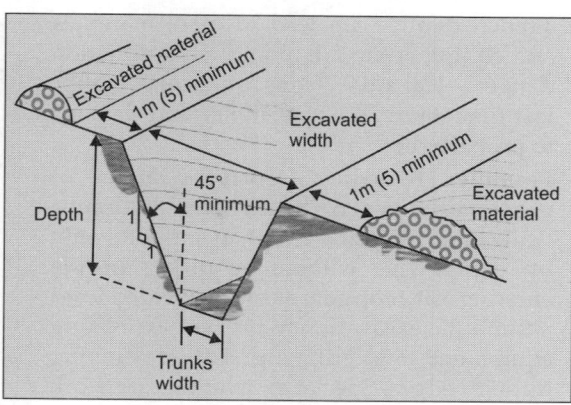

Benching   Slopping

Fig. 3.23

sloped, rather than vertical. The amount of slope needed depends upon the types of soil present, moisture content and loads on the nearby ground surface. This method is only suitable for excavation to 3 meters deep.

3. **Shoring**—must be installed along all of the excavation faces when battering or benching is impractical. Various types of timber and steel sheeting systems are used with support frames and hydraulic rams, to stop faces caving in.

Even if the excavation is shored you should frequently check for any signs of faces collapsing.

Look for:

- Cracking of the faces
- Cracking of the ground near the edges of the excavation
- Bulging or distorted shoring.

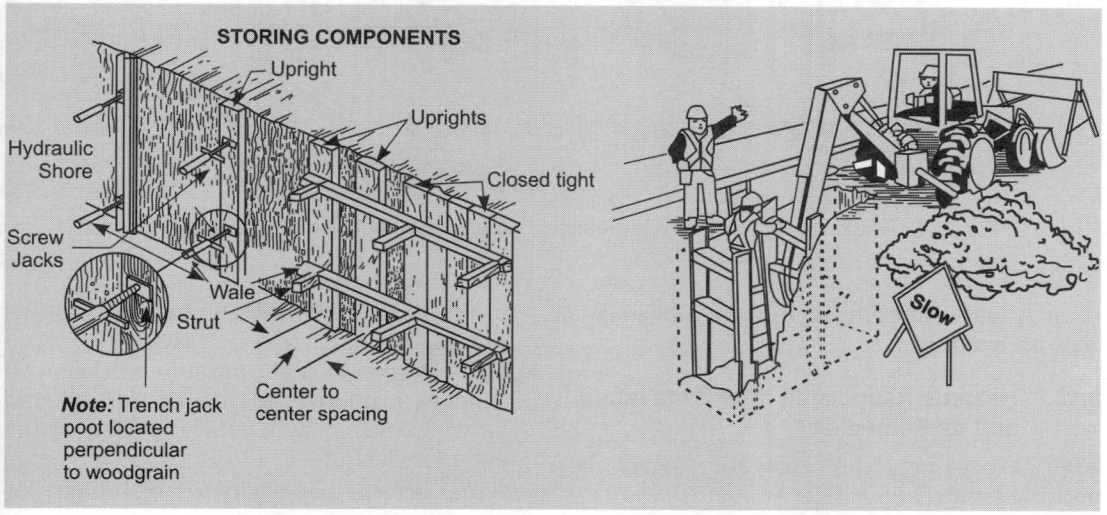

Shoring shutter   Excavating with shoring

Fig. 3.24

b. Trench depth—can lead to man, materials and equipment to fall in and make situation danger. **Hazards (consequence)**—fatality, lost time injury, fracture, injury and damage to property, etc.

Possible remedies: Correct range of proprietary trench boxes and hydraulic walls to put in place as trench supports only by equipment, without requiring people entering in it. Stop guards to be provided 1.5 m around the excavation to prevent excavating equipment and digging batter running through edges. Special excavation safe work procedure training shall be conducted for workers prior to start work about the hazards involved. Security personnel or banksman to be appointed for restricting unauthorized

time injury, fracture, injury and damage to property, etc.

Possible remedies: Solid barricading to provide around the excavation (1.5 m away from edges). Toe boards also to be provided (at least 900 mm height) on the bottom side barriers to prevent loose materials and tools to fall inside the excavation. Digging material has to be kept 5 m away from excavation edges. Machineries and equipment shall not work on edges while workers busy with excavation because the edges can collapse or machinery and equipment can fall in making big disaster. Daily Job Safety Analyses to be updated. Warning signage's "Deep excavation work in progress, keep away", "Danger of falling, keep away" to be place

Fig. 3.25: A safe way of erecting side protection walls without requiring people entering in it

entry and controlling equipment movements to excavating area.

3.18.3 Hazards Associated after Excavating and its Remedies

After excavating, the excavation shall be protected with:

a. Excavation Protection- man, materials and tools to fall in and strike the person working below. **Hazards (consequence)**—fatality, lost

on the barriers. Flashing orange lights to be provided on barricades for easy identification of excavation at night for shift workers. All small excavations (less than 1.5 m depths) to be well covered with solid materials at the end of each shift.

b. Ramp access—use walkways or bridges for access across the excavation.

Hazards (consequence)—fatality, lost time injury, fracture and injury.

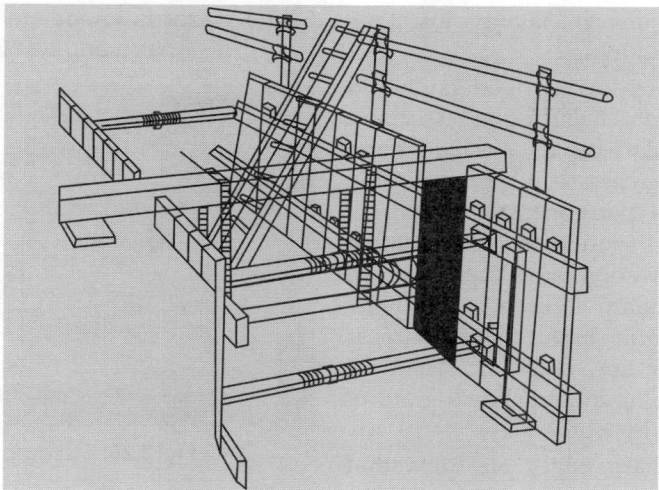

Fig. 3.26: Excavation edge protection

Possible remedies: Do not jump an excavation. Ramp access to cross excavation shall built with strong materials and to carry four times its intended load bearing capacity.

c. **Excavation Access**—unsafe access inside excavation can cause slip and fall to workers. If more than 10 persons working inside excavation provide with staircase access as a consideration for Emergency Evacuation Procedure.

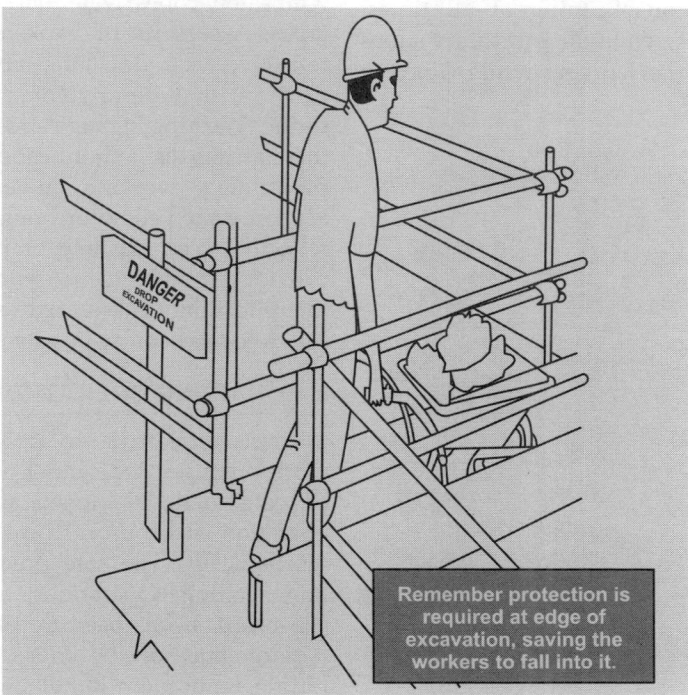

Fig. 3.27: Excavation ramp access to cross excavation

Hazards (consequence)—fatality, lost time injury, fracture and Injury.

Possible remedies—access ladders should fixed 1 m above zero level and to be secured firmly at both ends. Keep always ladder and stairways free from grease, oil and dirt. Only allow authorized personnel's to go inside the excavation. Ensure good illumination facilities in excavation area especially on access area. Standard PPE to be worn by workers at all time basic. Safety harness attached with life line or fall arrester to be used while stepping in and climbing out of deep excavations by workers.

d. **Protruding or sharp edges on Excavation Access**—can cause harm or injury to workers.

Hazards (consequence)—lacerations and deep cut.

Possible remedies: Regular inspection needs to carry-out for sharp edges and nails. If any thing found and cannot be removed, capping system to be adopted. Ensure good housekeeping at all times. Hard hat gloves, coverall and safety shoes to be worn by the workers while working on excavation.

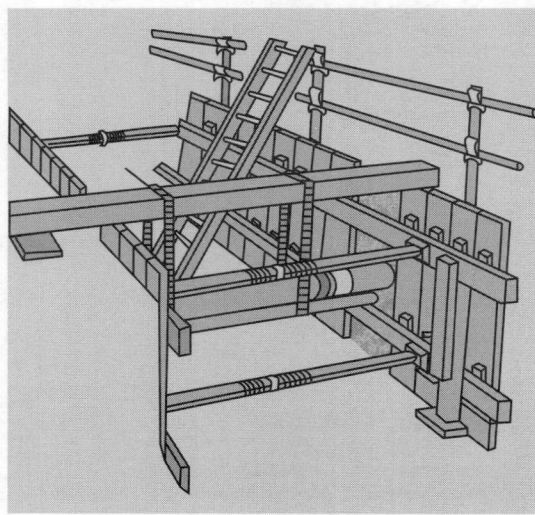

Fig. 3.28: Excavation access

3.18.4 Hazards Associated while Excavating on Roads and its Remedies

Fig. 3.29: Excavation on roads

Excavation on roads can cause man and vehicle to fall into the excavation.

Hazards (consequence): Fatality, lost time injury, fracture, Injury and damage to property, etc.

Possible remedies: Avoid excavations on roads, if possible. Shall be barricade or marked with safety cones, water barricade and warning tapes. A security or banksman wearing with reflective vest (for long visibility of vehicle driver's) shall be appointed for the control of traffic. Warning signage's to be placed around the barriers as a notification to public. The personnel's working in excavation should always wear yellow or orange color jackets or reflective vest. Flashing orange lights shall be provide around excavation on nights for easy identification. These excavations should be backfilled as soon as possible.

3.19 LIFTING OPERATION SAFETY

Lifting operation is an important activity on work premises and cannot be avoided in form of constructional organizations. The lifting operation shall proceed with manually and mechanically. The safe practice guide details how Lifting Operations are planned and organized. It has been developed to meet the requirements of the Lifting Operations and Lifting Equipment Regulations 1998 [LOLER], in particular Regulation 8, which states:

In this regulation "lifting operation", means an operation concerned with the lifting or lowering of a load. Every employer shall ensure that every lifting operation involving lifting accessories is:

a. Properly planned by a competent person
b. Appropriately supervised
c. Carried out in a safe manner.

In lifting operation company shall

a. Nominate formally in writing a Competent Person who is suitably trained and experienced to ensure safe lifting operations
b. Provide adequate resources to enable lifting operations to be carried out safely
c. Ensure that all appointees, i.e. supervisors, crane operators and operatives with duties under this procedure are trained, competent and aware of those duties
d. Confirm by regular monitoring that lifting procedures are being properly implemented.

The competent person

The Person planning the operation must have the necessary experience, skills and knowledge of the particular type of lift to be carried out so as to be able to discharge the duties required by Lifting operations and Lifting Equipment Regulations (LOLER) 1998. The necessary experience, skills and knowledge will include those of the Appointed Person in accordance with the Code of Practice for the Safe use of Cranes.

3.19.1 Manual Lifting

Most of the countries have their own manual lifting standards concern a single person can lift. Lifting and moving loads by hand is one of the most common causes of injury at work. Many manual handling injuries result from repeated operations, but even one bad lift can cause a lifetime of pain and disability. The Manual Handling Operations Regulations require employers to avoid the need to carry-out manual handling which creates a risk of injury. Where avoidance is not reasonably practicable, employers have to make an assessment, reduce

the risk of injury as far as reasonably practicable and provide information about the weight of loads. Back injuries are mainly occurring due to unsafe manual lifting practice.

Fig. 3.30: Safe manual lifting

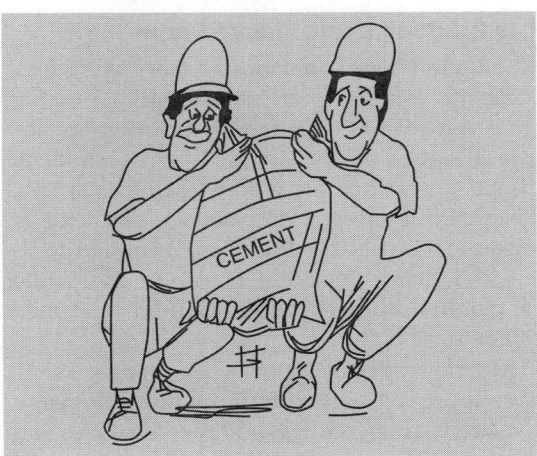

Fig. 3.31: Team lift

The thinking to be made before manual lifting is planning

• Do not attempt to lift or move objects that are obviously too heavy or bulky for one or which require getting into an awkward position. Get assistance
• Ensure you have a firm grip on the object before lifting it
• Watch out for nails, and sharp ends when handling objects. If possible, these will be removed from the object
• Ensure that you have a clear view of your route when carrying materials

- Ensure the traveling route to be taken clear of obstructions. Housekeeping to carry-out frequently to clear travel route for avoiding slip and trip hazards
- Ensure PPE's available are suitable and withstanding.

While lifting

- Ensure that you can see over the object you are carrying and Watch where you are going
- Keep your back as nearly upright as possible
- Use leg muscles instead of back or stomach muscles
- Avoid twisting motions
- Always carry lifted load close to the body.

3.19.2 Lifting Using Mobile Crane

Mobile cranes are sophisticated machines which is design for lifting efficiently. Mobile cranes are a versatile, reliable means of lifting on site. It is designed small enough to get on and off the site and to operate within it. Cranes are not design for pulling. Complacency can lead to serious accidents. No lift is small enough to be left to a chance for accident. Every lift should be planned and carried out under a trained competent Engineer.

Mobile cranes using on work premises shall have approved test certificates with a competent and licensed operator. Wire ropes, shackles and slings shall have approved certificates and must be visually inspected prior to use. A rigger or banks-man wearing reflective jackets or vest shall be appointed for giving signal to crane operator.

Planning a Lift

(a) Select the right crane for the job

It shall be able to lift the heaviest load at the required radius with capacity to spare. The maximum load a crane can lift decreases the further the load is from the crane, so a crane rated at 20 tons may be needed to lift a 1 ton load. The log chart of the cranes tells its safe working capacity at different radius or boom length.

(b) Checking to be conducted

Ensure the crane has a current approved certificate and that all accessories for lifting has approved certificate. Visual checks to be conducted on crane parts and checklist has to be updated by a competent Engineer. Make sure an automatic Safe Load Indicator is fitted and is in good working order. Ensure crane operator is well trained and experienced in the operation of the type of crane being used. Ensure that man machine language (MML) systems of the crane shall work in order.

(c) Mobilization of mobile crane

Site the crane in a safe place, so that the operator has a clear view. The crane shall place well away From excavations or overhead power lines. It shall place on level ground which can take its full weight and its load. Crane is not design to lift standing on tyres (otherwise manufacture design cranes to lift standing on tyres), so that the outriggers to be fully extended and shall placed on a firm surface or mud sills or timber packing. The crane shall be positioned to ensure adequate clearance from the exiting structure or buildings.

(d) Co-ordinating the lift

The load has to be properly slung. Chains and slings may be damaged by the load which has sharp edges, so packing is required. The center of gravity of the load may not be in the middle of the load causing it to shift or slip out of its slings when it is raised. It is important that loads are slung so that they are in balance with their center of gravity beneath the hook. Competent and certified slingers shall be appointed for above purpose. Ensure adequate clearance is maintained so that people are not struck or trapped by the load, counterweight or body of the crane. If traps are unavoidable, fence or barricade them off. None of the workers are allowed to work under the suspended load path, when the crane is carrying a load. Tag line has to be fixed on load for the control of the load. Signage's such as "Heavy Lifting in progress, Keep Away", "Falling Flying Objects Keep Away" to be place for the notification of other workers or visitors.

(e) Principles of rigging signals

Crane operator is allowed to lift any objects, if a certified and competent rigger or banksman wearing reflective jackets or vest signals to do so. The signals shown by rigger or banksman shall be clear and visible for crane operator. If the lifted object is erected in operator's unviewed place or long distance, a secondary banksman or rigger to be appointed and facilitate with whistle, flag and walky-talkies as the part of communication program between them.

Safety personnel duties while lifting a load using mobile crane

1. Inspect all records, license and text certificates of mobile crane, slings, chains, gears, operator and banksman and make sure that it is valid and up to date.
2. A special induction talk shall be conducted among the entire group which is participating in lifting operation and shall aware them the instructions, limitations, rules and regulation to follow while lifting is in operation.
3. Aware the fellow workers that the crane is design only for lifting not for draging or pulling a load.
4. Never allow to lift a load which is higher than cranes safe working load (SWL) capacity.
5. Make sure that the crane is mobilized on a level ground and outriggers fully extended. Also make sure that the automatic "safe load indicator" indicate the correct safe working mode.
6. Do not allow lifting people and never allow anyone to ride to the hoisting load.
7. Make sure that Banksman is wearing a reflective vest for operator's clear visibility. Also make sure that the operator is only obeying the signals given by the banksman.
8. Do not allow to lift load over people and no one shall be allowed under the suspended load. Barricade the whole area were the lifting operation is planned.
9. Always check that the correspondence and signals between the operator and banksman are in correct order and both can understand.
10. Never allow to leave the suspended load unattended.
11. Make sure that the sling is well balance and stable. Avoid trip loading. Never allow improper rigging of the load and using damaged lifting tackles or gears.
12. Never allow over speeding, sudden stops and quick reversal operation while lifting is on going.
13. Clearly inspect before beginning a lift that the entire proposed path of the crane is free from the overhead obstacles (electric lines and existing structures) and existing buildings.
14. Make sure that the fellow workers are not controlling the suspended load with their bear hand. The suspended load can be controlled

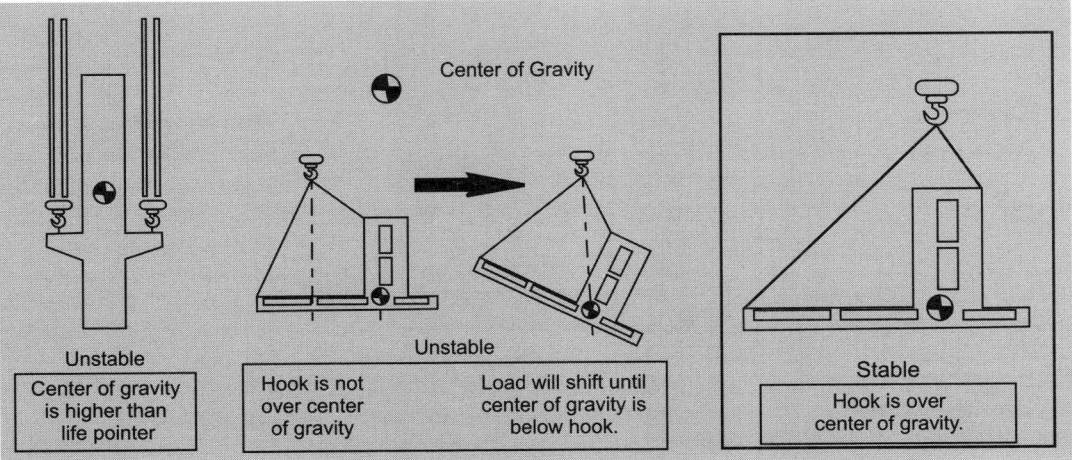

Fig. 3.32: Safe lifting chart

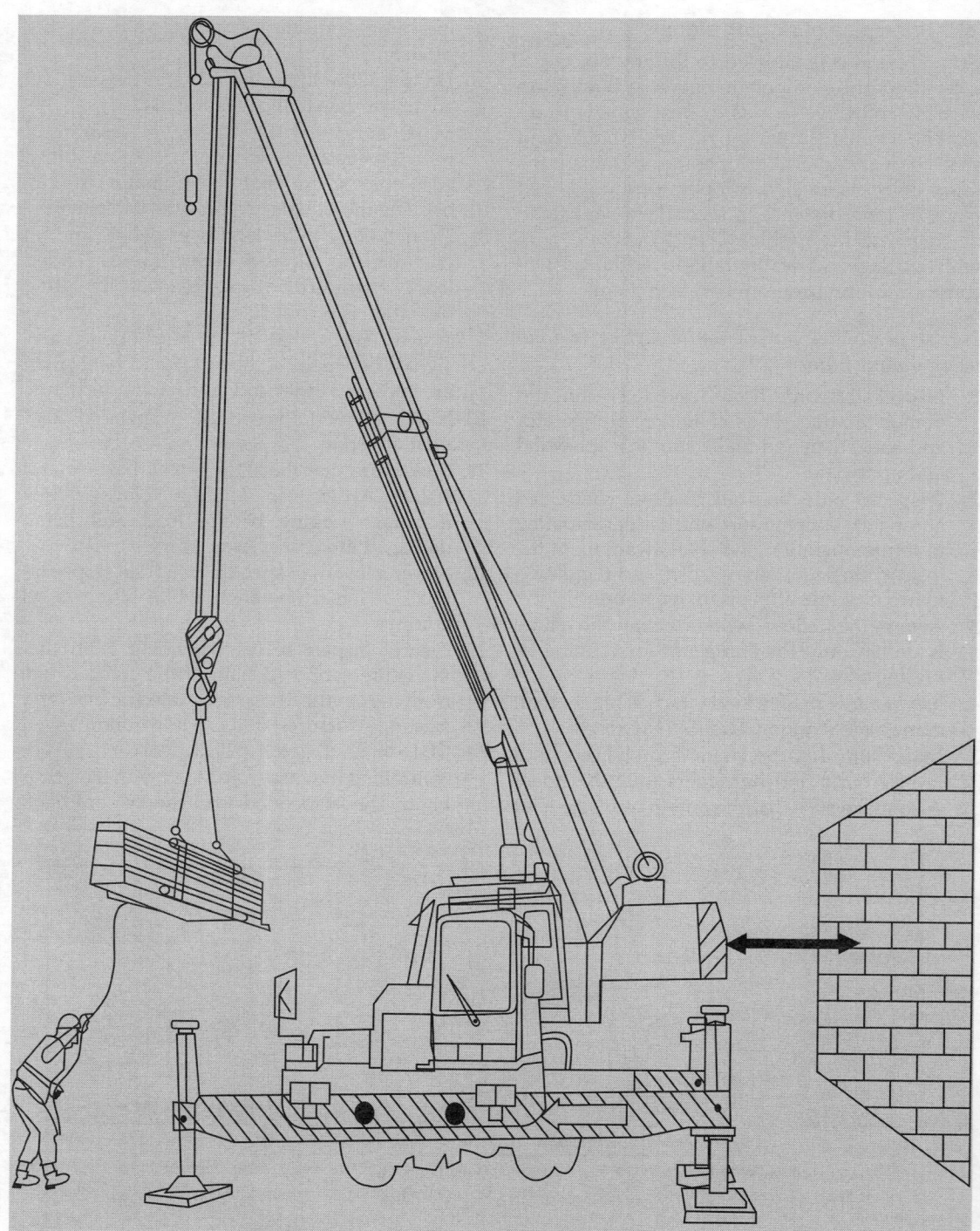

Fig. 3.33: Safe lifting operation by means of mobile crane

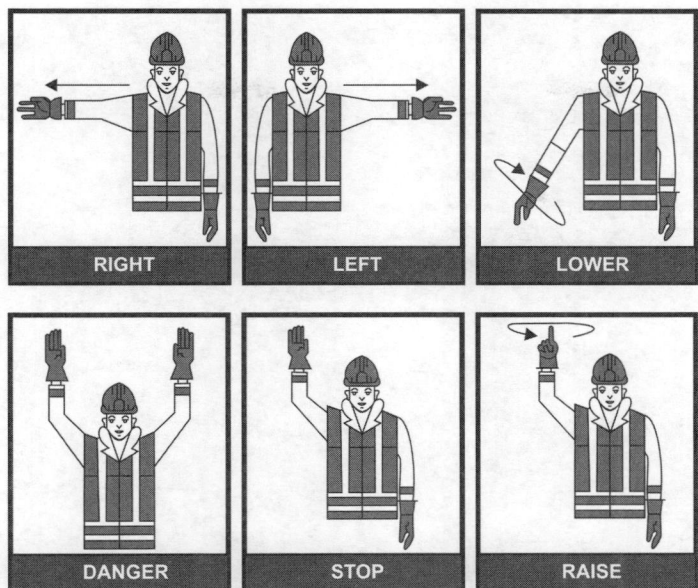

Fig. 3.34: Examples of rigging signals

only by means of a rope which should be attached to the suspended load.

15. Make sure that, after the completion of lifting program the operator shall shut down the mobile crane and park the vehicle on construction vehicle parking area by keeping all doors locked.

16. Make sure that the lifting operation is conducting under the supervision of a competent engineer.

(f) Tandem lifts

Tandem lift means two or more cranes lifting a single load at a time. Avoid tandem lifts, if possible. If cannot avoid tandem lifts, use similar weight capacity cranes. Both cranes shall fitted with anticollision sensors for avoiding collision accidents. Special safe lifting talks to be conducted by site engineer among the operators and banksmen prior lifting operations. All persons engaged in lifting operation shall facilitate with walky-talkies for the safe operation purpose. The load shall be lifted in a way that the center gravity of the load to be in the middle, stable and lifted uniformly as per lifting standards by both cranes.

3.19.3 Tower Crane Safety

Tower crane is the fixed crane heavily used for lifting purpose in construction sites. The tower crane is a modern form of balance crane. Fixed to the ground (and sometimes attached to the sides of structures as well), tower cranes often give the best combination of height and lifting capacity and are used in the construction of tall buildings. The jib (the 'boom') and counter-jib are mounted to the turntable, where the slewing bearing and slewing machinery are located. The counter-jib carries a counterweight, usually of concrete blocks, while the jib suspends the load from the trolley. The hoist motor and transmissions are located on the mechanical deck on the counter-jib, while the trolley motor is located on the jib. The crane operator either sits in a cabin at the top of the tower or controls the crane by radio remote control from the ground. In the first case the operator's cabin is most usually located at the top of the tower attached to the turntable, but can be mounted on the jib, or partway down the tower. The lifting hook is operated by using electric motors to manipulate wire rope cables through a system of sheaves.

Fig. 3.35: Tower crane

In order to hook and unhook the loads, the operator usually works in conjunction with a signaler (known as a 'rigger' or 'banksman'). They are most often in radio contact (walky-talkies), and always use hand signals. The rigger directs the schedule of lifts for the crane, and is responsible for the safety of the rigging and loads.

A tower crane is usually assembled by a telescopic jib (mobile) crane of greater reach and in the case of tower cranes that have risen while constructing very tall skyscrapers, a smaller crane will often be lifted to the roof of the completed tower to dismantle the tower crane afterwards.

Self-erecting tower crane

A type of tower crane, also called self-assembling or "Kangaroo" cranes, lift themselves off the ground using jacks, allowing the next section of the tower to be inserted at ground level or lifted into place by the partially erected crane itself. They can thus be assembled without outside help, or can grow together with the building or structure, when they are erecting.

Safety alert on the use of tower cranes

a. Tower cranes are erected and dismantled by competent people who have the necessary training and experience.

b. A thorough examination of the crane is undertaken after its erection by a competent Engineer.

c. Only competent and certified person is allowed to operate the tower crane.

d. Prestart checks are carried out by the crane operator at the start of each shift to ensure that the crane has not suffered any damage or failure and is safe to be used and records has to be filled.

e. A properly planned maintenance system is established and used. Competent people should undertake this maintenance. Any parts replaced should be installed in accordance with the manufacturers instructions.

Hydraulic section

Fig. 3.36: Self-erecting tower cranes

f. Further thorough examinations are carried out by a competent person at specified intervals, after major alterations or repair.

g. Lifting operations are properly planned according to the log-chart (of different load lifting capacity at different boom length) and appropriately supervised.

h. Operator should lift objects according to the instruction of rigger or banks-man wearing reflective jackets and also facilitate with walky-talkies and whistle for communication purpose.

Lifting tackle or gear safety

Lifting tackles or gears are used for the lifting operation purpose by means of mobile or tower crane. Wire rope slings, web slings, Chains rope and slings, man-made fiber slings (which has approved tested certificate), Shackles, eye bolt, etc. are some commonly using types. All lifting tackle shall be tested and stamped with its Safe Working Load (SWL).

Safe working load (SWL)

The maximum load (as determined by a competent person) which an item of Lifting Equipment may raise, lower or suspend under particular service condition.

Laws for lifting tackles or gear

• All lifting equipment shall be of good mechanical construction, manufactured from sound materials with adequate strength, free from defects and shall be properly maintained.

• Competent personnel (slinger) who have undergone training and certification by a third party shall operate power operated lifting equipment.

• A competent person as per the regulations shall inspect all items of lifting tackle on weekly or monthly wise.

• The SWL shall be clearly marked on each lifting gears.

- The SWL, any item of lifting equipment designed to lift persons shall clearly indicate the safe number of person for its use. In addition, any item of equipment that is not designed for lifting persons but could be construed to do so shall be clearly marked: "Not For Lifting Persons".
- Any item of lifting equipment which itself forms a significant part of the load (5 kg or more) shall be marked with its weight.
- Any accessory to an item of lifting equipment shall be marked to indicate what item it belongs to and the configuration that it is safe to be used in.
- Prohibit usage of damage equipment. Place as scrap or cut it down, the damage lifting equipment.

The following rules apply specifically to forklift operation

- Materials and equipment are to be loaded on the forklift in a manner that prevents any movement of the load that might create a hazard.
- All loads that might be subjected to shifting during transportation are to be restrained if such shifting will result in the forklift becoming unstable
- Carry loads as low as possible
- Must drive with arms, head or legs inside the confines of the forklift
- Operator shall clearly see the load which is carrying or off-load points and the full path of travel must use a signal person

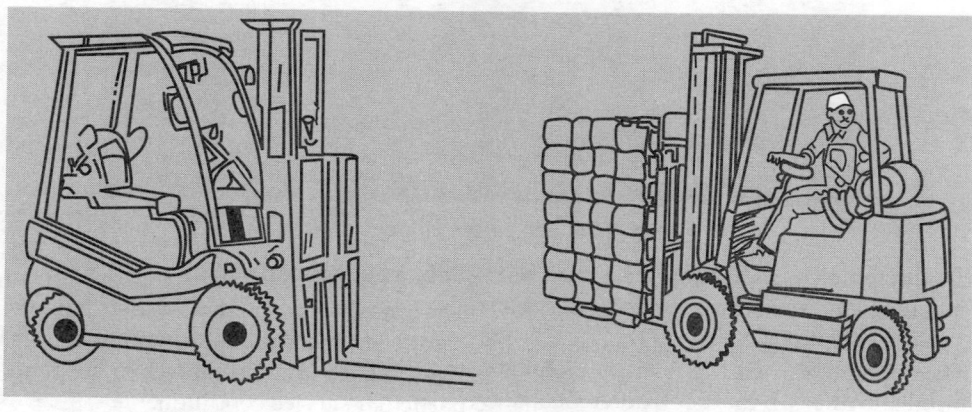

Forklift   Forklift carrying load

Fig. 3.37

- Checklists as per daily or monthly of lifting equipment shall be updated by a competent site engineer.

3.19.4 Forklift Safety

Forklifts are widely used on construction site for transporting heavy loads from one place to another, or from a, or to store on height or racks. Loads are loaded on a pallet and forklift tines (fork) carry it to one place to another. While carrying the load the operator has to turn the tines to locked position for the protection of load from falling down.

- Sound horn and slow down when approaching pedestrians, ramps and other forklifts or vehicles
- Must not exceed the load capacity of the forklift
- When shutting a forklift down, level and lower the forks, apply the parking brake and put the controls in neutral
- Do not elevate anyone on the forks unless they are in a registered man cage that is secured to the forklift
- Operator shall not move or carry any workers as standing on its forks.

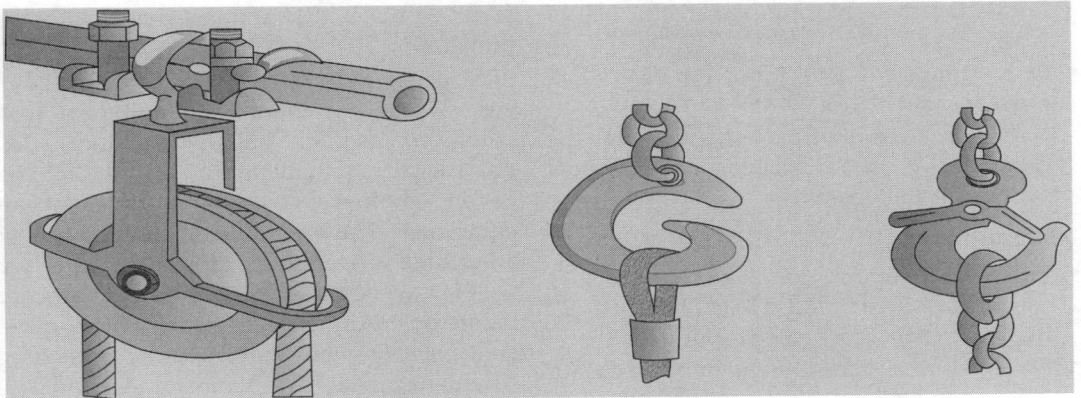

Fig. 3.38: Components of small lifting equipment

3.19.5 Small Lifting Equipment

Gin wheels or pulley and similar equipment provide a very convenient way of raising loads. Though simple pieces of equipment, care is needed when assembling and using them if accidents are to be avoided. If a pulley or similar is to be used, ensure it has:

- There shall has a strong solid materials or beams from which the hook of pulleys can be locked or secured.
- Pulleys been securely fixed to a anchorage, to prevent out of its position.
- A proper hook designed to prevent displacement of the load or a hook fitted with a safety latch. The safety latch will retain the load in case it snags.
- Do not use bent reinforcing rods or other makeshift hooks to load materials.

- There shall has a safe working platform or basement from which the hook can be loaded and unloaded.

3.19.6 Chain Block Safety

This equipment commonly used in construction, mining and rigging industries for lifting purpose. While pulling the chain or its turning lift gear (as per manufacture's regulation), the object hooked to be lifted as per its SWL. For better efficiency of the equipment certain points shall be checked:

- Ensure that the chain block gain the test certificate from concern authorities
- Ensure that the chain block is in sound condition and good working order
- Use a qualified person to lubricate and maintain the chain block and its brake system

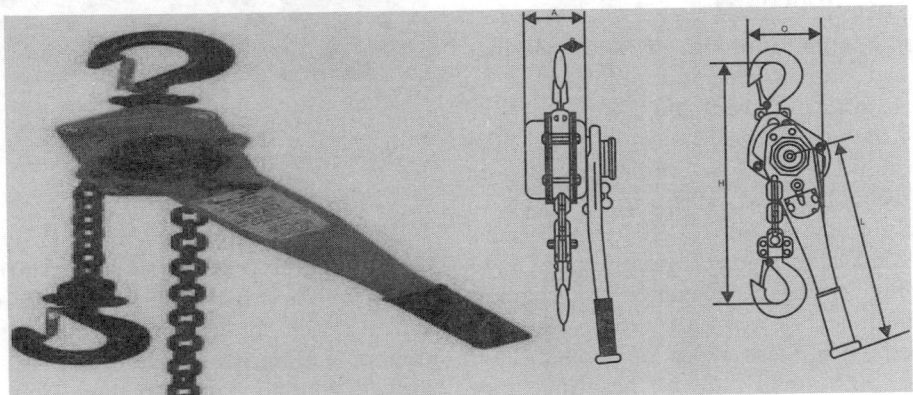

Fig. 3.39: Chain block

- Locate the chain block in a suitable, well lit work area
- Only competent and trained person shall operate chain blocks
- Carry-out a visual inspection every time the chain block is used and listen for any unusual sounds during operation.

While operating

- Ensure the support for the chain block is capable of withstanding a load of at least IV2 times the chain block safe working load
- Do not allow anyone to stand or pass beneath the suspended load
- Do not lift more than safe working load capacity (SWL) of the chain block
- Ensure that slings/chains/ropes around the load are adequate and in good condition and that the load is directly below block
- Start to raise load and check that it is level with no possibility of the load tilting and/or slipping from its restraints. If necessary, lower the load and readjust slings to obtain a safe, level lift
- Continue to raise load to required height in a slow and controlled manner. Do not rise so far that load hook comes into contact with the block
- Checklists as per daily or monthly of Chain Blocks shall be updated by a competent site Engineer.

3.20 SAFE OPERATION OF SITE PLANT AND EQUIPMENT

For the smooth running of site activities with speed and efficiency, employer has to provide with several plant and equipment (like shovel, JCB, mechanical breakers, etc). These sophisticated machines are built according to modern technology and which are very powerful and stronger beyond human efficiency because it is designed, calibrated and strengthened to do job with high speed, performance and accuracy. In modern technology all work activities are linked with these types of sophisticated machines. So a small fault to these machines can be deadly dangerous and can proceed with fatalities, disabilities and lost time accidents to humans.

It is strongly advised that the company ensures that the plant and equipment to be employed has public liability Insurance. This is not a legal requirement, but is in the interests of all employees working with plant and equipment. The basic rule on a site, whenever a machine is operating, is for everyone to keep well away because these machines are very dangerous. Any worker who will not keep well back or who appears to be particularly "accident-prone" by temperament, will be sent right away or to another site if necessary. Similarly all visitors must be excluded from the site while machinery is being used.

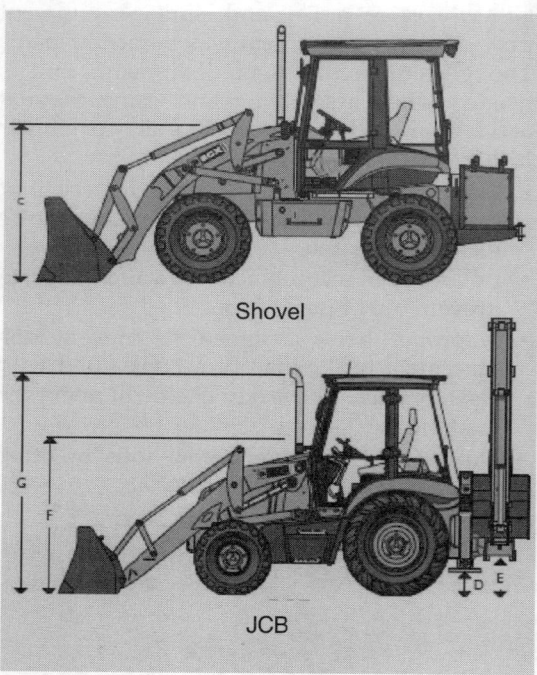

Shovel

JCB

Fig. 3.40

While working a "banksman" who shall stands at a safe point, to assist the operator and watch for important features. The "banksman" also watches for unforeseen hazards in the ground, and for any overturn, overadventure workers who approach too close. He must have a clear hand-signal code and well understood by the machine operator to transmit instructions. He must wear an orange reflective vest and safety helmet (with "banksman" pasted on it). While working with machinery only competent persons specifically nominated by the Site Supervisor may work within NEVER WORK behind a working machine. Workers or working with machines should always be visible to the operator.

While operating

a. All operators of vehicles and mobile equipment must possess the appropriate driver's license, induction and training programs completed

b. Every worker as well as operator of mobile equipment must have read the applicable safety rules and shall attended the prestart Toolbox talks

c. Workers must not operate company vehicles while impaired (e.g. alcohol, fatigue, sickness or drugs)

d. Seat belts are to be worn by operator while operating

e. Operators are to use running lamps or illuminated head lamps during day and night work time hours

f. Operation shall only be conducted by providing adequate supervision

g. No person shall be on any part of powered mobile equipment not intended for operator or passenger transport while the equipment is in motion

h. Vehicles and powered mobile equipment must be driven and operated at safe speeds according to safe work procedure

i. A banksman to be appointed wearing reflective vest to give clear hand-signal and it to be and well understood by the machine operator

j. Workers must not get on or off a moving vehicle except in an emergency

k. Operators must not leave the controls unless the equipment or vehicle has been secured against movement by setting parking brakes and transmission locks, lowering any blades, buckets or forks to the ground and chocking wheels where necessary

l. Tools and equipment carried in any part of a vehicle or piece of mobile equipment where workers are working must be placed or secured to prevent injury to workers

m. Where the equipment operator's vision is obstructed and the motion is in reverse, an audible warning device (MML systems) is required and it should work properly

n. A competent Engineer shall check the plant and equipment on dailywise. Checklists to be updated as per safe work procedure

o. Never allow working near or overhead charged electric lines without accessing Permit-to-work systems.

3.21 SAFE USE OF HAND AND PORTABLE POWER TOOLS

Definitions

Hand tools—tools that are powered manually. Hand tools include anything from axes to wrenches.

Pneumatic tools—tools that are powered by means of compressed air and include chippers, drills, hammers, and sanders.

Fuel powered tools or liquid fuel tools—tools that are typically operated by means of gasoline.

Electric power tools—tools that are powered by means of electricity.

Five basic safety rules

• Keep all tools in good condition with regular maintenance

• Use the right tool for the job. If not sure, ask what the right tool is

• Examine each tool for damage before use and do not use damaged tools

• Operate tools according to the manufacturers' instructions

• Provide and use properly the right personal protective equipment.

Personnel should be trained in the proper use of all tools. Employees should be able to recognize the hazards associated with the different types of tools and the safety precautions necessary. Personnel who use hand and power tools and are exposed to the hazards of falling, flying, abrasive, and splashing objects, or to harmful dusts, fumes, mists, vapors, or gases must be provided with the appropriate personal protective equipment.

3.21.1 Hand Tools Safety

Incidents with hand tools are usually the result of misuse or using the wrong tool for the work to be conducted. Hand tools are precision tools capable of performing many jobs when used properly. Training and proper use of the tools can prevent accidents. Supervisors and employees shall frequently inspect all hand tools used in the work being performed. Defective tools shall be immediately removed from service and tagged red color coding or "Do Not Use".

Fig. 3.41: Examples of hand tools

Safety precautions for preventing hazards while using hand tools

- Tools should be of good quality and adequate for the job. All tools should be kept in good repair and maintained. Never allow to use any wear and tear tools, e.g. cracked, split head, bended, mushroom type or make shift handle type, etc.
- Racks, shelves, or tool boxes shall be provided for storing tools which are not in use.
- When using tools while on ladders, scaffolds or platforms the worker shall tie tools on his hand by means of a piece of rope for preventing tools from falling down. Carry bags or Toolbox shall be used for tools not in use.
- Operators shall wear appropriate PPE while using the tool. Use right tools for right jobs.

3.21.2 Pneumatic and Compressed Air Tools Safety

Compressed air has the appearance of a relatively harmless gas. However, to avoid accidents, compressed air must be used correctly. The maximum air pressure approved for general use is 30 psi (pounds per square inch). This pressure is sufficient for most operations and is not significantly hazardous. Use discretion and good judgment when using compressed air, even at this low pressure. It is dangerous to pressurize any container not designed for that purpose. Safety precautions for preventing hazards while using compressed air tools:

- All personnel assigned with air compressors shall be familiar with compressor operating and maintenance instructions. Training to be conducted to make operators competent
- Compressed air is not to be used to blow dirt, chips, or dust from clothing
- Never apply compressed air to any part of a person's body

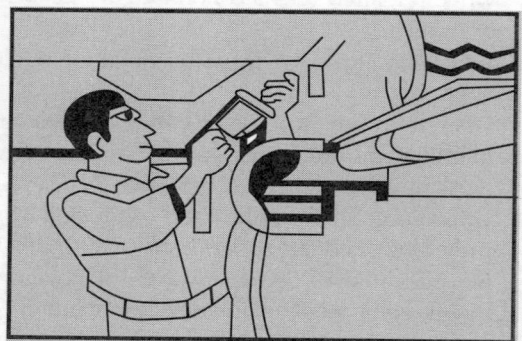

Fig. 3.42: Compressed air tool

- Air compressors shall be maintained strictly in accordance with the manufacturer's instructions
- Never use compressed air where particles can be accelerated by the air stream
- Do not use compressed air to clean machinery or parts unless necessary. Where possible, use a brush. Wear goggles to protect your eyes
- Do not use a compressed air line that does not have a pressure regulator for reducing the line pressure
- The maximum working pressure of compressed air lines shall be identified in psi. Pipeline outlets shall be tagged or marked showing maximum working pressure adjacent to the outlet

- Keep the hose length between tool housing and the air source as short as possible
- Inspect air supply and tool hoses before using. Repair or replace hoses where applicable
- Turn valve off and vent pressure from a line before connecting or disconnecting it. Never work on a pressurized line
- Do not use compressed air to transfer materials from containers when there is a possibility of exceeding the safe maximum allowable working pressure of the container.

3.21.3 Fuel-powered or Liquid Fuel Tool Safety

Fuel-powered tools are usually operated with gasoline. The most serious hazard associated with the use of fuel-powered tools comes from fuel vapors that can burn or explode and also give off dangerous exhaust fumes.

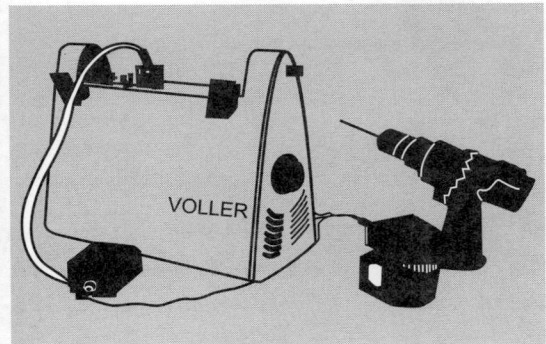

Fig. 3.43: LPG gas-powered portable tool

The following precautions shall be taken

- Be careful to handle, transport, and store gas or fuel only in approved flammable liquid containers, according to proper procedures for flammable liquids. Wear appropriate PPE while using these tools.

Hydraulic Tools

The fluid used in hydraulic power tools must be an approved fire resistant fluid and must retain its operating characteristics at the most extreme temperatures to which it will be exposed. The tools of hydraulic fluid shall be of the insulating type. While using tool do not exceed the manu-

facturer recommended safe operating pressure for hoses, valves, pipes, filters, and other fittings.

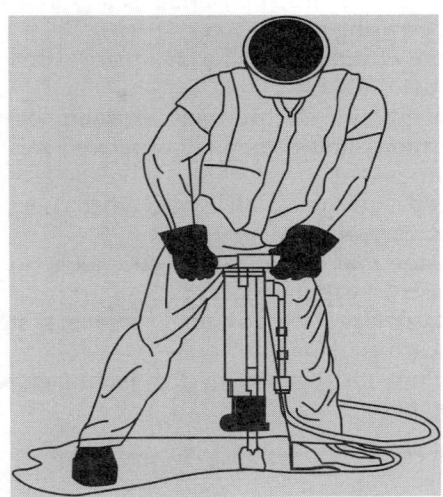

Fig. 3.44: Hydraulic tack hammer

All jacks including lever and ratchet jacks, screw jacks, and hydraulic jacks, must have a stop indicator, and the stop limit must not be exceeded. Manufacturer's load limit shall be permanently marked on the jack, and the load limit must not be exceeded. Never be using a jack to support a lifted load. Once the load has been lifted, immediately place a block under the base of the load where the foundation is not firm.

3.21.4 Electric Power Tool Safety

Employees using electric tools must be aware of several dangers, the most serious of which is electrical burns and shocks. Safe operation shall be consistent with the manufacturer's instructions. Supervisors are responsible for ensuring that personnel are authorized to use portable power tools, have met the requirements for authorization and are well train in safe operation of tool.

The following general precautions shall be followed when using electric tools

- Operate electric tools within their design limitations

- Only allow to use double insulated tools. Squire symbol is marked on every double insulated tools
- Use gloves and appropriate safety footwear when using electric tools
- Store electric tools in a dry place when not in use
- Do not use electric tools in damp or wet locations unless they are approved for that purpose
- Keep work areas well lighted when operating electric tools
- Ensure that cords from electric tools do not present a tripping hazard
- Use tools protected by circuit breaker systems or earthed enough
- Follow instructions in the manufacturer's manuals.

- If the wheels sound cracked or dead, do not use the tool because wheel could burst apart in operation and endanger seriously the operator or fellow workers. Change wheel with new one, immediately
- Be sure to keep good footing and maintain good balance when operating power tools
- Wear proper apparel for the task. Loose clothing, ties, or jewelry can become caught in moving parts
- Power tools must be fitted with guards and safety switches. Safety guards must never be removed when a tool is being used. The codes or switches shall check for any damages for avoiding electrocution
- Never allow to use trigger-locked switch type power tools on work premises.

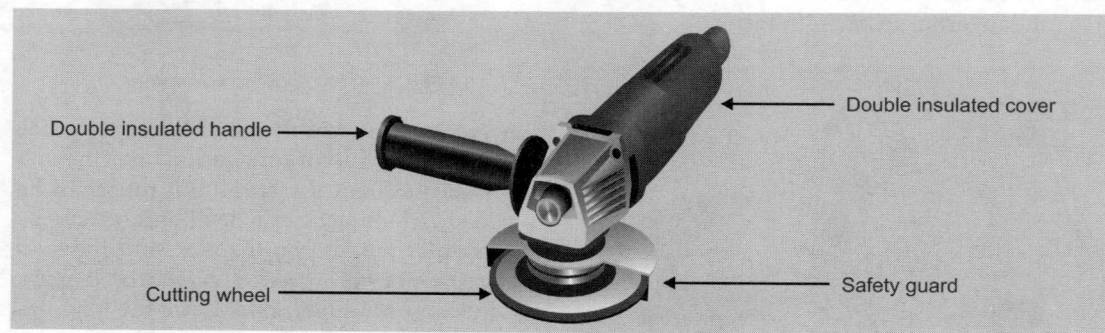

Fig. 3.45: Portable electric cutter

Portable abrasive wheel tool safety

Portable abrasive grinding, cutting, polishing, and wire buffing wheels create special safety problems because they may throw off flying fragments that can endanger eyes and respiratory system Wear appropriate PPE while using the tools.

When using the portable electric tool, ensure that

- The tool is inspected for any evidence of damages. Remove all damaged portable electric tools from use and tag them: "Do Not Use"
- Only allowed authorized or competent personnel, using the tool
- Conducted sound or ring test to ensure that it is free from cracks or defects

3.22 ELECTRIC SAFETY

Electricity is a very useful energy when it is used under control. It is widely used in the construction premises. At the same time it is a serious source of potential danger if fails to ensure safe design, operation, work practice and maintenance. Therefore great care shall be given to ensure that electricity is used in a proper way by taking all necessary precautions and safety measures.

Electricity can kill employees, being human body is the conductor of electricity. It can effect as electric shock and burn on human body. Injuries can vary from slight shock to fatal. Injuries are usually less severe when the electrical current passes nerve center and vital organs. Electrical

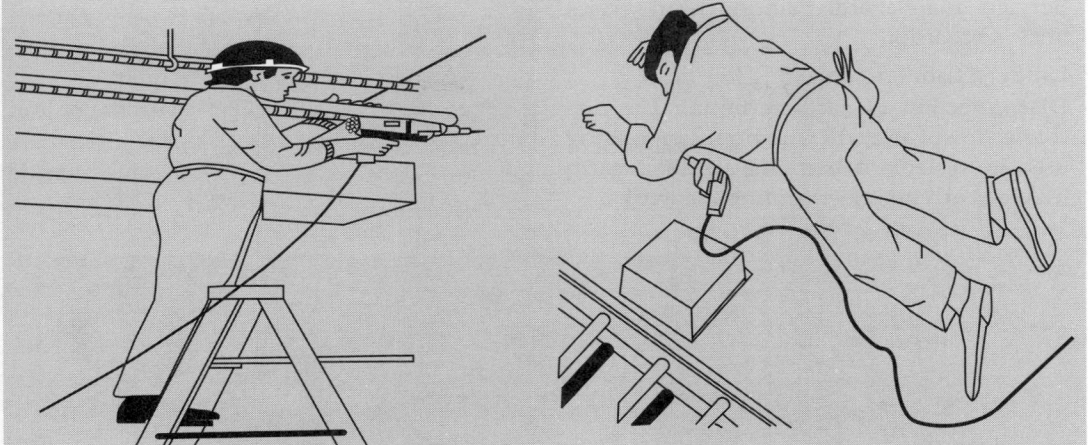

| Person working near charged electric line | Death can happen if he touches the electric line |

Fig. 3.46

current that flows through the body from limb to limb usually passes through the heart and lungs causing serious injury, possibly death. When the human body is wet, the resistance will reduce and the person is in the danger of receiving a shock of great severity. Once the skin resistance is broken down, the current flows rapidly through the blood and tissues. High voltage currents tend to pass over the body surface causing burns and flash injuries.

3.22.1 Physical effects from an electrical shock most likely to cause death or serious injury

a. Contraction of the chest muscles, resulting in an inability to breathe, and asphyxiation

b. Paralysis of the nerves, usually temporary, resulting in an inability to breathe

c. Interference to the heart rhythm, causing accelerated heartbeat, uncoordinated twitching of the heart muscles or total arrest

d. Hemorrhage in the brain or other vital organs

e. Destruction of tissues, nerves, or muscles.

Burns are another result of electrical injuries. Electrical burns usually look less severe than they are because the burn is deep-seated. Side effects, such as gangrene from embedded molten metal or flash burns to the eyes, often mean hospitalization and prolonged convalescence.

3.22.2 Common Causes of Electric Accidents on Construction Sites

- **Inattention:** A worker is engrossed in ground level work, and fails to notice overhead lines.

- **Inadequate Line Clearance:** This is really a version of inattention, when a crane carrying a load "walks" into an overhead line. Both the operator and dog-man are watching the load. Each expects the other to watch out for live lines.

- **Inadequate structural clearance:** Adjacent aerial cables that will pose a threat at some time during the building or demolition project must be removed or re-routed at commencement, otherwise they are overlooked due to familiarity.

- **Careless use of mobile scaffolding:** Many times mobile scaffolds have been pushed into overhead lines.

- **Fencing hazards:** Straining wire fences across gullies can bring them into contact with aerial power lines.

- **Survey hazards:** Similar to fencing hazards above but involving the surveyor's steel tape.

- **Demolition site hazards:** Where disconnection of power has been overlooked, or an unexpected second supply line is left alive.

- **Service line hazard:** Painting near service lines is hazardous.
- **Ladders:** Ladders contacting live lines.
- **Disconnection of earth terminal:** The disconnection of an earth wire from a pipe being used as an electrical earth may result in earth leakage currents, electrocuting the worker.

- Switch where material may fall on and operate it
- Outer insulation of lead pulled back
- Switch mounted upside down

Mobile equipment in contact with over head electric line

Workface contact with broken electric line can lead to electrocution

Fig. 3.47

3.22.3 Common Faults on Construction Sites

Gear	Defect
Skill—saw, drills, grinders, sanders, etc.	– On/off does not always function – Switch permanently on, a very dangerous fault – Exposed terminals and wiring insulation pulled out of housing
Extension leads	– Damaged insulation – Outer insulation pulled back, showing colored insulation of conductors – Incorrect connections (transposed wires)
Saw benches, electric welders, etc.	– Switches inconveniently placed
Site lighting	– Lights and fittings set at head level or lower – Plugs broken or cracked, conductors exposed – Leads coiled when in use (this may result in overheating and insulation damage) – Permanent wiring cable used in place of flexible leads – Cables that impede access – Empty sockets exposing live terminals – Trouble lamps without bulb cages

Only authorized by persons may carry-out electrical installation, repair or maintenance work on construction sites. Registered electricians make up the bulk of those authorized. However, persons holding a certificate shall fit plugs and connectors and replace fuse links. Temporary installations must be installed by a registered electrician and examined at least once every 3 months by that electrician.

PTW systems to be follow while working near charged lines. Training shall be conducted among the workers in the connection of sockets to the same voltage lines. For this purpose color coding of the sockets come into part (i.e. Yellow color sockets mark for 110 V, Blue color sockets mark for 220 V, Red color sockets mark 440 V, etc). Overcrowd connection from a socket line for different purpose to be prohibited on work premises.

3.22.4 Electrical Accident Safeguards

There are several safeguards used on construction sites to reduce the possibility of electrical accident.

(a) Fuses

When electricity flows along a conductor some heat is generated. If twice as much electricity flows, then four times as much heat can be given out. When the electric current is greater than intended the fuse melts breaking the circuit, effectively turning the power off. If the circuit is broken, something is wrong. If it is simply an overloaded circuit, it will be cured by unplugging some appliance. If the circuit breaks a second time, an electrician is needed.

(b) Direct earthing

Appliances with metal parts that could become live are earthed by a conductor that is usually incorporated in the flexible lead. This protection is only suitable for dry indoor situations, and then not if earthed metal is being worked on. By itself this is not a satisfactory safeguard on construction work.

(c) Double-insulated appliances

A double-insulated tool may be used inside or outside, in wet or dry conditions or on earthed metals, but it is prudent to use an isolating transformer with double insulated tools. Double insulation will not protect the user from risks caused by faulty leads or plugs. The international symbol for double insulation is a square and is always labeled in equipment.

(d) Isolating transformers

A worker who touches a live wire becomes part of a closed circuit and is at risk. The isolating transformer removes that risk. The worker will receive a shock on the outlet side of an isolating transformer only when in contact simultaneously with both outlet terminals.

Always position the transformer as near as possible to the switchboard or point of supply. This ensures that both the tool and lead share the protection provided by the isolating transformer. Use only one tool per transformer unless the exposed metal surfaces of the appliances are all effectively bonded together.

(e) Earth leakage circuit breakers (ELCBs)

ELCB is usually fitted at the switchboard or distribution box. It continuously compares the electric current passing into the work site with that leaving the site. If less current leaves than enters, then there must be a fault of some sort that is letting the current "leak" to earth. The device switches off all current when a very small difference occurs by trip.

Each time it is reset unless the fault is found and rectified, or the faulty equipment is disconnected. Being electromechanical devices, they need regular checking to ensure that their action has not been impaired by dust, rust or mechanical fault.

(f) Monitored earth supply

Monitored earth protection should be fitted to those machines and tools that use two- or three-phase power. Saw benches, buzzers, welders, and hoists are machines that need this form of protection. The integrity (or continuity) of the earth wire is monitored by passing a low voltage current along it. If the earth wire is broken, a relay switches off the power.

3.23 HOT WORKS

"Hot Work" is an operation involving open flame operations, abrasive grinding and cutting, welding, thermal or oxygen cutting or heating and other related heat-producing or spark-producing operations. The common fires occurs on site premises is due to conducting the hot works without taking additional safety precautions.

The Contractor's Site Manager or Supervisor has overall responsibility for the safe execution of all operations relating to the hot work. The Contractor shall nominate a "Responsible Officer" who will directly supervise the work. This person shall have satisfactory knowledge of the fire, explosion and toxicity hazards associated with hot work and be adequately trained and experienced in the testing procedures and precautions necessary for the elimination of any risk involved. This nominated person shall be approved from the part of client and has the authority to issue Permit-to-Work systems.

Any hot works can be done on work premises after accessing the Permit-to-Work systems. This document is a permit or authorization given in writing after a detailed checking, describing that the concern area is safe to enter, can start work, way in which the sequence of work is to be done, the duties and responsibilities of the persons involved in work activities, the safety checks and precaution to be taken while work in progress and the time when work must stop.

3.23.1 General Precautions

(1) Removing combustible materials

All combustible materials shall be removed from the hot work area. If any combustible materials cannot be removed, then it shall be well covered with fire blankets or with non-combustible sheets.

(2) Off-cuts and electrode stubs

Before any hot work commences, arrangements shall be made to prevent any work off-cuts, hot metal, slag or electrode stubs from lodging in places where there is a possibility of starting a fire.

(3) Timber

Where any hot work is to be carried out adjacent to or above timber, the timber shall be protected by fire safe non-combustible blankets or other suitable means from the direct heat of any flame or arc and from sparks, slag and hot metal particles.

(4) Rope

During hot work, ropes shall be protected from heat of any flame or arc and from sparks, slag and hot metal particles. Particular care shall be exercised with respect to ropes supporting loads, guy ropes and scaffolding ropes.

(5) Dusty work areas

In buildings there are often large quantities of dust present. This dust may be combustible, especially in below floor and ceiling areas. Loose dust shall be cleared from the hot work area for a distance of at least 2 meters. The preferred cleaning process is by vacuum cleaning.

(6) Grass fires and bush fires

Before hot work commences near grass or bush, the immediate area shall be cleared or wetted sufficiently to prevent the hot work from starting a grass or bush fire.

3.23.2 Procedure

(1) Inspection of site

Before hot work commences, the site shall be thoroughly inspected by the Contractor and made safe, or alternative methods of carrying out the work shall be adopted.

On completion of hot work, a thorough inspection of the site shall be carried out by the Contractor to ensure that the site is safe.

(2) Before performing hot work

Prior to commencing hot work, the following precautions shall be taken, to prevent fire, explosion, injury or other danger developing during the performance of the hot work

a. Identify and control any fire hazard (including the presence of flammable or combustible liquids, gases, vapors, dusts, fibers or substances) within 15 meters from the hot work.

b. Consider relevant hazards that may exist outside the 15 meters area.

c. Properly ventilate the hot work area.

d. Suitably locate equipment, including emergency fire fighting equipment.

e. Fire watchers to be allocated in search of any evident of starting a fire.

f. Arrange for temporary isolation of Smoke Detection Zone.

g. Appropriately isolate the area with safety barricades where the hot work is to be performed.

h. Ensure passers by are adequately protected from welding flashes.

i. Proper signage's to be placed for the notification to coworkers.

j. Do not commence the hot work, until complying with all of the above requirements.

(3) Conduct of work

While carrying out hot work the following requirements shall apply:

a. A current hot work permit shall cover the work and shall be prominently displayed. *Note: A Hot Work Permit is only valid for the date or job it has been issued for.*

b. Each person associated with the hot work shall be conversant with the precautions to be taken as specified on the hot work permit and with the safety requirements of the site.

c. Workers shall not work alone.

d. Supervisors shall be present at all times.

e. Fire watchers to be stand there at least one hour after the accomplishment of work in search of any evident of starting a fire.

3.24 CONFINED SPACE

Confined space means an enclosed or partially enclosed space that is not primarily designed or intended for human occupancy, except for the purpose of performing work and has a restricted means of entrance and exit (e.g. tanks, vessels, silos, storage bins, hoppers, vaults, storage tanks, reaction vessels, enclosed drains, sewers and pits).

3.24.1 Definitions

Confined Space—it means an enclosed or partially enclosed space that:

i. Is not primarily designed or intended for human occupancy, except for the purpose of performing work.

ii. Has a restricted means of entrance and exit.

iii. May become hazardous to any person entering it owing to

    a. Its design, construction, location or atmosphere,

    b. The materials or substances in it, and

    c. Any other conditions relating to it

Isolate—it means to physically interrupt or disconnect pipes, lines and sources or energy from a confined Space.

3.24.2 Regulatory Requirements

- Identify the confined space and the associated hazards
- Avoid entry into a confined space if possible
- Ensure there is a safe entrance and exit
- Hazards are assessed and monitored by a competent person
- Entry permit must be completed
- Equipment necessary for a rescue is readily available and personnel are properly trained.

3.24.3 Hazard Identification

(a) Oxygen deficiency

Oxygen deficiency occurs in confined space when the oxygen has been consumed by the oxidation process of metals, bacterial action, combustion or the displacement of the oxygen by other gases.

Normal air contains 20.9% oxygen. An oxygen deficient atmosphere is one that contains less than 19.5% oxygen. Ventilation can be used to

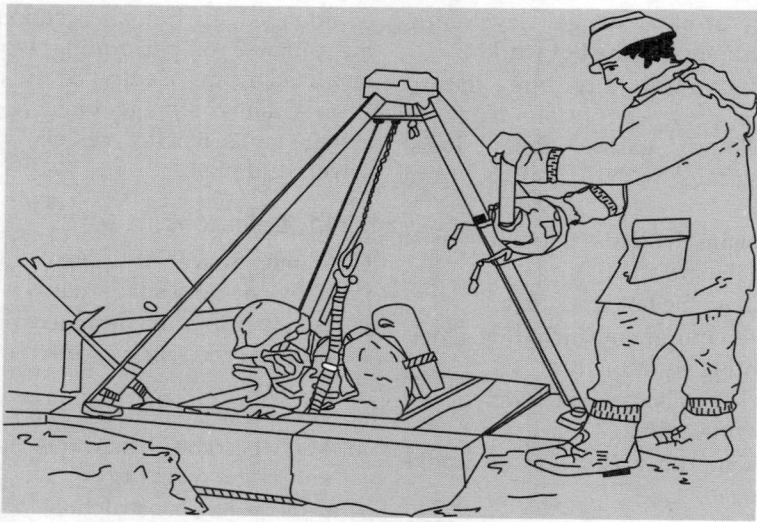

Fig. 3.48: Worker safely entering confined space wearing breathing apparatus and lifeline attached

help maintain a normal atmosphere. Workers shall not be permitted to work in any atmosphere that is less than 19.5% oxygen without an air supply. If the oxygen concentration is above 23% oxygen then there is an oxygen rich atmosphere and the area must be ventilated to reduce the concentration of oxygen.

(b) Combustible gases and vapors

Combustible gases and vapor hazards include naturally occurring gases and vapors used as fuel and solvents. Some of these materials vaporize quickly. When mixed with air they will burn or explode when ignited. This will cause a fire hazard and the gases could also be toxic. Purging the confined space and ventilating with fresh air will keep the gases and vapors below the lower explosive limits. The Material Safety Data Sheets (MSDS) will provide these limits. Ignition sources must also be eliminated.

(c) Toxic atmosphere

In confined spaces, toxic atmospheres can cause serious health effects and even death. The poisonous effects can be immediate (acute), delayed (chronic) or a combination of both. Contaminants of these kinds usually occur from material previously stored in the space or as a result of the use of coatings, solvents or preservatives. Decomposing organic matter can produce gases such as methane, carbon monoxide, carbon dioxide and hydrogen sulfide. Some have color and odor but human senses cannot be relied upon for detection of these gases.

When evaluating the toxic hazard of a chemical or a contaminant it is important to know the route of entry into the body. Materials can enter the body through inhalation, ingestion, absorption and injection. Knowing the route of entry will assist with deciding the type of PPE required for the worker, e.g. gloves and respirator.

(c) Other

There are numerous other hazards that can be associated with a confined space. Some of these include: radiation, vibration, noise, extreme temperatures, slippery surfaces, cramped quarters, electrical shock, moving equipment and sharp edges. The presence of these hazards must be evaluated and safe guards put in place to protect the worker.

3.24.4 Hazard Controls

(a) Isolation and lockout

Any confined space must be isolated from the possible release of hazardous substances into the work area and the start up of any equipment associated with the confined space. There are only two acceptable methods of isolating the flow of material from pipes into a confined space. The first is using a Blank in the pipeline and the second is the Double Block and Bleed Method. If there is a possibility of a startup of equipment then the equipment must be locked out using the "Lock-out/Tag-out Procedure".

(b) Ventilation

Before any work starts, consideration must be given to ventilation. This may be through natural ventilation through clean-out doors or mechanical means. Ventilation is used to maintain a safe breathing atmosphere while workers are inside. It can be used to displace or dilute potentially hazardous conditions before they become dangerous.

Natural ventilation might not be sufficient alone because air will travel along the path of least resistance and the air might not reach all areas in the confined space. Air must be forced into the confined space to be effective. The ventilation of the confined space must continue until the work is complete and must not be interrupted while the worker is in the area. If the ventilation of the area stops then the workers must leave the area until ventilation resumes.

(c) Entry permit

The Entry Permit is an administrative tool used to document the completion of a hazard assessment for each confined space entry. All work done in a confined space must be done under a "Confined Space Entry Permit". The purpose of the permit is to ensure that all necessary precautions are taken before any confined space entry is made. The type of operation to be performed in a confined space will determine the controls necessary to do the work safely and protect the health of the workers. The permit outlines the precautions necessary before entering the confined space. The permit will contain job specific information and is only valid on the day that it was issued. The entry permit is unique for each confined space to be entered and must remain at the confined space entrance until the completion of the job.

(d) Personal protection equipment (PPE)

Each confined space has its own particular hazards. There could be any number of physical hazards like slipping or physiological hazard such as a toxic gas. The most common PPE worn are for: eye protection, head protection, foot protection, respiratory protection, safety belts and lanyards, lifelines and fall arrest devices. The PPE could vary from one confined space to another.

(e) Accident/Incident reporting

Reporting accidents or incidents is an important key to allow the prompt correction of hazardous situations that could have serious affect on the health and safety of individuals. If an accident or incident occurs the worker should follow the "Accident/Incident Reporting Procedure".

(f) Stand-by person

A stand-by person has a critical role to play in confined space work. The stand-by person is the only one who will be able to help or begin a rescue operation from outside the confined space.

Some things a stand-by person should do

- Know the hazards associated with the confined space
- Barricade the entrances to the confined space
- Ensure the permit has been completed and posted at the entrance
- Be aware of the preentry air monitoring results
- Know the workers and how many will be in the confined space
- Have the knowledge of proper PPE is being used by the workers
- Ensure there is a means of communication with the workers inside
- Ensure all rescue equipment is at the confined space.
- Have accessed the training of rescue procedure
- Have the well knowledge of using lifelines
- Never leave the station unless the workers are out of the confined space.

3.24.5 Entry and Rescue Procedures

(a) Entry procedure

Everyone working in a confined space must be properly trained in safe entry and rescue procedures and must have a knowledge and understanding of their equipment and potential hazards that exist. A "Confined Space Entry Permit" must be completed and issued prior to entry into the confined space. The work space must be made safe for the workers. The workers entering the confined space must have the proper PPE as indicated by the Entry Permit. A system of communication must be established between the worker and the stand-by person.

(b) Rescue procedure

Rescue procedures must be established prior to any worker entering the confined space. This is done so that injured workers may be quickly and safely removed. The time required for the rescue shall calculated before entry. Doing this ensures workers know their assigned tasks and have the equipment ready if needed to perform an emergency rescue. The type of rescue equipment and procedures will vary with each confined space. The workers must be trained on the proper use of all safety and rescue equipment.

Rescuers must not enter a confined space unless others are standing by in the event of additional help is required in the rescue.

3.24.6 Managers and Line Management Responsibility

- Identify the types of confined space that a worker may be required to enter
- Identify the types of hazards that are or may be present at each confined space
- To facilitate or provide proper instruction to their workers on protection requirements including confined space entry and emergency Egress Procedures
- Identify alternative means to perform the work that will not require the worker to enter the confined space
- Ensure there is a safe entrance and exit from the confined space

- Prevent any unauthorized entry into a confined space
- Before a worker enters a confined space, a competent person is appointed to assess the hazards and test the atmosphere for hazards
- Ensure a Confined Space Entry permit is completed for each entry
- Post the permit at the work area
- Provide training on the hazards and the proper use of PPE.

3.24.7 HSE Department Responsibility

- Assist in the identification of the confined space and the hazards
- Perform atmospheric testing to determine the presence of hazards in the confined space upon request from the Manager or Supervisor
- Train the workers engaged in the confined space work according to the safe work procedure and Emergency Egress Procedures.

3.24.8 Workers Responsibility

- Must be competent in confined space entry to identify the work procedures required to enter the confined space
- Ensure there is reasonable means exit from all parts of the confined space
- Ensure that ventilation and purging is established and allows acceptable air levels to be achieved and maintained
- Establish method of communication to allow immediate contact with necessary personnel if rescue or assistance is required, confirm alarm system
- Shall have trained in different gas concentrations (if required)
- Before entry, the vessel or confined space must be tested by a competent worker wearing breathing apparatus, for oxygen content, combustible gas (LEL) and hydrogen sulfide
- Continuous monitoring may be required of the vessel or confined space atmosphere
- Must be conversant with Rescue Procedures.

3.25 DEMOLISHING WORK

Demolition is the dismantling, wrecking, pulling down or knocking down of any building or structure or part thereof, but does not include such work of a minor nature which does not involve structural alterations.

A company carrying out demolition work shall

a. Develop and implement safe work procedures for demolition work including plant and equipment used
b. Ensure that workers involved in demolition work are trained in those safe work procedures
c. Ensure that workers involved in demolition work comply with those safe work procedures.

An employer must ensure that before demolition work begins the following are removed from the building or structure, or part of the building or structure, being demolished

a. Glass, metal cornices or other material that may shatter
b. Hazardous substances, including asbestos
c. Any tanks, wells, windows, doors, piping systems, flammable or explosive materials or gas cylinders
d. Any live electric power supply connections
e. Any unauthorized personnel. For the purpose barricade the surrounding area with solid materials.

An employer must ensure that a competent person

a. Inspects the demolition site and confirms that subsection has been complied with before proceeding with demolition work
b. Supervises the demolition work at all times when demolition work is in progress.

When carrying out demolition work, an employer must ensure that

a. Adequate ventilation is provided for a machine powered by an internal combustion engine that operates in an enclosed area
b. Every floor, roof or other surface is of sufficient strength to safely support any of the following loads:

i. The load of a worker who is required or permitted to be on it
ii. The load of any equipment, including powered mobile equipment, placed on it.

c. Material or debris is removed promptly and is not allowed to accumulate:

i. In an area that might result in the collapse of all or part of the building or structure due to overloading
ii. On the ground immediately outside of the building or structure being demolished.

d. Unless it is being demolished at the time, no wall or other part of the building or structure is left unstable or in danger of collapsing
e. An employer must ensure that a chute provided under subsection:

i. Is properly secured
ii. Has all of its openings adequately guarded and empties into an area that is appropriately barricaded to prevent access by any person
iii. Has placed adjacent to its outlet, in a conspicuous location, a clearly visible and legible sign bearing the wording "DANGER, NO ENTRY" in a minimum of 5 cm letters.

f. Workers and equipment are located clear of any falling material or debris
g. When carrying out demolition work, an employer must ensure that no crane boom is used for pushing or pulling a building or structure.

When a part of a building or structure is demolished by pulling and pushing

a. This has to be done under the supervision of a competent Engineer
b. The horizontal distance from the machine used for the pulling to the face of the part being pulled is at least 20% greater than the height of the part
c. No person stands between the machine and the building or structure
d. At least two-thirds up the height of the part, measured from the base of the part being pushed.

When an explosive is to be used to carry-out the demolition work, an employer must ensure that

a. A competent person develops a demolition procedure to protect the safety and health of workers

b. The worker who uses the explosives does so in accordance with the requirements.

3.26   SAND BLASTING

Abrasive blasting involves forcefully projecting a stream of abrasive particles onto a surface or metals, usually with compressed air or steam. Silica sand is commonly used in this process. Workers who perform abrasive blasting are often known as sandblasters.

Health Affects

When workers inhale the crystalline silica used in abrasive blasting, the lung tissue reacts by developing fibrotic nodules and scarring around the trapped silica particles. This fibrotic condition of the lung is called silicosis. If the nodules grow too large, breathing becomes difficult and death may result. Silicosis victims are also at high risk of developing active tuberculosis. The silica sand used in abrasive blasting typically fractures into fine particles and becomes airborne. Inhalation of such silica appears to produce a more severe lung reaction. This factor may contribute to the development of acute and accelerated forms of silicosis among sandblasters. These hazards can be avoided by a great extend by wearing proper personal protective equipment.

Type of PPE's required

a. Abrasive blasting gloves
b. Appropriate respirators
c. Body shields
d. Aprons
e. Non-slip and steel-toed shoes
f. Full eye protection
g. Full-body jump suits for dust protection
h. Hard hats
i. Ear protectors
j. Hairnets
k. Foot guards.

Training element requires for sandblasters

1. Information about the potential health effects of exposure to crystalline silica
2. Material safety data sheets for silica, masonry products, alternative abrasives, and other hazardous materials
3. Instruction about the purpose and set-up of regulated areas marking the boundaries of work areas containing crystalline silica
4. Information about safe handling, labeling, and storage of toxic materials
5. Discussion about the importance of substitution, engineering controls, work practices, and personal hygiene in reducing crystalline silica exposure
6. Instruction about the use and care of appropriate protective equipment (including protective clothing and respiratory protection)
7. Safe use of sandblasting equipment or machineries
8. In emergency egress procedures.

General requirements

1. Sandblasting shall be conducted according to the approved safe work procedure and manufacturers specifications
2. Sandblasting work shall be started after accessing the Permit-to-work systems
3. Sandblasting work shall be strictly supervised by a competent person
4. Sandblasting work shall be done by competent persons
5. Operatives shall wear all required PPE's
6. Unauthorized entries to sand blasting areas have to be strictly prohibited. For the purpose signage's and securities to be placed
7. Nets to be placed around the sandblasting area for the prohibition of chemical penetration to the surrounding areas
8. Precautions shall be activated for noise and vibration hazards
9. Ensure proper sized air supply with regulated pressure and functional shut off readily accessible
10. In extremely hot conditions a ventilation air hose may be needed attached to the blaster's hood

11. Ensure proper size and length of air hose and 'o' clips to secure twist lock connectors

12. Follow sandblasting safe work procedure step by step

13. The operative shall well aware of emergency procedure to follow if any emergency situation arrives

14. Any open inlets to drains shall be covered to prevent ingress of chemicals to the drainage systems

15. Waste must be collected and removed from the site at regular intervals.

3.27 PAINTING WORKS

Paints are chemical mixed liquids or solvents. Some may contain a wide range of toxic, corrosive and flammable substance. The harmful effects are the paints can entry into the body through inhalation, skin absorption and ingestion. Some paint solvents can penetrate the skin and exert harmful effects or diseases. The repeated use of white sprit or other solvent often leads to dermatitis. Water based paints and emulsions are much less hazardous unless they contain toxic fungicide or similar preparations. Paints or coatings containing led or zinc or any other metal shall not to be used in industrial, domestic or administrative buildings because it can lead to serious respiratory diseases to the people belong there. So special care has to be maintained while stacking, transporting, handling and working with paints.

MSDS of these chemical tells exactly what its hazards is. Wearing PPEs like overalls, gloves, eye protector and respiratory protector shall minimize most of its hazards. Safe work procedures for the use of these chemicals shall be updated and to be followed. Workers have to be trained according to the safe work procedure prior starting work activities and to enforced to wear all relevant PPE. Smoking to be prohibited in paint work or stacking areas. Workers shall not allow eating and drinking with paint contaminated hands or in the areas where painting or spraying is underway. Proper ventilation is required to reduce levels of solvents vapor, paint powder or dust, so as to eliminate the problem of health hazards, fire or explosion.

General requirements on paint stacking or working area

1. "No Smoking" and "Flammable Material, Keep Away" signage's shall be displayed

2. Prestart safety talks shall be conducted among the worker for remembrance of hazard and precautions to be taken

3. All combustible materials shall be kept away and fire watchers to be appointed

4. Adequate charged Fire extinguishers shall be placed in these locations

5. Adequate ventilation shall be provided to prevent vapors to build up above Lower Explosive Limit

6. Ensure strict supervision at all time basis

7. Ensure workers are wearing PPE at all times

8. All painting booth shall be built with fire-resistance material and provided with adequate ventilations

9. Housekeeping to be carried out on regular basis

10. Good hygienic practice to be built up among the workforce for avoiding diseases

11. MSDS shall be displayed on stacking areas

12. If a worker faces any problem, immediately report to the concern supervisor or safety department.

3.28 PRECAST CONCRETE STRUCTURE ERECTION WORKS

Precast concrete beams or slabs or stairs are prefabricated or made according to the length and breath in workshop or off-site and fix to the concern position on constructing buildings by means of mechanical aid or manually. In our current environment, speed is a critical aspect of any construction project. The use of precast allows not only the speedy erection of the structure, but also flexibility and overall program shortening. This is achieved by allowing the production of components at the same time the footing system is being prepared. Precast structures have been shown to be extremely cost effective, durable, stable, and of the highest quality and strength.

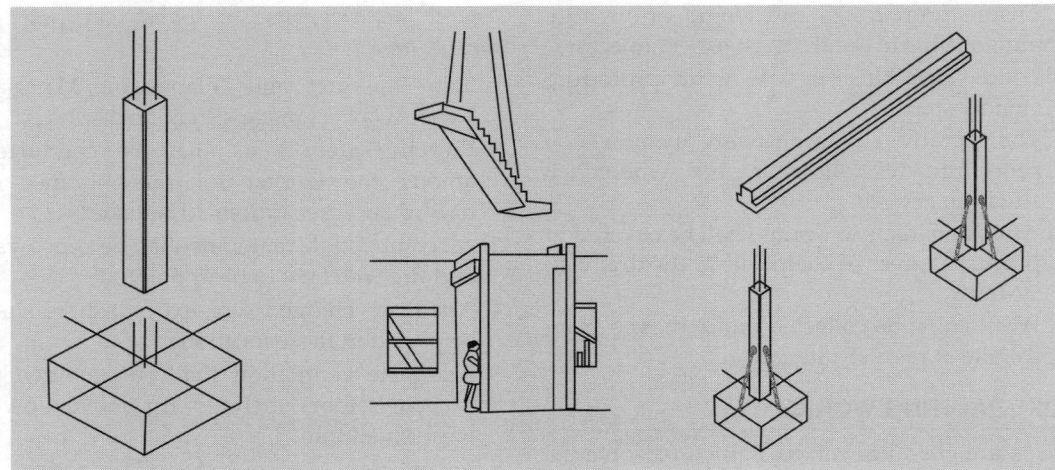

| Precast slab erection | Precast stair erection | Precast erection details |

Fig. 3.49

General requirements

1. Prefabricated parts shall be designed and made as they can be safely transported and erected

2. Lifting plate and equipment used shall have valid certificate and have a licensed and competent operator

3. Use only the concern lifting capacity plat and equipment as per lifting regulations.

4. Lifting shall be done, if a banksman signals the operator to do so

5. Safe work procedures for the erection of precast structures shall be updated and to be followed

6. Before lifting any precast structure a QA. QC Engineer has to certify that it is safe for lifting

7. Prestart safety talks shall be conducted among the worker for remembrance of hazard and precautions to be taken

8. All workers has to wear appropriate PPEs such as helmet, gloves, safety shoes, eye protectors and safety harness where required

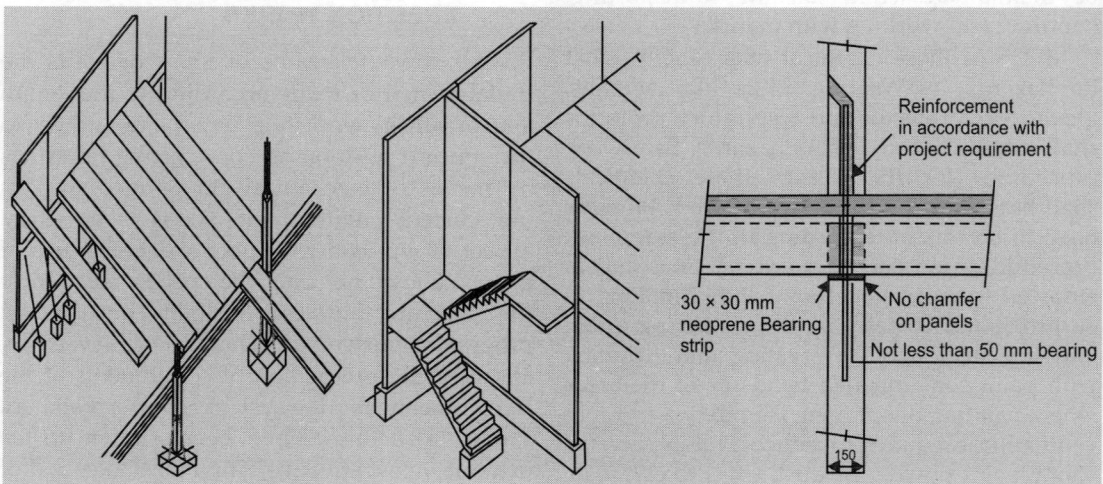

Reinforcement in accordance with project requirement

30 × 30 mm neoprene Bearing strip

No chamfer on panels

Not less than 50 mm bearing

150

Fig. 3.50

9. Erection shall be done under strict supervision (Engineers or Supervisors standing on work phase at all time bases)

10. No person is allowed to work under the suspended precast structures and the area below to be barricaded

11. Gangways and scaffolding as the access for the worker to the erection point to be build with strong materials and as per safe working standard

12. While lifting or erecting precast structures all protruding parts to be removed. If cannot be removed then it shall be caped or safely bended for avoiding laceration or cuts to fellow workers

13. While in heavy windy condition the lifting operation must be stopped for avoiding the load from spinning or overturning which can turn to major accident.

3.29 FORMWORK AND STEEL REINFORCED CONCRETE WORKS

Steel fixing on the surface where beams, columns and slabs are planned before pouring concrete is steel reinforcement. Formwork means erecting the shutters as a mold for the concrete of structures as planned and designed after steel reinforcement.

The main risks are

- People falling during steel-fixing and erection of formwork
- Collapse of the formwork or false work
- Materials falling while striking the form-work
- Manual handling of shutters, reinforcing bars, etc.
- Silica dust from scrabbling operations
- Arm and back strain for steel-fixers
- Cement burns from wet concrete.

While on steel reinforcement and formwork ensure that

- A health and safety method statement has been agreed before work starts and that it is followed.
- Guard rails or other suitable barriers to prevent falls are put in place as work progresses.

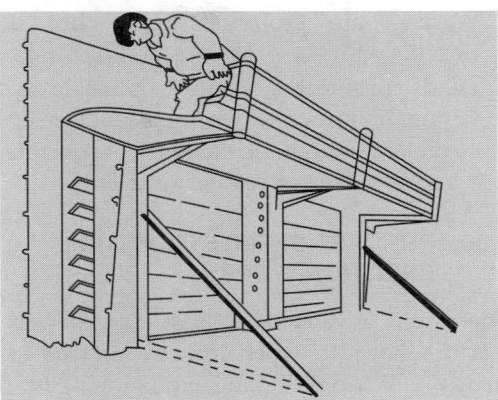

Formwork of a vertical wall

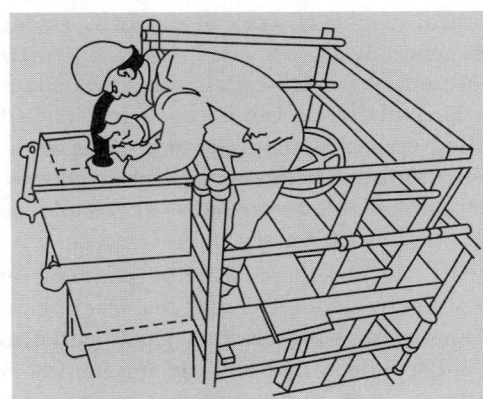

Formwork of vertical beam

Fig. 3.51

- Workers have safe access to the work - it is not safe to stand on primary or other open timbers.
- A ladder or a tower scaffold is used for access.
- Ladders are tied for climbing up vertical sections of reinforcement or the wedges of column forms should not be permitted.
- Equipment is in good order before use. Do not use substitutes for the manufacturers pins in adjustable props.
- The formwork, false work and temporary supports are checked, properly tied, footed, braced and supported before loading and before pouring walls or columns.
- Workers are protected from wet concrete provide with gloves and Safety boots.

- Workers are protected from silica dust provide with eye protectors and respirators.
- Loads are spread as evenly as possible on the temporary structure. Do not place large loads of timber, reinforcing bars or wet concrete in a localized area, spread loads evenly.

3.30 PRECAST STEEL STRUCTURE ERECTION WORKS

There will always be a high risk of falls if erectors have to work directly off the steel structure. When designing and planning for the erection of steel frames the first consideration should be to eliminate or reduce the need to work at height. If work at height is essential it is generally safer if other access, e.g. mobile platforms or tower scaffolds or other suitable working platforms can be used for access for bolting-up and similar operations. Prepare and level the ground before work starts to allow safe use of mobile platforms or tower scaffolds and safe standing for a crane.

This work requires careful planning and execution by a competent Mechanical Engineer. Whenever possible erectors shall work from an independent platform or mechanical aid platform. If it is not possible erectors shall move through steel beams by sitting on it. But the erectors should wear a full body harness attached to either the steel or a running life line.

There is also much potential to reduce risks during design and planning by

1. Ensuring erection is sequenced so that stairs and handrails can go in as early as possible to provide safe access to high levels of the structure.
2. Designing connection joints to make bolting-up easy.
3. Adding bracing into the design to ensure integral stability of the structure through all stages of erection.
4. Ensuring adequate information is passed on to alert erectors about special sequences which need to be followed to ensure stability.

Fig. 3.52: Safe movement through steel beams

The main hazards to be controlled on site while erecting steel structures are

- Falls when working at height
- Erectors being hit or knocked off the steel by moving steel members or decking packs being craned into position
- The structure collapsing before it is fully braced
- Materials dropping onto people working below
- The manual lifting of heavy steel members, causing back and other strains and injuries
- Cranes overturning.

Before work starts, plan safe working methods

- Plan for good access onto the site and proper standing areas for delivery vehicles, cranes, mobile platforms and tower scaffolds.
- Arrange for materials to be stored safely.
- Agree a safety method statement and ensure it is followed.
- Erection shall be done under strict supervision (Engineers or Supervisors standing on work phase at all time bases).
- Lifting plat and equipment used shall have valid certificate and have a licensed and competent operator.
- Lifting steel structures shall be done, if a banksman signals the operator to do so.

- Program work to make sure other trades do not have to work beneath the erectors, so avoiding the risk of them being injured by dropped materials.
- Arrange for safe working at height using mobile work platforms, tower scaffolds or another form of independent access if possible.
- Where erectors work directly from the frame, make sure they wear a harness and lanyard which is connected to the steel or provide other means (such as nets) to arrest their fall.
- Make it clear that workers must not walk on the top flange of steel beams.
- Ensure that all necessary materials (including braces and fixings) are delivered to the site in the correct sequence.
- Check temporary bracing to ensure stability has been provided.

Workers fall not only during the erection of the frame, but also when decking sheets are being handled. People often fall

- During the landing and splitting of decking packs
- When decking sheets are being moved around the frame
- During the laying of decking sheets
- From edges of decked areas (including leading edges).

To prevent these falls

- Position guard rails at all fixed edges and openings.
- Where possible, store and split packs of decking sheets at ground level or on a previously decked area of the frame provided with suitable guard rails.
- Position decking sheets from mobile access platforms or tower scaffolds, or develop a safe system of work to prevent falls from the leading edge. Harnesses and lines may be required.

3.31   ROOF WORK

Almost one in five workers killed in construction accidents are doing roof work. Most of these are specialist roofers, but some are simply involved

in maintaining and cleaning roofs. Some of these workers die after falling off the edges of flat and sloping roofs. Many other workers die after falling through fragile materials. Many roof sheets and roof lights are, or can become, fragile. Asbestos cement, fiber-glass and plastic generally become more fragile with age. Steel sheets may rust. Sheets on poorly repaired roofs might not be properly supported. Any of these materials could give way without warning. Do not trust any sheeted roof. Do not stand directly on any of them. On a fragile roof, never try to walk along the line of the roof bolts above the purling, or along the roof ridge, as the sheets can still crack and give way.

Roof openings and roof lights are an extra hazard. Some roof lights are difficult to see in certain light conditions and others may be hidden by bituminous paint. Openings and lights can be protected by barriers or with covers either secured in place or labeled with a warning.

Guard rails and toe boards or suitable barriers erected at the edge or eaves level of a roof are usually needed to stop people and materials from falling off. On fragile and most sloping roofs, purpose made roof ladders or crawling boards to spread the weight of workers and materials will be required.

If work is going to be done on any roof, make sure there is

- Safe access onto and off the roof, for example, a general access scaffold
- A safe means of moving across the roof is roof ladders or crawling boards
- A safe means of working on the roof is a guard railed platform
- Do not throw materials such as old slates, tiles etc from the roof or scaffold may strike someone passing by. Use enclosed debris chutes or lower the debris in containers
- Agree a safety method statement and ensure it is followed
- Strict supervision shall be provided at all time bases

Fig. 3.53: A safe scaffold platform on a pitched roof

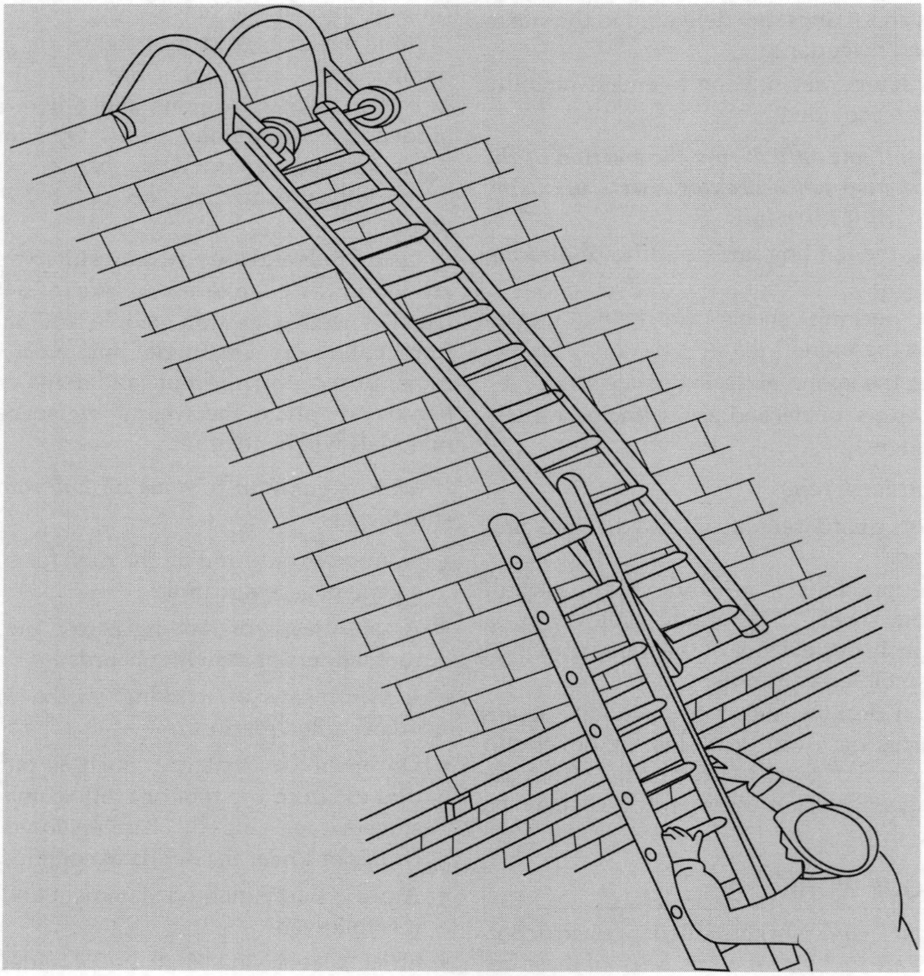

Fig. 3.54: Ladder with handrail access on roof

- All workers have to wear appropriate PPE's where required
- If a worker faces any problem, immediately report to the concern supervisor or safety department.

3.32 GAS WELDING SAFETY

3.32.1 Main Components of Oxygen/fuel Gas Equipment

- Cylinders of oxygen and fuel gas (propane or acetylene)
- A means to shut off or isolate the gas supply, usually the cylinder valves
- A pressure regulator fitted to the outlet valve of the gas cylinder, used to reduce and control gas pressure
- A flashback arrester to protect cylinders from flashbacks and backfires
- Flexible hoses to convey the gases from the cylinders to the blowpipe
- Non-return valves to prevent oxygen reverse flow into the fuel line and fuel flow into the oxygen line
- A blowpipe or other burner device where the fuel gas is mixed with oxygen and ignited.

3.32.2 Main Hazards

The main hazards are from fire and explosion. These are caused by:

- Careless handling of a lighted blowpipe resulting in burns to the user or others using the blowpipe too close to combustible material
- Cutting up or repairing tanks or drums which contain or may have contained flammable materials
- Gas leaking from hoses, valves and other equipment
- Misuse of oxygen
- Backfires and flashbacks.

3.32.3 Fire and Preventing Fire

The flame from an oxygen/fuel gas blowpipe is a very powerful source of ignition. Many fires have been caused by the careless use of oxy/fuel blowpipes. The flame will quickly ignite any combustible material it comes into contact with: wood, paper, cardboard, textiles, rubber, plastics. Many processes also generate sparks and hot spatter which can ignite these materials.

Preventing fire

The following precautions will help to prevent fire:

- Allow to start work only after accessing permit to work systems
- Move the work place to a safe location for carrying out the hot work process
- Remove any combustible materials (such as flammable liquids, wood, paper, textiles, packaging or plastics) from within about 10 meters of the work
- Ventilate spaces where vapors could accumulate, such as vehicle pits or trenches
- Protect any combustible materials that cannot be moved, from close contact with flame, heat, sparks or hot slag. Use suitable guards or covers such as metal sheeting, mineral fiber boards or fire retardant blankets
- Check that there are no combustible materials hidden behind walls or partitions which could be ignited, particularly if prolonged welding or cutting is planned. Some wall panels contain flammable insulation materials
- Use guards or covers to prevent hot particles passing through openings in floors and walls (doorways, windows, etc.)
- Maintain a continuous fire watch during the period of the work, and for at least an hour afterwards
- Keep fire extinguishers nearby.

3.32.4 Preventing Injury

The following precautions will help to prevent injury:

- Work in a safe location away from other people
- Wear protective clothing, gauntlets and eye protection
- Shut off the blowpipe when not in use. Do not leave a lighted blowpipe on a bench or the floor as the force of the flame may cause it to move

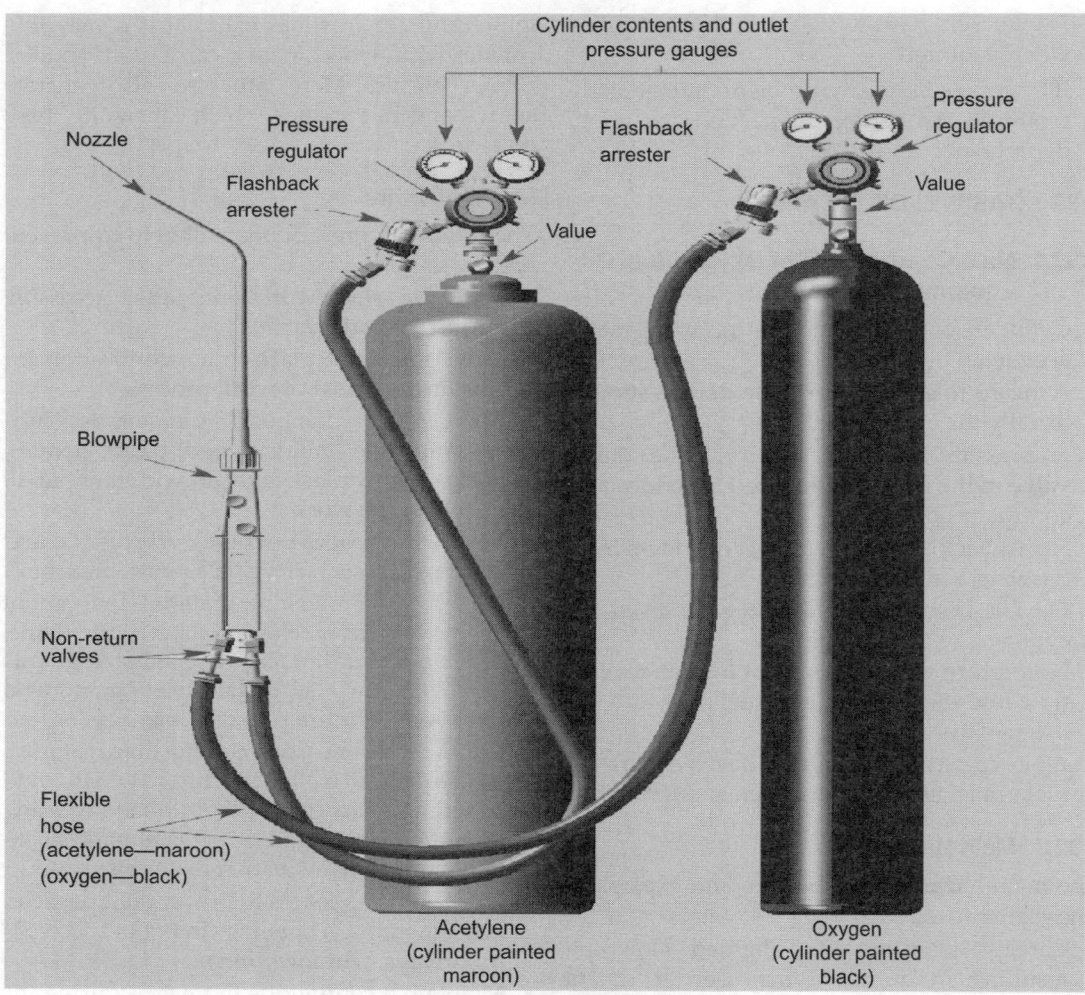

Nozzle

Pressure regulator

Flashback arrester

Cylinder contents and outlet pressure gauges

Flashback arrester

Pressure regulator

Value

Value

Blowpipe

Non-return valves

Flexible hose (acetylene—maroon) (oxygen—black)

Acetylene (cylinder painted maroon)

Oxygen (cylinder painted black)

Fig. 3.55: Components of gas welding set

- Clamp the work piece, do not hold it by hand
- Keep hoses away from the working area to prevent contact with flames, heat, sparks or hot spatter.

3.32.5 Preventing Explosion

Flammable liquids and vapors such as petrol, diesel, fuel oil, paints, solvents, glue, lacquer and cleaning agents are found in many places of work. If a welding blowpipe or burner is used on a tank or drum containing flammable material (solid, liquid or vapor), the tank or drum can explode violently. People have been killed and seriously injured by such explosions.

Tanks and drums that are 'empty[1] usually still have residues in the bottom, and in seams and crevices. Just a teaspoon of flammable liquid in a drum can be enough to cause an explosion when heated and turned into vapor.

Similarly, you must never weld or flame cut wheels to which tires are fitted. The heat may generate flammable vapor from any oil or lubricating fluid on the inner rim of the wheel. This vapor, confined by the tire may be enough to cause an explosion, if ignited. These explosions are very violent and can kill. Always remove the tire.

You must never use an oxy/fuel gas blowpipe on a drum or tank that has contained or may have contained flammable material unless you know it has been made safe. If it contains flammable material, it will need thorough cleaning. It may be safer for a specialist company to carry-out the work.

3.32.6 Preventing Gas leaks

There is a risk of fire and explosion if oxygen/fuel gas equipment is allowed to leak. Acetylene and other fuel gases are highly flammable, and form explosive mixtures with air and oxygen. Even small leaks can have serious consequences, particularly if they are leaking into a poorly ventilated room or confined space where the gases can accumulate. A leak of flammable gas could cause a flash fire or explosion. Gas leaks are often the result of damaged or poorly maintained gas control equipment, hoses, blowpipes and valves, poor connections and not closing valves properly after use.

Preventing leaks

The following precautions will help to prevent leaks:

- Keep hoses clear of sharp edges and abrasive surfaces or where vehicles can run over them,
- Do not allow hot metal or spatter to fall on hoses,
- Handle cylinders carefully. Keep them in an upright position and fasten them to prevent them from falling or being knocked over. For example, chain them in a wheeled trolley or against a wall,
- Always turn the gas supply off at the cylinder when the job is finished,
- Maintain all equipment and keep in good condition,
- Regularly check all connections and equipment for faults and leaks.

Oxygen leaks also increase the fire risk. In particular, if clothing is contaminated with oxygen, it will catch fire easily and burn very fiercely resulting in severe injury. Even fire retardant clothing will burn if contaminated with oxygen. Also oxygen can cause explosions if use d with incompatible materials. In particular, oxygen reacts explosively with oil and grease.

You should always take the following precautions

- Never allow oil or grease to come into contact with oxygen valves or cylinder fittings,
- Never use oxygen with equipment which is not designed for it. In particular, check that the regulator is safe for oxygen and for the cylinder pressure.

3.32.7 Ventilation

Small leaks may not be detected immediately. If they leak over a period of time into a poorly ventilated room or confined space, a dangerous concentration of gas may accumulate.

To prevent gas accumulating

- Always provide adequate ventilation during welding and cutting operations
- Store gas cylinders outside whenever possible or in a well-ventilated place
- Avoid taking gas cylinders into poorly ventilated rooms or confined spaces.

3.32.8 Back Fires, Flashback and Arresters

A backfire is when the flame burns back into the blowpipe often with a sharp bang. This may happen when the blowpipe is held too close to the work piece, or if the nozzle is blocked or partly blocked. The flame may go out or it may re-ignite at the nozzle. Sometimes the flame burns back into the blowpipe, and burning continues at the mixing point. Backfires do not usually cause serious injury or damage but they indicate a fault in the equipment.

If a backfire does occur

- Shut off the blowpipe valves, oxygen first and then the fuel gas
- Shut off the oxygen and fuel gas cylinder valves
- Cool the blowpipe with water, if necessary
- Check the equipment for damage or faults, particularly the nozzle.

Flashbacks

Flashbacks are commonly caused by a reverse flow of oxygen into the fuel gas hose (or fuel into the oxygen hose), producing an explosive mixture in the hose. The flame can then burn back through the blowpipe, into the hose and may even reach the pressure regulator and the cylinder. The consequences of a flashback are potentially very serious. They can result in damage or destruction of equipment, and could even cause the cylinder to explode. This could end in serious injury to personnel and severe damage to property.

To protect a cylinder, you should fit flashback arresters onto the regulator, on both the fuel and oxygen supply. Arresters may be fitted on the blowpipe but these do not give protection from a fire starting in the hose. For long lengths of hose, you should fit arresters on both the blowpipe and the regulator.

If a flashback does occur

- Immediately close the cylinder valves, both fuel gas and oxygen, if it is safe to do so. The flame should go out when the fuel gas is shut off. If the fire cannot be put out at once, evacuate the area and call the emergency fire services.
- The blowpipe, hoses, regulators, flashback arresters and other components may have been damaged. Check carefully and replace if necessary before reuse. If in doubt, consult the supplier.

3.32.9 Gas Cylinder Storage

After welding work the cylinders shall be return to the cylinder stacking area. Cylinder stacking area shall be:

- Cylinder stacking area shall be in separated compartments and shall built with the requirements of local authorities and even client. Re-filling shall not allow on work premises.
- Separate racks shall be provided for different gas cylinders. Empty cylinders shall be keep in different racks.
- Empty or full cylinders shall be stacked in up-right position and shall chain them to the existing Structure or beams for arresting cylinder fall. Unauthorized entries shall be prohibited.

- Shall paste safety signage's and place appropriate fire extinguishers on entrance. Prohibit smoking on these area.

3.33 RADIOGRAPHY TEST

Radiography involves the use of penetrating X or Gamma radiation to examine parts and products for imperfections. An X-ray machine or radioactive isotope is used as a source of radiation. Radiation is directed through the inspection object and onto film or an electronic detector producing a shadow graph that shows the internal structure of the inspection object. Possible imperfections show up as density changes in the image, the same way a medical X-ray can show broken bones. All personnel performing radiographic tests or interpreting the films under this Specification shall be qualified and authorized. The qualification record of each radiographer must be complete. It must include the name of the authorized entity participating in such qualification and a description of all the procedures for which it has been approved. Such certificate shall be valid for 3 years.

Radiography experts and any other personnel involved in radiographic activities shall carry a dosimeter film and a pocket dosimeter. Care shall be taken to prevent personal radiation from exceeding the limits established by the regulations in safe work procedure. A Geiger meter shall be available for all gamma ray or X ray equipment. When gamma ray source is not being used, it shall be kept in a metallic locked container with radioactive signs and access shall be permitted to radiography personnel only. The Radiography Contractor shall have an Emergency Procedure for accidents occurred during the performance of radiographic tests. Signs as flashing lights, flags, etc will be established at safe intervals to warn all personnel when radiography is in progress.

General safety requirements are

- Radiography test shall be conducted according to the approved safe work procedure.
- Radiography test shall be carried out by taking approval from client and consultation by taking proper precautions.
- Only authorized and certified personnel shall be allowed to carry radiography activities.

- Technician operating the radiography equipments shall fully aware of radiography rays radiation precaution.
- Radiography sources shall be stored in a designated place and protected and to barricade properly.

3.34.2 Site Tidiness

Plan how the site will be kept tidy. In particular, walkways and stairs should be kept free of tripping hazards such as trailing wires and loose materials. This is especially important for emergency routes. Remove nails from loose

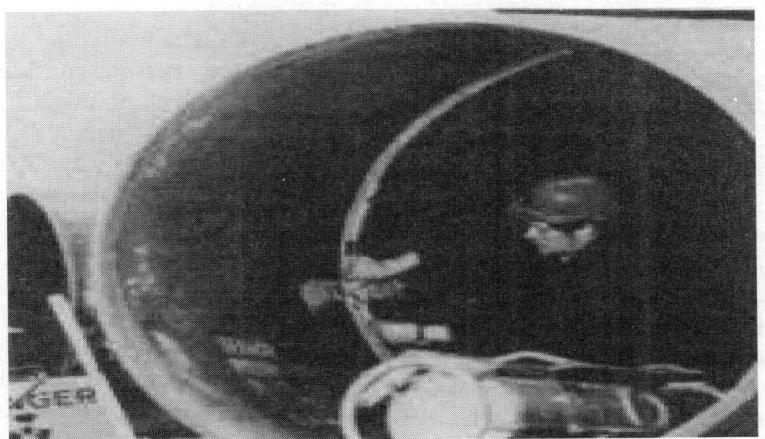

Fig. 3.56: Authorized person conducting radiography test

- Safety signage's and precaution procedures shall be placed of radioactive work premises.
- Area shall be barricaded for restricting unauthorized persons. For the purpose security person shall be appointed.

3.34 COMMON MATTERS ON CONSTRUCTION PREMISES

3.34.1 Falling Materials

It is important to provide toe boards on working platforms to stop materials being kicked, or rolling off platforms. Fans, netting, hoardings or protected walkways may also be needed to protect workers, site visitors and the public where there is particular risk from falling materials. Some processes such as demolition and cutting operations may throw materials out from the work area. Again, fans, hoardings, covered walkways, etc. may be needed to protect against this risk. Make sure there is a system to maintain necessary protection.

timbers to prevent foot and other injuries. Clear paper, timber off-cuts and other flammable materials to reduce fire risks.

3.34.3 Doors and Gates

As part of providing safe access and egress, it may be necessary to provide doors and gates. These can range from the existing doors within a house under refurbishment to the installation of a new door or gate as part of a new building. It may be necessary to fit devices to doors and gates which allow them to be used properly. Doors which open onto traffic routes may need vision panels or windows. Sliding doors and gates will need stop ends to prevent them coming off their runners or tracks. Rising doors or gates will need a means of preventing them from falling back.

3.34.4 Site Rules

Clients may insist on certain safety precautions, especially where their business continues at the premises while construction work is in progress. It may assist everyone if site rules are applied.

Site rules might cover, for example, safety helmets, safety footwear, site transport, fire prevention, site tidiness, hearing protection or permit-to-work systems. Make sure everybody knows and follows any rules relevant to them.

3.34.5 First Aid

First aid can save lives, reduce pain and help an injured person make a quicker recovery. The number of a qualified first adder needed depends on the risk of injury and ill health on site. But first-adder is prohibited to prescribe medicines for any diseases and taking injection or stitching to any injured employees. Any site needs:

- A first aid box with enough equipment to cope with the number of workers on site
- An appointed person who knows how to contact the accident and emergency services
- Information telling workers the name of the appointed person or first aider and where to find them. Paste the names on site main notice boards is a good way of doing this.

3.34.6 Reporting Incidents, Accidents, Near-miss and Diseases

Supervisors must report to HSE department any accident, incident, near-miss and diseases which happen to work force for proceeding follow-up actions. However the control of the site must also report those accidents which involve a self-employed worker or member of the public. Details of requirements for accident or incident reporting are to investigate it and to find direct, indirect and root causes of event happen and to take remedial action to prevent the reoccurrence. Investigating accidents and near misses can help identify circumstances where more precautions are required or where procedures are not being followed.

3.34.7 Working Alone

If possible avoid working alone because of the cosideration of human failures, human errors and the unexpected disorders of physical and physiological functions of human body. It is strongly recommended that never work alone in confined spaces, over or near edges like lift shaf openings, near deep water source or wells and on live electrical installations or equipment.

If it is unavoidable to work alone these precaution shall be taken:

- Supervisor and the person engaged shall do themselves risk assessment for prepare and implement concern person rescue safe work procedure in case of any incident
- Plan the work carefully that one person can manage it
- Supervisor shall allocate the work to an experienced person considering his age, mental and physical capability
- Training and instruction shall be given to perform the job safely by supervisor
- Ensure that the work area got safe access, emergency escape routes and maintain proper time management system
- Ensure proper communication system is in place at all times with the alone working person either by phone or walky-talkies
- Ensure periodic supervisory check
- Provide the alone worker with manually operated alarms for alerting others if he is in any danger.

3.34.8 Working Near or over Water

Never work alone, always work with a team ensuring some person in the team are well aware of swimming and can rescue others incase of any incident originating. Provide the edges with guardrails, toe-boards and adequate warning signages. Ensure that HIRA and safe work procedure are in place prior committing the work activities. Supervisor shall conduct training and to give strict instruction to concern workers according to safe work procedure.

If the work area is high accidental nature, provide workers with safety life jackets, nets to arrest fall and rescue life lines. Work is planed near or over tidal water or fast flowing rivers, a power driven rescue boat shall be provided. Supervision shall be provided at all times basis in these areas.

3.34.9 Overhead Power lines

Contact with overhead electric lines is a regular cause of death and injury. Any work near electric distribution cables or railway power lines must be carefully planned to avoid accidental contact.

The most common operations leading to contact with overhead lines are:

- Handling long scaffold tubes
- Handling long metal roof sheets
- Handling long ladders
- Operating cranes and other lifting plant
- Raising the body or inclined container of tipper trucks
- Using MEWP.

Where work is to take place close to overhead lines, detailed precautions shall be discussed with client and legal authorities. Isolate the power supply by taking permission from authorities. PTW system to be accessed prior committing work activities. While crane, shovel or MEWP are working close to over head lines, goal post system to erected for preventing machine contact with the power lines. Strict supervision shall be provided at all times.

3.34.10 Risk Works

The hazard while committing risk work is that, the concern person can face fatality, disability, multiple fracture, lost time accident and severe injuries. Steps to be taken for avoid taking risk works is the overall responsibility of top management and line management. Client and consultant authorities shall take strict disciplinary action, if they found any employees committing risk works.

All work activity shall plan and implemented according to the safe work procedure. Supervisors shall be strictly instructed by concern engineer, to avoid entertaining his workers for taking short cuts. Always provide strict supervision to the area where inexperienced workers are working.

Safety personnel shall be enforced to give training according the safe work procedure among the work force for avoiding ignorance and for building up the expertness, professionalism and competency in their level of work. Supervisors shall conduct Toolbox talk to their work force at the beginning of every shift for the remembrance of hazard and precaution to be taken to each group of workers.

3.34.11 Safety Documentation

This is a record of information for the client, in the time of audits or end user. The planning supervisor ensures it is produced at the end of the project and is then passed to the client. It gives details of health and safety risks that will have to be managed during maintenance, repair, renovation or demolition.

Contractors should pass information on these matters which becomes available during the construction phase to the planning supervisor for inclusion within the file. The client should make the file available to those who will work on any future design, construction, maintenance or demolition of the structure. As per OSHA standards every safety file that made to be stored two years from the date of the document has been originated.

3.34.12 Monitoring Health and Safety

With any business activity checks need to be made from time to time to make sure that what should be happening is actually being carried out in practice, this includes health and safety. Make sure that everyone is fulfilling their duties. If a supervisor is nominated or a safety adviser is employed to visit sites and review safety, then make sure that:

- Do they report problems to the site manager and to the employer?
- Are matters put right?
- Do the same problems keep recurring?
- If there are problems, find out why?

Keeping a record of accidents, illnesses and treatments given by first adders will help to identify trends. New instructions may need to be issued or extra training provided. Act before there is an accident or Some ones health is damaged. If an accident happens, find out what happened and why. Minor accidents and near misses can give an early warning of more serious problems. Consider whether the accident would have happened if the work had been better planned or managed or employees had been better trained. Could site or company rules have been clearer or could plant and equipment have been better maintained. Do not just put the blame on human error or other people without thinking why the error was made.

3.35 WASTE MANAGEMENT SYSTEM

In Construction work waste is generated whenever the activities like constructing tower buildings, roads, bridges, subways, remodelling and demolition take place. The first motive of the waste management system is to reduce waste that can generate in the construction site. For the purpose a team of qualified professionals experienced in environment shall identify potential wastes early in the design process and shall design the construction using limited materials.

The waste that generated on construction site is often heavy and bulky and occupy considerable storage place. The waste consists of mainly:

1. Inflammable materials like wood, plastic, paper, chemicals, fuels, etc.
2. Biodegradable materials like food waste and toilet waste, etc.
3. Inert materials like concrete, plaster, Iron and other metal waste, etc.

Accumulation of inflammable materials may cause large fires on site and biodegradable materials can cause deceases to the staff and workers. Inert materials scattering on site may cause trip and fall hazards which may lead to injury, fracture or fatality to the fellow workers. For avoiding these hazards on site waste management plan is to be implemented. The plan shall elaborate the production and management of waste, including accurate details showing quantities that collected, stored, reused and disposed to the recycling yard.

3.35.1 Collection of Waste

The collection of waste is the combined responsibility of top, line and low management. The top management shall give strict orders to line management to train workers how to identify waste whether it can be reused or disposed, to the specified storage area. The management shall input a plan to dispose waste after the completion of specific work activity or at the end of each shift. The supervisors of concerned areas shall take initiative action for this purpose.

3.35.2 Reuse of Waste

Identify waste that can be salvaged for reuse on current project, on another project or can be donated. For the purpose the construction company shall develop a reuse and salvage plan. The plan shall include a list of items being reused in place or elsewhere on site, a list of items for reuse off-site through salvage, resale or donation, a plan for protecting, dismantling, handling, storing and transporting them. This plan shall be communicated to line management to train their low management for knowledge and work practice.

3.35.3 Storage, Transportation and Disposal of Waste

First issue is to identify how many type of waste are produced from construction site. For each waste different skip shall be placed for storage because each skip needs to be transported to different yards for recycling. For example Metal scrap needs to be transported to metal melting units for recycling. For the purpose the skips shall be colour coded according to the identification of waste to be dumped. The waste management plan shall define clearly that in each colour coded skip which waste is to be dumped. Supervisors shall train his work force, how and which skip the generated waste of their area is to be disposed.

For waste transportation process, the top management shall select and appoint a competent person or company for transportation and disposal of fully dumped waste skip. From the part of construction company also needs to assign a competent person to supervise and eliminate hazards while on transportation and disposal process. For example penetration from the toxic waste while transporting through road can cause big hazard.

Chemical Handling and
Hazardous Substance Safety

4

4.1 CHEMICAL AND HAZARDOUS SUBSTANCE SAFETY

Any hazardous substances that are going to be used or processes which may produce hazardous materials should be identified. The risks from work which might affect site workers or members of the public should then be assessed. Designers should eliminate hazardous materials from their designs. Where this is not possible, they should specify the least hazardous products which perform satisfactorily.

Contractors often have detailed knowledge of alternative, less hazardous materials. Designers and contractors can often help each other in identifying hazardous materials and processes and suggesting less hazardous alternatives. If workers use or are exposed to hazardous substances as a result of their work, the Control of Substances Hazardous to Health Regulations (NI) 1995 (COSHH) 1995 make it a legal duty to assess the health risks involved and to prevent exposure or else adequately control it.

4.1.1 Identification

People may be exposed to hazardous substances either because they handle or use them directly, (for example, solvents in glues and paints), or because the work itself results in the creation of a hazardous substance, (for example, scrabbling concrete generates silica dust). Identify and assess both kinds of hazard. If hazardous substances are going to be used, manufacturers and suppliers of such substances have a legal duty to provide information. Read the label on the container or the material safety data sheet (MSDS). Approach the manufacturer or supplier directly for more information if necessary.

Also, some hazardous substances may be on site before any work starts, for example, sewer gases or ground contaminants. Assess these risks in the same way as for other hazardous substances. Information to help identify these risks may be available from the municipality, client or the design team.

4.1.2 Assessment

Look at the way people are exposed to the hazardous substance in the particular job that is about to be done. Decide whether it is likely to harm anyone s health. Harm could be caused by:

- Breathing in fumes, vapors: Does the manufacturers information say that there is a risk from inhaling the substance? Are large amounts of the substance being used? Is the work being done in a way which results in heavy contamination of the air, for example, spray application? Is the work to be done in an area which is poorly ventilated, for example, a basement? Does the work generate a hazard, for example, hot cutting metal covered with lead causes lead fumes to be given off?

- Direct contact with skin or eyes: Does the manufacturer's information say there is a risk from direct contact? How severe is it, for example, are strong acids or alkalis being used? Does the method of work make skin contact?

- Swallowing or eating contaminated material: Some dusty materials can contaminate the skin and hands. The contamination can then be passed to a person's mouth when they eat or smoke. This is a particular problem when handling lead and sanding lead-based paint. Make sure people do not smoke or eat without washing first.

Once a full assessment has been completed and where the same work is being done in the same way under similar circumstances at a number of sites, the risk assessment does not have to be repeated before every job. Review the assessment from time to time, but every few years will probably be enough. However, look out for new products which could be safer substitutes.

If, however, there are many processes which result in different hazardous substances being used in a wide range of circumstances, a fresh assessment may be needed for each job or set of similar jobs. This will make sure the assessment is relevant to the job being done and the circumstances in which it is being carried out. Remember to assess both immediate risks and longer term health risks. Materials like cement can cause dermatitis. Sensitizing agents like isocyanides can make people using them have sudden reactions even though they may have used the substance many times before.

4.1.3 Prevention

If harm from the substance is likely, the first step to take is to try and avoid it completely by not using it at all. This will mean either:

- Doing the job in a different way, for instance, instead of using acids or caustic soda to unblock a drain, use drain rods
- Using a substitute substance, for example, instead of using oil or spirit-based paints, use water based ones which are generally less hazardous. However, always check one hazard is not simply being replaced by another.

4.1.4 Control

If the substance has to be used because there is no alternative, or because use of the least hazardous alternative still leads to significant risk, the next step is to try and control exposure. Some of the ways this could be done include the following:

- Ensuring good ventilation in the working area by opening doors, windows and skylights. Mechanical ventilation equipment might be needed in some cases

- Using as little of the hazardous substances as possible, Do not take more to the workplace than is needed
- Rather than spraying solvent-based materials, use a roller with a splash guard or apply by brush
- Transferring liquids with a pump rather than by hand. Keep containers closed except when transferring
- Using cutting and grinding tools fitted with exhaust ventilation or water suppression to control dust
- Using blasting equipment fitted with exhaust ventilation or water suppression to control dust.

4.1.5 Personnel Protective Equipment

If exposure cannot be adequately controlled by any combination of the measures already mentioned, also provide personal protective equipment (PPE). This might take the form of:

- Respirators which can protect against dusts, vapors and gases. Make sure the respirator is of the correct type for the job. It is essential that respirators fit well around the face. Make sure the user knows how to wear the equipment and check for a good face seal.
- Protective clothing, such as overalls, boots, gloves. Protection may be needed against corrosive substances.
- Eye protection, such as goggles or face shield. Protection of the eyes is important. If the protection needed is against corrosive splashes, face shield can protect the whole face.

Select PPE with care. Choose good quality PPE which is ISO marked. Let the user of the equipment help choose it, they will be more willing to wear it. Explain to the user why the equipment has to be worn and the hazards the equipment protects against. Users need to know how the equipment shall be operated and what maintenance checks they should carry-out. Supervise the user to make sure the equipment is being used properly. Regularly maintain the equipment and check it for damage. Store it in a dry, clean place and have replacement and spare equipment to hand. Make sure the PPE does not become a source of contamination by keeping the inside of dust masks and gloves clean. Store them in a clean box or cupboard and Do not leave them lying around in the work area.

4.1.6 Personnel Hygiene

Substances can also be a hazard to health when they are transferred from workers hands onto food, cigarettes, etc. and so taken into the body. This can be avoided by good personal hygiene. For example:

- Washing hands and face with soap and anti-infection lotion before eating, drinking and smoking and before using the toilet
- Eating, drinking and smoking only away from the site of exposure.

Make sure people exposed to the substances are separated by excluding people not directly involved in the work from the contaminated area while eating or using toilet. Make sure those at risk know the hazards. Provide good washing facilities and somewhere clean to eat meals. Good clean welfare facilities can play an important part in protecting the health of everyone involved in the work.

4.1.7 Health Surveillance

Sometimes workers health can be protected by checking for early signs of illness. Such surveillance is a legal duty of the employer while employees working with hazardous chemicals. Employer safe work procedure and local rules can give advice on when and how to carry-out health surveillance or medical checks. Employer shall conduct these checks as stipulated on approved HSE plan without any inconvenience and within a reasonable time frame. If the surrounding of the work premises has occupancy of the people which is not connected to work or neighbors, health checks shall be conducted to them also at regular intervals. These medical consultations will be provided without loss of salary to the employee as well as neighbors without any fees.

4.2 CHEMICAL SAFETY

Chemicals can be extremely useful for a wide range of applications, but they can also be quite dangerous at times, especially when they are handled and processed improperly. The practice of chemical safety is designed to identify chemicals which are hazardous, to use clear marking systems to label hazardous chemicals, and to have protocols in place for handling such chemicals with the goal of preventing accidents. Occupational health and safety laws often specifically address chemicals and the way in which they should be handled, stored, and destroyed.

One innovation in the field of chemical safety is the development of Material Safety Data Sheets (MSDS), informational sheets which clearly describe various chemicals, their effects on the human body, the ways in which they should be handled, and unique hazards, such as exposure to heat which could cause spontaneous combustion for some chemicals. Chemical labeling systems shall be displayed on chemical containers are also a form of chemical safety which are designed to clearly convey relevant safety information.

The focus of chemical safety is on identifying potential hazards so that incidents with chemicals can be avoided, and developing protocols for dealing efficiently and correctly with incidents such as spills. Many companies which deal with chemicals have their own safety protocols and guidelines in place, with these documents being formulated in response to recommendations from professional organizations and the government. It is critical for people to comply with guidelines for their own safety and the safety of others.

This guideline of chemical safety shall contain these procedures

- Details and safety plan about the chemicals used
- MSDS of the chemicals used
- PPE's to be worn and signage's to be placed
- Training Programs
- Chemical Injury or Exposure Response
- Hazardous Chemical/Substance Spills
- Toxicity of the chemical used
- Inflammable level of chemicals
- Chemical Labeling and housekeeping

- Fire fighting details
- Emergency response/evacuation procedure
- Storage, transportation and handling of the chemicals used
- Waste disposal systems, etc.

4.2.1 Chemical Hazards and its Safety

Chemical hazards represent potential for illness or injury due to single acute exposure or chronic repetitive exposure to toxic, corrosive, sensitizing or oxidative substances. They also represent a risk of uncontrolled reaction, including the risk of fire and explosion, if incompatible chemicals are inadvertently mixed.

Chemical hazards can most effectively be prevented through a hierarchical approach that includes

- Replacement of the hazardous substance with a less hazardous substitute.
- Implementation of engineering and administrative control measures to avoid or minimize the release of hazardous substances into the work environment keeping the level of exposure below internationally established or recognized limits. Keeping the number of employes exposed, or likely to become exposed, to a minimum.

nication should be in an easily understood language and be readily available to exposed workers and First Aid personnel
- Training workers in the use of the available information (such as MSDSs), safe work practices, and appropriate use of PPE.

4.2.2 Characteristics, Storage Rules and Hazard Identifying Signages of Different Chemical Properties

1. Flammable and combustibles

These chemicals are easily ignited and may present a serious fire and explosion hazard. Flammable liquids have a flash point below 100 F. Combustible liquids have a flash point of 100 F to 200 F. Flammable solids have an ignition temperature below 212 F. Flammable solids include finely divided solid materials which, when dispersed in air, could ignite. Other classes of chemicals with a high fire hazard include oxidizers, pyrophoric chemicals, and water reactive chemicals.

Keep flammables away from all ignition sources from open flames, hot surfaces, direct sunlight, spark sources. Keep flammable liquids that require cold storage in laboratory

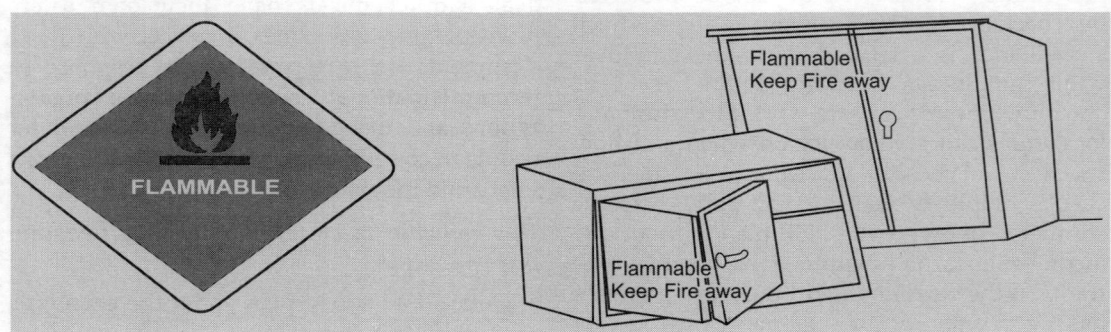

Flammable chemical identification signage   Storage area signage's

Fig. 4.1

- Communicating chemical hazards to workers through labeling and marking according to national and internationally recognized requirements and standards, including the International Chemical Safety Cards (ICSC), Materials Safety Data Sheets (MSDS), or equivalent. Any means of written commu-

safe flammable material refrigerators or freezers to avoid ignition of the materials by sparks or static electricity. Keep flammable gases from oxidizing gases with an approved non-combustible partition or by a distance of 20 feet. Store flammable liquids in approved safety containers or cabinets.

2. Corrosives

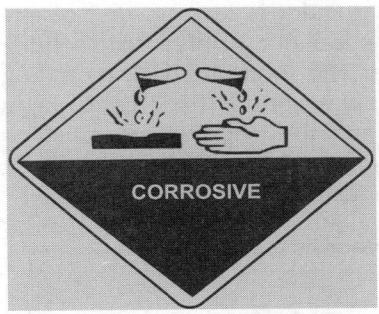

Fig. 4.2: Corrosive chemical identification signage

Strong acids and bases can destroy human tissue and corrode metals. Acids and bases are incompatible with one another and may react with many other hazard classes. Segregate acids from bases. Segregate inorganic oxidizing acids (e.g. nitric acid) from organic acids (e.g. acetic acid), flammables, and combustibles. Segregate acids from chemicals that could generate toxic gases upon contact (e.g. sodium cyanide and iron sulfide).

Segregate acids from water reactive metals such as sodium, potassium, and magnesium. Use tight-fitting goggles, gloves, and closed-toe shoes while handling corrosives. Store solutions of inorganic hydroxides in polyethylene containers. Store corrosives on lower shelves, at least below eye level and in compatible secondary containers. Do not store corrosives on metal shelves. Although ventilation helps, chemicals will still corrode the shelves. Store containers in plastic tubs or trays as secondary containment.

3. Toxics

Fig. 4.3: Toxic chemical identification signage

Overexposure to toxic chemicals can cause injury or death. Toxics are chemicals with a lethal dose (LD50) of more than 50 and less than 500 milligrams per kilogram body weight or a lethal concentration (LC50) in air of more than 200 and less than 1,000 parts per million. Segregate toxics from other hazard classes and store in a cool, well ventilated area, away from light and heat. Containers should be tightly sealed to minimize exposure to personnel and contamination of other chemicals.

4. Highly toxics

Fig. 4.4: Highly toxic chemical identification signage

These chemicals can cause serious injury or death at low concentrations. Highly toxics are chemicals with a lethal dose (LD50) of less than or equal to 50 milligrams per kilogram body weight or a lethal concentration (LC50) in air of less than or equal to 200 parts Chemicals per million. Maintain the lowest possible quantities of highly toxics. Segregate highly toxic chemicals from other hazard classes and store in an area that is cool, well ventilated, and away from light and heat. Use highly toxic chemicals in a designated area or laboratory. Highly toxic chemicals that produce fumes or dusk should always be handled within a chemical fume hood.

5. Oxidizers

Oxidizers are a fire hazard. They will readily decompose under certain conditions to yield oxygen or react to promote or initiate the combustion of flammable or combustible materials.

Segregate oxidizers from flammable and combustible materials (paper, wood). Segregate oxidizers from reducing agents (zinc, alkaline metals, formic acid). Segregate inorganic oxidizers from organic peroxides. Take care not to contaminate oxidizers.

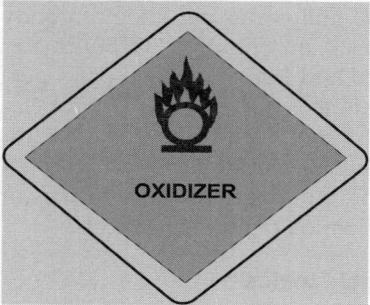

Fig. 4.5: Segregate oxidizer

Some oxidizers, such as perchloric acid, can become explosive mixtures if contaminated with trace amounts of organic materials or metals. Store oxidizers in cool and dry place. Do not store under sink. Remember that perchloric acid, nitric acid, and hydrogen peroxide are oxidizers and shall not be stored on wooden shelves or in cardboard boxes.

6. Compressed gases

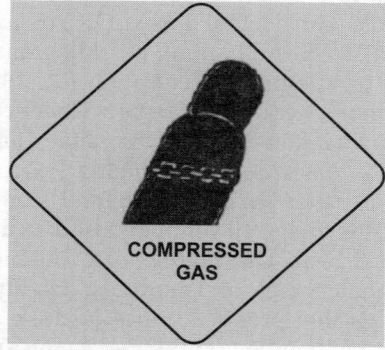

Fig. 4.6: Oxidizer identification signage

What all compressed gases have in common is the large amount of energy stored in the cylinder from the compression of the gas. Dropping or knocking over a cylinder can cause the energy to be rapidly released. It may even propel a cylinder like a rocket. Additional hazards can arise from the toxicity, flammability, corrosivity, or reactivity of the gas.

Secure cylinders so they will not fall with chains. Keep cylinders away from heat and

open flames. Leave the valve protection cap on the cylinder unless it is in use. If you suspect that a cylinder is leaking, do not attempt to sniff the leak out. Apply a soap solution to the cylinder and locate the leak by noting where the bubbles appear. Toxic gases, highly toxic gases, and pyrophoric gases must be managed in accordance with toxic gas program requirements.

7. Cryogens

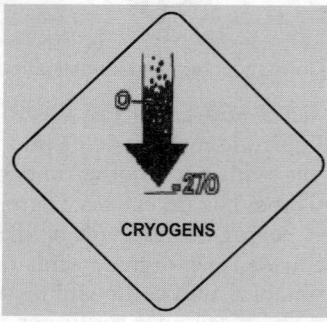

Fig. 4.7: Cryogens identification signage

These materials are extremely cold (–100 C to –270 C). Upon contact with cryogenic materials, living tissue can freeze and become brittle enough to shatter. Additional hazards include rapid pressure buildup, oxygen enrichment and asphyxiation. Rapid pressure buildup could lead to an explosion if cryogen is improperly contained. Cryogenic liquids and gases have many properties and hazardous characteristics in common with compressed gases.

Store and handle in a well-ventilated area. When liquid cryogens are converted to the gaseous phase, they may create an oxygen deficiency. Do not use cryogens in small enclosed spaces. Use only approved storage vessels (i.e., thermos-like evacuated, double-walled containers) with pressure-relief mechanisms. Non-approved vessels may explode.

Liquid nitrogen and liquid helium are capable of liquefying oxygen from air. This form of oxygen enrichment can become a strong fire or explosion hazard. Use appropriate protective equipment for handling cryogens (insulated holders for carrying vessels, eye protection, goggles, or face shields and aprons).

Use cryogenic gloves or leather gloves when handling super cold surfaces.

8. Pyrophorics (air reactive)

Fig. 4.8: Air reactive identification signage

Pyrophorics substances that ignite spontaneously upon contact with air. Store in a cool, dry place. Prevent contact with air. Take extreme care to prevent containers of pyrophorics from leaking or breaking. For additional protection, consider keeping the chemicals in the manufacturer's original shipping package (i.e. surrounded by vermiculite inside a metal can). Manage pyrophorics gas as toxic gases and highly toxic gas, shall be treated in accordance with toxic gas program requirements.

9. Water reactive

Fig. 4.9: Water reactive identification signage

These substances often react violently with water and may ignite or generate toxic, flammable, or corrosive gases. Store in a cool, dry place. Keep away from water. In case of fire, do not use water. Use a dry chemical extinguisher.

10. Explosive and potentially explosive chemicals

Explosive chemicals can rapidly release tremendous amounts of destructive energy. Explosive chemicals can cause death, serious injury, or severe property damage. Heat, shock, friction, or even static electricity can initiate explosions of these chemicals. The family includes pure chemicals (e.g. TNT) and mixtures (e.g. ammonium nitrate/fuel mixtures).

Fig. 4.10: Explosive identification signage

In addition to explosive chemicals, which constitute a known high hazard, there are chemicals that may become explosive, depending on how they are handled. This category is commonly referred to as potentially explosive chemicals and includes:

- Pure chemicals or mixtures that may become explosive through contamination (e.g. perchloric acid contaminated with organic compounds or metals
- Pure chemicals or mixtures that may degrade over time and become explosive (e.g. hydrated picric acid, which becomes explosive upon drying). This category also includes certain alcohols and ethers that may accumulate explosive levels of peroxides by interacting with air.

4.2.3 Chemical Procurement

- The decision to purchase a chemical shall be considered a commitment client to handle, store, and use the chemicals properly from initial receipt to ultimate disposal.
- Requests for procurement of new chemicals shall be submitted to the Chemical Hygiene Officer. Information on proper handling, storage, and disposal shall be known to all involved personnel prior to the procurement of the chemicals. Chemicals utilized in the premises shall have appropriate ventilation facilities to be installed.

- When chemicals are decide to purchase, effort shall be made to hand over MSDS from the part of manufacture to know concern hazards and precautionary measures co-link with it.

4.2.4 Chemical Hygiene Safety Plan

A chemical hygiene plan (CHP) has been created that sets forth procedures, equipment, personal protective equipment, and work practices that are capable of protecting employees from the health hazards presented by hazardous chemicals used in the workplace. Components of the CHP include standard operating procedures for safety and health, criteria for implementation of control measures, measures to ensure proper operation of engineering controls, provisions for training and information dissemination, permitting requirements, provisions for medical consultation, designation of responsible persons and identification of particularly hazardous substances.

In order to ensure the safety of employees and maintain a safe working environment all personnel must know and follow the procedures outlined in this plan. All operations performed must be completed in accordance with the enclosed procedures. In addition, all employees are expected to develop safe, personal chemical hygiene habits, aimed at the reduction of chemical exposure to themselves and co-workers.

This policy applies to all persons trained and then only authorized to work. The CHP will be updated and reviewed on an annual basis to ensure that the provisions and procedures described are consistent with the proper requirements and standards. The Chemical Hygiene safety Officer is responsible for maintaining this standard.

4.2.5 General Chemical Storage Regulation that follows on Industrial or Workface

Materials should always be segregated and stored according to their chemical family or hazard classification. Do not store chemicals alphabetically unless they are compatible. Accidental contact between incompatible chemicals can result in a fire, an explosion, the formation of highly toxic or flammable substances, or other potentially harmful reactions:

- Oxidizers mixed with flammable solvents can cause a fire.
- Acids mixed with metal dust can produce flammable hydrogen gas.
- Alphabetical storage can bring incompatibles together. For example, if chromic acid (an oxidizing acid) and chromium powder (a combustible metal) were stored together and an accident broke their containers, the chemicals could mix and react with explosive violence.
- Each chemical family should be separated from all other chemical families by an approved non-combustible partition or by a distance of twenty feet. Incompatible chemicals within the same hazard class should also be separated from one another.
- Keep oxidizers away from other chemicals, especially flammables or combustibles.
- Storage area shall be build with non-combustible material while storing inflammable chemicals. While in the case of toxic chemicals, shall build leak proof compartments. These storage compartments shall meet all rules and regulation of local authorities and even client.
- The storage area will be well illuminated, with all chemical storage maintained below eye-level. Large bottles shall be stored no more than two feet from ground level.
- Chemicals shall be segregated by hazard classification and compatibility in a well-identified area, with local exhaust ventilation.
- Mineral acids will be separated from flammable and toxic materials and shall be stored in different Compartments. Acid-resistant trays shall be placed under bottles of mineral acid.
- Danger signage's to be paste outside the storage areas. Unauthorized entries to the storage area shall be prohibited. All person deals with storage area shall wear appropriate PPEs.

- Acid-sensitive materials such as sulfides shall be separated from acids or protected from contact with acids.
- Highly toxic chemicals or other chemicals whose containers have been opened shall be stored in unbreakable and leak proof secondary containers.
- The storage area shall not be used as a preparation or repackaging area. The storage area shall be accessible during normal working hours without any obstruction.
- Storage of chemicals at the work areas shall be limited to those amounts necessary for one operation or shift. The container size shall be the minimum size that is convenient for work. The amount of chemicals at the work area shall be as small as possible.
- Chemicals in the storage area as well as workplace shall not be exposed to sunlight or heat.
- Stored chemicals shall be examined at least monthly by the store keeper or material controller for replacement, deterioration, and container integrity. The inspection will determine whether any corrosion, deterioration, or damage has occurred to the storage facility as a result of leaking chemicals.
- Periodic inventories of chemicals outside the storage area shall be conducted by the material controller or store keeper. Unneeded items shall be properly discarded or returned to the storage
- area. Unwanted or rejected chemicals shall be cleared from the site through proper waste disposal system.
- MSDS of the chemical stored shall be pasted on the entrance of the storage area. Appropriate charged fire extinguishers shall be keep outside chemical storage area for meeting fire emergencies.
- Gas storage areas should be secured and separated from other areas of chemical storage.
- Confine chemical storage areas so that leaks or spills are controlled. Prevent chemicals from running down sink, floor, or storm water drains. Clean up spills and drips immediately.

- Keep emergency equipment such as fire extinguishers handy and in good working order outside chemical storage area. Signages shall be pasted according to the chemical characteristics stored.

4.2.6 Record Keeping

All records relating to the chemicals shall be collected and filled at least two years from the date of document originated. These documents shall be audited periodically for finding any defects occurred from implementing actual chemical hygiene plan. The following records shall be maintained by the Chemical Hygiene Safety Officer:

- MSDS of all chemicals in use
- Accident reports will be retained for life of project
- Exposure record for hazardous chemicals will be retained for life of project
- Medical records for employees exposed to hazardous chemicals will be retained for life of project
- Inventory and usage records for high-risk substances
- Records of inspections of equipment will be maintained for life of the project
- Records of employee training will be maintained for the life of the project.

4.2.7 Chemical Handling

Each employee shall be trained, educated, and resources provided by safety department shall develop and implement work habits consistent with this plan to minimize personal and co-worker exposure to the chemicals. Based on the realization that all chemicals are dangerous under certain conditions, exposure to all chemicals will be minimized. General precautions that will be followed for the handling and use of the chemicals are:

- Skin contact with the chemicals will be avoided
- All employees shall wash all areas of exposed skin prior to leaving the premises
- Eating, drinking, smoking, gum, tobacco, or the application of cosmetics in the areas

where chemical are present will be avoided. All such areas will have a sign clearly posted

- Storage, handling, and consumption of food or beverages shall not occur in chemical storage or working area
- Any chemical mixture shall be considered as toxic as its most toxic component
- Substances of unknown toxicity shall be assumed to be toxic
- Employees shall be familiar with symptoms of exposure for chemicals with which they work and the precautions necessary to prevent exposure
- In all cases of chemical exposure, immediately report to first aid center and send for medical attention
- Engineering controls in the premises shall be utilized and inspected in accordance with manufacturer's recommendations
- Specific precautions based on the toxicological characteristics of individual chemicals shall be implemented as deemed necessary by the safety department.

4.2.8 MSDS Availability

Material safety data sheets (MSDS) must be readily available for each chemical. MSDS must be available at all times and near the hazard source. Prior to the use of a new chemical the MSDS must be provided for review to the chemical user and those persons in the work face.

MSDS copies shall be filled or displayed on

- On central documentation room as a accessibility to all top management and line management personnel's for better planning, prior committing work with concern chemicals.
- On safety office to know concern hazards and precautionary measures co-link with concern chemicals for inspection purpose.
- On First Aid center for knowing what type of First Aid treatment shall be given by the first-adder in case of concern chemical injury or exposure.
- Displayed on chemical stacking area to know the first aid treatment shall be given by a

co-worker in case of concern chemical injury or exposure by reading its MSDS.

4.2.9 PPE's and Signages

All relevant PPE's according to the safe work procedure, CHP and MSDS shall be worn by the work force while transporting or handling concern chemicals. The common PPE's that can use while chemical transporting or handling are:

- Respirators which can protect against dusts, vapors and gases
- Protective clothing, such as overalls, boots, gloves
- Appropriate protection like leather, PVC or rubber gloves and overall may be needed against corrosive Substances
- Eye protection, such as goggles or face shield or hood
- Breathing apparatus, if chemicals using are toxic
- Safety shoes for leg protection.

Appropriate signage's shall be displayed on work areas as well as storage areas for the identification of dangerous chemicals and precautions to be taken by the workers and visitors. Unauthorized entry to the chemical storage and working area shall be prohibited by placing securities.

4.2.10 Training programs

All employees will be apprised of the hazards presented by the chemicals in use. Each employee shall receive training at the time of initial assignment to the job, prior to assignments involving new exposure situations, and semi-annually thereafter.

This training shall include methods of detecting the presence of a hazardous chemical, physical and health hazards of chemicals, and measures employees can take to protect themselves from those hazards.

The training shall present the details of this plan and include:

- Location of work
- Permissible exposure limits and precaution shall be taken

- Signs and symptoms associated with exposure to chemicals present in the work or storage area
- Toxic condense in chemicals
- PPE's to be worn
- Inflammability of the chemicals
- Type of fire extinguisher to be used
- Emergency evacuation procedure
- Location and availability of reference materials on chemicals.

This training will be conducted by the Safety officer. Daily written prestart safety talks shall be conducted by the supervisor for the remembrance of chemical safety for work force in the presence of safety personnel.

4.2.11 Chemical Injury or Exposure Response

When an injury has occurred general response guidelines are as follows:
- Protect yourself from exposure and stabilize the injured person. When possible wash your hands prior to and after giving first aid. Use gloves whenever possible.
- Call emergency numbers when medical attention is required or when not sure how to respond.
- Utilize the safety shower available. Clothing must be removed to prevent prolonged chemical contact with the skin. Wash the exposed area for at least 15 minutes.
- Use the emergency eyewash stations to wash harmful chemicals from the eyes when appropriate. Eyes must be washed for a minimum of 15 minutes.
- Offer the injured person medical attention by calling emergency numbers.
- All injuries shall be reported to Safety department for the investigation of accident happen. An accident report must be completed within 24 hours of the incident happened.

4.2.12 Hazardous Chemical/Substance Spills

Determine the extent and type of spill. Chemical spill cannot be cleaned immediately, report the spills to your supervisor.

Immediately alert area occupants and evacuate the area, if necessary and posted signs for location information. If a volatile flammable material is spilled immediately warn everyone. Control sources of ignition. Ventilate the area by turning on the fume hoods with the sashes completely open and open all windows. Use the appropriate personal protective equipment for the hazard involved. Refer MSDS or other available references for the concern chemical information. Do not enter a contaminated atmosphere without protection or use a respirator or without training. When respiratory protection is used for emergency purposes there must be another trained person outside the spill area. This person must have communication abilities with the person in the spill area. Contact Safety department when no one is available for backup.

Cover or block floor drains or any other route that could lead to an environmental release. Use the appropriate media when cleaning spills. Begin by circling the outer edge of the spill with absorbent. Next, distribute spill control materials over the surface of the spill. This will effectively stop the liquid from spreading and minimize volatilization. Place absorbed materials in an appropriate container using a brush and scoop. Small spills can be placed in polyethylene bags. Larger quantity spills may require five-gallon pails or 20-gallon drums with polyethylene liners. Absorbent materials used on the chemical spill will most likely require disposal as hazardous waste. Place a completed hazardous waste label on the container. Clean the surface where the spill occurred using a mild detergent and water.

Companies should be prepared for chemical spills by having a spill kit or materials available and supervisory personnel trained to respond. The spill kit must be in an obvious location and all persons responsible for the activities conducted and shall have knowledge in the use of the spill kit. Spill kits are commercially available or you may request a listing on what items must be included in a spill kit.

Recommended items for a universal chemical spill kit are

- Spill absorbents and equipment
- One container of spill-X-A Acid neutralizer and absorbent
- One container of spill-X-B Base neutralizer and absorbent
- One container of spill-X-S Solvent absorbent,
- A small broom
- A plastic dustpan
- Personal protective equipment such as:
  a. Safety goggles
  b. Chemical resistant apron or lab coat
  c. Safety gloves
- Spill consumables:
  a. Chemical waste disposal bags
  b. Chemical waste labels
  c. 5-gallon pail

4.2.13 Chemical Labeling

All containers in the work premises shall be labeled. This includes chemical containers and waste containers. The labeling shall be informative and durable, and at a minimum will identify contents, source, and date of acquisition, storage location, and indication of hazard. All Portable containers shall also be labeled. The labeling program will be periodically inspected by the concern supervisor to ensure that labels have not been defaced or removed. If any thing is removed in condition then re-labeling to be enforced.

4.2.14 Housekeeping

- Each worker is responsible for the cleanliness of his or her workspace, and jointly responsible for common areas of the premises. Management shall monitor on the maintenance of housekeeping standards.
- All spills on benches shall be immediately cleaned and properly disposed of. Large spills may necessitate the implementation of the Emergency spill response plan. The work benches or platforms shall be kept clear of equipment and chemicals except for those chemicals or equipment that are currently being used.

- All floors, aisles, exits, fire-extinguishing equipment, eyewashes, showers, electrical disconnects and other emergency equipment shall remain unobstructed.
- All labels will face the front of container or portable cans.
- Chemical containers will be clean, properly labeled and returned to storage upon completion of usage.
- Hazardous chemicals shall be stored according to compatibility and in chemical safety cabinets or in the ventilated base cabinet of the fume hood.
- Toxic and flammable chemicals are color coded separately for ease of segregation and also keep in separate compartment as per client and local authority requirements.
- All chemical wastes will be disposed of in accordance with the waste disposal plan.

4.2.15 Emergency Response Procedure

- All employees shall be trained according to the emergency response procedure prior committing the work activity with chemicals.
- Emergency routes to assembly point shall be marked, signages shall be pasted on marked routes and emergency drills to be conducted regularly.
- Telephone numbers of emergency departments, personnel, supervisors, and other workers as deemed appropriate shall be posted in the work area on site notice boards or person gathering areas in sight.
- All personnel will be trained in the proper use of fire extinguishers when hired and annually thereafter. Prior to the procurement of new chemicals, the Supervisor will verify that extinguishers and other emergency equipment are appropriate for such chemicals.
- Any employee who might be exposed to chemical splashes shall be instructed in the location and proper usage of emergency eyewashes. Emergency eyewashes shall be inspected weekly by the safety office for any maintenance. Records shall be maintained with daily work area inspections.

- Location signs for safety and emergency equipment must be posted.
- All first aid medicines and equipment shall be kept in good condition for attending any emergency situations.

4.2.16 Chemical Fire, Explosion Prevention and Fire Procedure

Fires and or explosions resulting from ignition of flammable materials or gases can lead to loss of property as well as possible injury or fatalities to project workers. Prevention and control strategies include:

- Storing flammables away from ignition sources and oxidizing materials. Further, flammables storage area should be:
  a. Remote from entry and exit points into buildings
  b. Away from facility ventilation intakes or vents
  c. Have natural or passive floor and ceiling level ventilation and explosion venting
  d. Use spark-proof fixtures
  e. Be equipped with fire extinguishing devices and self closing doors, and constructed of materials made to withstand flame impingement for a moderate period of time.
- Providing bonding and grounding of, and between, containers and additional mechanical floor level ventilation if materials are being or dispensed in the storage area.
- Where the flammable material is mainly comprised of dust providing electrical grounding, spark detection and quenching systems. Defining and labeling fire hazards areas to warn of special rules (e.g. prohibition in use of smoking materials, cellular phones, or other potential spark generating equipment).
- Providing specific worker training in handling of flammable materials, and in fire prevention or Suppression.

Fire procedure

All employees shall be informed at the beginning of site of building evacuation routes. It is the supervisor's responsibility to provide this information. In the event of a fire, immediate evacuation is essential. On the way out of the building remember these safety precautions.

- Never enter a room containing a fire.
- Never enter a room that is smoke filled.
- Never enter a room in which the top half of the door is hot to the touch.

Small fires

1. Pull the fire alarm and call the emergency numbers.
2. Alert people in the area to evacuate. Assist those individuals with disabilities.
3. Turn off gas main.
4. If you have been trained to use a fire extinguisher, do so while maintaining a clear exit path behind you. If can't fight against fire, march to assembly point. On the way tell others to go to assembly point.

Large fires

1. Pull the fire alarm on the safe area and immediately dial emergency numbers of fire department
2. Shout fire for the attention to others to follow evacuation procedure
3. Alert people in the area to evacuate. Assist those individuals with disabilities
4. Turn off gas mains, only if time permits
5. Close the doors to confine the fire
6. Move to a designated assembly area away as quickly as possible
7. Report to the supervisor in assembly area to find any missing personals
8. Trained search party shall be ready for finding missing persons
9. Persons having knowledge about the incident and location must provide this information to emergency response personnel
10. Do not return to the work area until to do so.

4.2.17 Chemical Waste Disposal System

(a) Objectives of waste disposal system

- To collect, store and dispose of chemical wastes in an environmentally sound manner
- To comply with environmental and health legislation for disposal of chemical wastes

- To provide a chemical waste classification system as CHP
- To avoid risks to health, safety and the environment.

The following process is to be followed

- Only use containers that are compatible with the chemical waste and that are sealed so that they will not leak during road transport.
- Ensure the container is labeled accurately with the substances name along with approximate concentrations and relevant dangerous goods class.
- All empty chemical containers should have the label defaced so that the chemical name can't be identified, and thoroughly cleaned prior to being placed in general waste.
- Waste Oil must be sealed in an appropriate and compatible container. Where applicable, the container should be labeled with the type of oil and total quantity, waste generator's name and contact phone number.
- Any chemicals discharged to sewer shall be:
  a. Waste must be soluble with water
  b. Waste must not be toxic or hazardous to aquatic, marine and terrestrial life and environments
  c. Acceptable flammable liquids must be less than 10% of their explosive concentrations.
- Only one 'individually generated' chemical waste is permitted in each container, do not mix chemical waste from different processes even if they are of similar properties. The waste disposal contractor will assess which wastes may be combined and will perform this task.
- Persons engaged in waste disposal systems has be adequately trained and shall wear all required PPE's while in work.
- Provide strict supervision all times. All points that mention in CHP shall be implemented.
- Safety officer shall inspect the area and shall clear all deviations while the waste collecting process is ongoing.

4.3 SOLVENT SAFETY

Many chemical substances which are used to dissolve or dilute other substances and materials are called 'solvents'. Industrial solvents are often mixtures of several individual substances. They can be found under a variety of trade names. You are most likely to be exposed to solvents if working in the following industries where they are used extensively in engineering, footwear, construction, textiles, chemicals, foodstuffs, printing woodworking, rubber, dry cleaning, plastics, paint manufacture, pharmaceutical, ink manufacture, etc mainly as:

- Cleaning and degreasing materials
- Paints, lacquers and varnishes
- Inks and ink removers
- Pesticides Toiletries
- Paint removers, etc.

Different solvents can affect your health in different ways. Unconsciousness and even death can result from exposure to very high concentrations of solvent vapors. Some of the short-term effects are:

- Irritation of eyes, lungs and skin
- Headache
- Nausea
- Light-headedness, etc.

When you are affected by solvents, there may be an increased chance of having an accident. There can also be long-term effects on your health from repeated exposure to particular solvents. These may include dermatitis. Other possible effects on health vary according to which solvent you are exposed to.

Solvent can get into the body by breathing vapors and fumes, contact with your skin and get absorbed or swallow liquid solvents by bad hygienic practice.

The following precaution shall be taken while working with solvents

- Make sure employer provides details of the hazards of the particular solvents in use, the precautions to take while use them and the procedures to follow in an emergency
- Ensure MSDS of the solvents using are available and container has been labeled for identification
- Ensure employees have been trained according to the safe chemical handling.

- Make full use of any ventilation equipment that employer provides to remove vapors from work area.
- Wear all adequate respiratory protection. Keep protective equipment in a clean place. Make sure it is kept clean so it is fit to use.
- Prevent unnecessary evaporation of solvents by using the minimum amount for the job, keeping lids on containers and using sealed containers for solvent contaminated waste. Do not leave solvent contaminated rags lying around.
- Avoid skin contact with solvents and any products containing solvents by wearing suitable protective clothing (gloves, apron, goggles or face shield, etc.) where necessary. Do not use solvents to remove paint, grease , etc. from your skin.
- Do not eat or smoke in areas where there are solvents. Wash thoroughly after working with solvents before eating or smoking. Do not weld, burn or use any naked lights in areas which may contain the vapor of chlorinated solvents as very toxic gases may be given off.
- If any employee's health has affected with solvent hazard, immediately report to first aid center and seek medical attention as early as possible.

Working with solvents in confined space

When working in confined spaces special precautions are needed to prevent from being exposed to solvents. In a confined space like a tanks, pit, small rooms or inside a vehicle, solvent vapors cannot escape easily and can build up to dangerous and even fatal concentrations very quickly. Beware of a build-up of vapors and gases which could be poisonous, explosive or flammable.

The following precaution shall be taken while working with solvents in confined space

- Ensure HIRA and safe work procedure are completed and implemented.
- Ensure training has conducted for work force according to the safe work procedure.

- PTW systems shall be accessed and ensure strict supervision.
- Do not use any materials containing solvents unless the confined space area is adequately ventilated and, where necessary provided with suitable respiratory protection.
- Do not enter a confined space where solvent vapors are present, unless employer has ensured it is safe to do by ventilating the space and testing the atmosphere, or unless special precautions shall taken.
- Special plan for providing trained and armed rescue persons at the work face.
- Ensure flame proof tools and equipment are using where using inflammable nature solvents or chemicals.

4.4 CORROSIVE, OXIDIZING AND REACTIVE CHEMICAL SAFETY

Corrosive, oxidizing, and reactive chemicals present similar hazards and require similar control measures as flammable materials. However, the added hazard of these chemicals is that inadvertent mixing or intermixing may cause serious adverse reactions. This can lead to the release of flammable or toxic materials and gases, and may lead directly to fires and explosions. These types of substances have the additional hazard of causing significant personal injury upon direct contact, regardless of any intermixing issues.

The following controls should be observed in the work environment when handling such chemicals

- Corrosive, oxidizing and reactive chemicals should be segregated from flammable materials and from other chemicals of incompatible class (acids vs. bases, oxidizers vs. reducers, water sensitive vs. water based, etc.), stored in ventilated areas and in containers with appropriate secondary containment to minimize intermixing during spills.
- Workers who are required to handle corrosive, oxidizing, or reactive chemicals should be provided with specialized training and provided with, and wear, appropriate PPE (gloves, apron, splash suits, face shield or goggles, etc).

- Where corrosive, oxidizing, or reactive chemicals are used, handled, or stored, qualified First Aid should be ensured at all times. Appropriately equipped First Aid stations should be easily accessible through out the place of work. Eye-wash stations or emergency showers should be provided close to all workstations where the recommended First Aid response is immediate flushing with water.

4.5 BIOLOGICAL HAZARDS AND ITS SAFETY

Biological agents represent potential for illness or injury due to single acute exposure or chronic repetitive exposure.

Biological agents should be classified into four groups

Group 1

Biological agents unlikely to cause human disease, and consequently only require controls similar to those required for hazardous or reactive chemical substances.

Group 2

Biological agents that can cause human disease and are thereby likely to require additional controls, but are unlikely to spread to the community.

Group 3

Biological agents that can cause severe human disease, present a serious hazard to workers, and may present a risk of spreading to the community, for which there usually is effective prophylaxis or treatment available and are thereby likely to require extensive additional controls.

Group 4

Biological agents that can cause severe human disease, are a serious hazard to workers, and present a high risk of spreading to the community, for which there is usually no effective prophylaxis or treatment available and are thereby likely to require very extensive additional controls. The employer shall all times encourage and enforce the highest level of hygiene and personal protection, especially for activities employing biological agents of Groups 3 and 4 above. Work involving agents in Groups 3 and 4 should be restricted only to those persons who have received specific verifiable training in working with and controlling such materials. Areas used for the handling of Groups 3 and 4 biological agents should be designed to enable their full segregation and isolation in emergency circumstances, include independent ventilation systems, and be subject to requiring routine disinfection and sterilization of the work surfaces.

HVAC (heating, ventilation and air conditioning) systems serving areas handling Groups 3 and 4 biological agents should be equipped with high efficiency particulate air filtration systems. Equipment should readily enable their disinfection and sterilization, and maintained and operated so as to prevent growth and spreading of disease agents, amplification of the biological agents, or breeding of vectors, e.g. mosquitoes and flies, etc.

Biological hazards can be prevented most effectively by implementing the following measures

- If the nature of the activity permits, use of any harmful biological agents should be avoided and replaced with an agent that, under normal conditions of use, is not dangerous or less dangerous to workers. If use of harmful agents can not be avoided, precautions should be taken to keep the risk of exposure as low as possible and maintained below internationally established and recognized exposure limits.
- Work processes, engineering, and administrative controls should be designed, maintained, and operated to avoid or minimize release of biological agents into the working environment. The number of employees exposed or likely to become exposed should be kept at a minimum.
- The employer should review and assess known and suspected presence of biological agents at the place of work and implement appropriate safety measures, monitoring, training, and training verification programs.

- Measures to eliminate and control hazards from known and suspected biological agents at the place of work should be designed, implemented and maintained in close co-operation with the local health authorities and according to recognized international standards.

4.6 RADIOLOGICAL HAZARDS AND ITS SAFETY

Radiation exposure can lead to potential discomfort, injury or serious illness to workers. Prevention and control strategies include:

- Places of work involving occupational or natural exposure to ionizing radiation should be established and operated in accordance with recognized international safety standards and guidelines.

- Exposure to non-ionizing radiation (including static magnetic fields; subradio frequency magnetic fields, static electric fields, radio frequency and microwave radiation, light and near-infrared radiation, and ultraviolet radiation) should be controlled to internationally recommended limits.

- In the case of both ionizing and non-ionizing radiation, the preferred method for controlling exposure is shielding and limiting the radiation source. Personal protective equipment is supplemental only or for emergency use. Personal protective equipment for near-infrared, visible and ultraviolet range radiation can include appropriate sun block creams, with or without appropriate screening clothing.

Project Safety Analysis 5

5.1 MEP PROJECT SAFETY

MEP projects mean Mechanical Electrical Plumbing work projects. These works to be planned and committed according to Local regulation and stipulated standards. All employees and workers involved in these activities shall be certified, experienced and competent persons. Electricians engaged in MEP works shall have the license or competent certificate from approved agencies or local authorities. Safe work procedure for each activity in MEP works shall be originated and followed after detailed hazard identification and risk assessment. According to the approved safe work procedure training programs to be conducted among the employees and workers for building their expertness and safe work practice. Strict supervision shall be provided at all time bases. Safety personnel has to provide special attention for finding any deviation from actual safe work procedure and corrective action to be implemented in work premises.

5.1.1 Mechanical Ducting and HAVC Systems

Mechanical ducting is mainly fixed for air condition, fire line, ventilation, plumbing, etc. Usually an interior duct system is installed either in a fur-down chase below the ceiling insulation or in a fur-up chase and insulated with the ceiling. Though ducts in floor cavities between upstairs and downstairs are often thought of as being in conditioned space, this is rarely the case. Interior duct work in a fur-down system typically consists of metal or fiberglass

duct board. Space limitations generally rule out the use of flex duct.

Fur-down ducting

Fur-up ducting

Fig. 5.1

HVAC that stands for the closely related functions of "Heating, Ventilating, and Air Conditioning", the technology of indoor environmental comfort. All the HVAC installations works should comply with relevant ISO standards. The indoor environmental condition including the noise level in each zone should meet the functional requirement of the space. The requirements for ductwork are three performances criteria, i.e. Stability, integrity and Insulation. Contractor should consider the following procedures when designing/ modifying existing ducting or HVAC installation.

1. Contractor to substantiate with calculation the chilled air distribution for each zone
2. Chilled water system should not be tapped to introduce additional cooling equipment
3. System design brief and safe work method of statement shall be submitted wherever applicable
4. All the duct works should comply with ISO standards and flexible ducts should not exceed 1.5 m in length
5. Chilled water pipes should be drained for any modification
6. Proper access shall be provided for all the equipment, valves and controls located above false ceiling
7. Erection at height shall be performed as per height work regulations
8. Location of condensing units of additional A/C equipment and the refrigerant pipe routes shall be coordinated with existing services and building elevation
9. Toolbox talks to be conducted prior starting works
10. Any ducting maintenance shall be done after accessing Confined space entry permit
11. Welding works to be conducted after accessing Hot work permit systems
12. Works to be conducted under strict supervision
13. Mechanical aid equipment engaged in work premises shall have approved certificate and competent operators
14. All required PPE shall be worn by the workers
15. Tools and portable electric tools used shall be of ISO marked or approved voltage types.

5.1.2 Fire Protection Systems

All the Fire Protection installations should comply with relevant code of application and to be approved by legal Department. Contractor should consider the following procedures when designing/modifying existing Fire Fighting Installations.

1. All the rooms/spaces except IT rooms should be covered with sprinkler system if the building is provided with sprinkler system.
2. Computer server rooms and critical equipment rooms shall be provided with Dry Automatic Fire Extinguisher System (FM 200 or equivalent).
3. If the building is not provided with Sprinkler System then adequate numbers of hand held Fire Extinguisher should be provided at easily accessible locations.
4. Each room shall be provided with adequate numbers of smoke detectors and connected to the building fire alarm system as per regulations.
5. Kitchen extract hood shall be provided with wet chemical suppression system.

5.1.3 Electric Works

The design and execution of electrical installations must be in accordance with ISO regulations. The fit-out electrical contractor should comply with relevant standards during the design and installation of electrical services. The key points to be followed during the electrical services design are as given below:

1. Component and parts of the installation such as cables, apparatus, equipment and accessories shall comply with the relevant ISO or BS standards.
2. Switches and sockets shall generally be fixed in accordance with the mounting heights and the measurement recommended in relevant standards.
3. Engineer shall planned installation work according to the safe work procedure and local regulations.
4. Recommended illumination intensity shall be as per Standards.

5. All the electrical symbols and signage's used shall be as per ISO or BS standards.
6. All electric installation and connections shall be fixed by approved and licensed electrician.
7. Live line works shall be done only after accessing PTW systems.
8. Tag-out or Lock-out system to be implemented on work face.
9. All required PPE shall be worn by the workers.
10. Housekeeping to be done on regular basis.
11. Works to be conducted under strict supervision.

Tag-out or Lock-out Procedure

Whenever a person is working on an installation or equipment a safe system of work is required. When this work is other than the normal operation of the equipment, such as fault finding, maintenance or repair an increased level of risk is created. The lockout tag out system is to be incorporated in all work practices to help reduce this risk. Each person performing work is to be made aware of isolation points and is to effectively lockout or control all energy sources affecting their work, this allows for group lockout procedures, with an individually keyed lock and tags.

(a) Out of service (caution) tags

Out of Service tags are to be of a durable nature, marked with the workers name, contact (preferably mobile), company or section, date of placement, details of defect and signature. Tags will be yellow with black striping bearing the label "Caution" or "Out of Service".

Out of Service tags are placed for the protection of the general populace or plant and are to be left attached to the device until the defect listed is remedied. Once the defect is remedied an authorized and competent person may remove the tag, returning the device to service.

(b) Danger tags

Danger tags are to be marked with the workers name, contact (preferably mobile), company or section, current date of work and signature. This is a minimum requirement on the tag.

Tags will be white with red striping, prominently marked "Danger—Do Not Remove". Danger tags are to be placed only for the protection of personnel working on plant, they are not to be left on after that person completes their work, or finishes their shift. Changeover procedures may be required in certain situations, or an Out of Service Tag is to be placed if an item is unfit for use after the worker has left. Danger tags are to be used in conjunction with locks. Danger tags are to only be removed by the person who placed them.

5.1.4 Plumbing and Drainage Works

All the installation should comply with relevant industry standards and practices. The following shall be Considered when designing/modifying existing plumbing and drainage works.
1. Alternatively drain pump with pipe works connecting to existing drain pipe at higher level should be proposed where coring is prohibited
2. All the wet area should be water proofed and tested
3. All plumbing installation and connections shall be fixed by approved and licensed plumbers
4. Condensate drain from additional AC units (if any) should be routed with proper slope and connected to existing system
5. Proper access shall be provided for all the equipment and valves located above false ceiling
6. Works to be conducted under strict supervision
7. Mechanical aid equipment engaged in work premises shall have approved certificate and competent operators
8. All required PPE shall be worn by the workers
9. Tools and portable electric tools used shall be of ISO marked or approved voltage types.

5.2 POWER PROJECT SAFETY

Power projects are constructed to make electricity. The safety element concern mainly in:
1. Total Management Commitment and resources

2. Safety culture and awareness
3. Safety organization/structure
4. Documentation
5. Training and education
6. Hazard identification and risk management
7. Reporting and evaluation
   - Reporting hazards
   - Reporting incidents
   - Investigation and analysis of both
8. Process to produce recommendations
9. Processes to ensure actions are completed
10. Emergency response.

Specific health and safety issues in power projects include the potential for exposure to

- Hot pipe lines
- Confined spaces
- Toxic or flammable gas release
- Heat
- Noise.

(a) Hot pipelines

Workers may be exposed to physical hazards associated with the wells and related pipeline networks. Hazards may result from contact with hot components, equipment failure, or the presence of active and abandoned well infrastructure which may generate confined space or falling hazards. Recommended management techniques to mitigate these impacts include:

- Placement of access deterrents, such as fences and warning signs, to prevent access and warn of existing hazards
- Minimizing the length of necessary pipeline systems
- Consideration of the feasibility of subsurface pipelines or heat shields to prevent public contact with hot pipelines,
- Managing closure of infrastructure such as pipelines and access roads including cleaning, disassembly, and removal of equipment, re-vegetation of site and blockade, and reclamation of access roads where necessary.

(b) Confined spaces

Confined space hazards in this and any other industry sector are potentially fatal. Confined space entry by workers and the potential for accidents may vary depending on design, on-site equipment, and presence of groundwater or thermal fluids. Specific and unique areas for confined space entry may include:

- Turbines
- Condensers
- Cooling water tower (during maintenance work)
- Boilers (during maintenance work)
- Monitoring equipment sheds (during sampling).

Power projects should develop and implement confined space entry procedures as described in the general HSE guidelines or safe work procedure. PTW systems to be accessed prior to entry in any confined spaces by concern supervisor.

(c) Toxic or flammable gas release

Occupational exposure to thermal gases mainly hydrogen sulfide gas, may occur during non-routine release of thermal fluids (for example, pipeline failures) and maintenance work in confined spaces such as pipelines, turbines, and condensers. The significance of the hydrogen sulfide hazard may vary depending on the location and geological formation particular to the facility. Where there is a potential for exposure to hazardous levels of hydrogen sulfide or any gasses Power projects should consider the following management measures:

- Installation of hydrogen sulfide or dangerous gasses monitoring and warning systems. The number and location of monitors should be determined based on an assessment of plant locations.
- Development of a contingency plan for gas release events, including all necessary aspects from evacuation to resumption of normal operations.
- Provision of facility emergency response teams and workers in locations with high

risk of exposure, with personal gas monitors, self-contained breathing apparatus and emergency oxygen supplies and training in their safe and effective use.

- Provision of adequate ventilation of occupied buildings to avoid accumulation of gas.
- Providing workers with a fact sheet or other readily available information about the chemical composition of liquid and gaseous phases with an explanation of potential implications for human health and safety.

(d) Heat exposure

Occupational exposure to heat occurs during construction activities, and during operation and maintenance of pipes, wells, and related hot equipment. Non-routine exposures include potential blowout accidents during drilling as well as malfunctions of the steam containments and transport installations.

Recommended prevention and control measures to address heat exposure include:

- Reducing the time required for work in elevated temperature environments and ensuring access to drinking water.
- Shielding surfaces where workers come in close contact with hot equipment, including generating equipment, pipes, etc.
- Use of personal protective equipment (PPE) as appropriate, including insulated gloves and shoes.

(e) Noise exposure

Noise sources in Power projects are mainly related to steam flashing and venting. Other sources include equipment related to pumping facilities, turbines, and temporary pipe flushing activities. Temporary noise levels may exceed 100 d6 (decibel) during steam venting activities. Noise abatement technology includes the use of rock mufflers, sound insulation, in add-ition to silencers on equipment in the steam processing facility. Further recommendations for the management of occupational noise and vibration, such as the use of appropriate PPE as described in the general HSE guidelines or safe work procedure.

5.2.1 Power Project Construction Safety

Contractor shall comply with the rules and regulations set forth by OSHA or any national or local governing authority. Work shall done carefully as possible at all times with the knowledge that each employee is responsible for his own safety. Safety on the project is the prime responsibility of the contractor and its subcontractors and suppliers. OSHA requires the Contractor to have a project safety plan or program. During the progress of the Project, the client and consultant shall monitor the contractor's compliance with the safety requirements of the contract documents as well as the contractor's own project safety plan or program. Prior to the Construction activities, the client and consultant shall request and obtain from the contractor a copy of the contractor's written project safety plan/program for the project file.

Regular or at least weekly inspections shall be carried out by client safety personnel on construction area in search of any deviations from actual safe work procedure. If anything found, NCR's to be draft and ensure corrective action implemented. NCR's and explanation copy to be filed for future clarification. The client shall periodically audit the Contractor's file for compliance with its Hazard Communication Standard. The client safety authorities shall conduct Weekly Construction Coordination Meeting for safety issues, including but not limited to:

- Observations of non-compliance with the Project Safety Plan/Program or a job site accident or incident,
- Outstanding safety issues not resolved by the Contractor,
- New activity safety alerts,
- Current Emergency Preparedness Plan.

General Safety Requirements of Contractor are

1. Heavy lifting shall be carried out under a competent Engineer.
2. In cranes or chain blocks or vehicle or hoisting equipment other than the operator is prohibited. Lifting cables, slings, shackles, etc. shall be inspected regularly. Employees shall be raised only in equipment designed

to carry personnel. While lifting a banksman shall be placed to give signals to the operator. Workers are not allowed under any suspended load, the areas to be barricaded immediately.

3. Hot works and confined space entry shall be carried out only after accessing concern PTW systems.

4. Radiography test shall be conducted frequently in search of any harmful ray's presence.

5. Obey all Safety and Warning signs and posters.

6. Hard hats, safety shoes and safety glass are to be worn as full time basis on work premises. Dark mirror or clear glasses may be worn outside on the jobsite and clear safety glasses are to be worn inside of any structure.

7. A full-face shield shall be worn by any employee who cut, grind, saw, burn or weld.

8. Fall protection systems shall be worn by any employee who is working above 2 meters.

9. All other forms of PPE such as coverall, hearing protection, dust masks, gloves, full body harnesses, lifelines, lanyards, etc. necessitated to perform the work safely and properly must be provided and maintained in good condition.

10. Scaffold erection and dismantling shall be supervised by a competent person. Scaffold planking and runways shall be kept clear of obstructions at all times. Scaffold tagging (red, yellow and green) shall be used in the erection and dismantling of the scaffold. All work platforms, scaffolds, and floor openings must be equipped with guardrails on all sides that comply with OSHA requirements. Any scaffold found not to be in compliance shall be red tagged and entry to be prohibited. Scaffold inspector shall inspect scaffold and to provide green tag for scaffold entry.

11. All excavations and floor openings shall be barricaded (1.5 meters away from edges) and shored as required by the work being performed. Toe-boards shall be erect underneath the barricading.

12. Ladders must be kept in good condition and to secure at both ends. Any ladder that is damaged must be removed from the construction site.

13. Hand tools and small power tools shall be kept in good repair and used only for their intended purpose. Defective tools, tools with frayed or damaged cords should be removed from service and tagged "OUT OF SERVICE" until such time the tool is repaired or replaced. Double insulated tools must also be marked.

14. All protruding rebar steel or parts on which employees shall be guarded by approved rebar guards (caps) to eliminate this hazard.

15. Good general housekeeping is to be maintained and practiced at all times. Work areas are to be kept free of debris.

16. Empty containers, papers or debris of any kind should not be allowed to accumulate in areas used by personnel on the jobsite. Trash cans are provided for disposal of such debris.

17. Compressed gas cylinders must be maintained in an upright position, secured and capped when not in use. Stored gases must be kept in a designated area away from the immediate work area. Oxygen and Acetylene bottles in storage are to be kept separated by a distance of no less than 20 feet.

18. All injuries, no matter the extent, are to be reported immediately to your supervisor or safety department.

19. Provide information and training for employees handling or potentially exposed to hazardous material. MSDS sheets are to be updated as new materials are brought onto the construction site.

20. All materials shall be stacked on material stacking area or workshop and only material needed for a shift to be bring to the work premises.

21. Contractor shall take all steps to eliminate unsafe condition and to provide a safe work place. Strict supervision shall be provided at all time bases under competent persons for preventing unsafe acts.

22. Pressure test on erected pipelines shall be conducted under the supervision of competent Engineer as per safe work procedure. Whole area shall be barricaded with warning tape and signage's of "No Entry, Pressure test in Progress, Keep Away" to be placed.

5.3   CIVIL AVIATION SAFETY

Aviation or airport and flight safety element concerns mainly with:

1. Total Management Commitment and resources
2. Sound Aviation Safety Culture and Awareness
3. Safety Organization/Structure
4. Documentation
5. Training and Education
6. Hazard identification and Risk Management
7. Reporting and Evaluation
   - Reporting Hazards
   - Reporting Incidents
   - Investigation and analysis of both
8. Process to produce recommendations
9. Processes to ensure actions are completed
10. Emergency response.

5.3.1   Responsibilities and Requirements

1.   Center management

1. Ensure that an aviation safety program that meets applicable requirements is in place.
2. Allocate aviation resources to meet objectives/programs safely, promulgate safety awareness, conduct mishap investigations, and develop/implement corrective actions.
3. Appoint a Mishap Investigation Board for all mishaps involving aircraft.
4. Attend or send a representative to selected program Airworthiness and Flight Safety Reviews.
5. Appoint competent Aviation Safety Officers.
6. Ensure that current policies/procedures that address safety in flight, flight test, airfield, and aircraft operations are in place.

2.   Aviation safety officer

1. Foster aviation safety measures and use all resources available to promote mishap prevention
2. Assist with aircraft mishap investigation
3. Attend all Airworthiness and Flight Safety Reviews
4. Assist with the identification, evaluation, and control of ground safety hazards
5. Perform safety surveys, at least annually, in all facilities including aircrafts.

3.   Aviation personnel

Aviation personnel include ground crews, support staff, researchers, and pilots.

1. Determine and observe the safety requirements for the work being performed.
2. Report any potential or obvious safety hazard to management.
3. Use specified personal protective equipment (PPE).
4. Do not use damaged equipment that could pose a hazard.
5. Take all required safety training and adheres to all medical monitoring protocols.

4.   Aviation safety subcommittees

The two Aviation Safety Subcommittees are:

- The Aircraft and Ground Safety Subcommittee is responsible for reviewing ground operations of aircraft to detect and reduce safety hazards to persons, aircraft, and ground support equipment. The scope of committee activities includes hangar and ramp safety, personnel activities, and employment of equipment in support of aircraft operations.
- The Aviation Management Office is responsible for the review of the safety aspects of pilot training and proficiency requirements and flight procedures and practices.
- The coordination and integration of these two subcommittees is provided by the Aviation Safety Officer, who chairs them both. Responsibilities for the committees include:

1. Attend all Airworthiness and Flight Safety Reviews
2. Identify and evaluate hazards that impact flight and ground safety
3. When hazards are identified, implement the necessary safety controls
4. Ensure that personnel are trained and have adequate skills to perform their work safely
5. Resolve safety issues reported by workers
6. Maintain documentation of hazard assessments, mishap reports, and personnel training
7. Perform periodic visual inspections of work areas to ensure that safety controls are in place, are being used, and are effective.

5.3.2 Civil Aviation Construction Safety

Safety on the project is the prime responsibility of the Contractor and its subcontractors and suppliers. OSHA requires the Contractor to have a Project Safety Plan/Program. During the progress of the Project, the Airport authorities or Safety personnel shall monitor the Contractor's compliance with the safety requirements of the Contract Documents as well as the Contractor's own Project Safety Plan/Program. Construction area shall be barricaded separately from existing aviation premises for avoiding obstructions on runways or any security measures. Separate gates and security systems shall be provided for the construction premises. If construction activity is conducting near runways, all persons engaged in construction activities shall wear a reflective vest as the visibility for pilots while the flights are landing.

Prior to the Construction activities, the aviation authorities shall request and obtain from the Contractor a copy of the Contractor's written Project Safety Plan/Program for the project file. The aviation authorities shall become familiar with the project Safety Plan/Program, as this will serve as the primary benchmark for monitoring the Contractor's safety performance. The aviation authority shall request from the Contractor the names of the Competent Persons for all shifts of the project. A file on the named Competent Persons shall be maintained by the airport authorities and updated as the Competent Persons changes. At each Weekly Construction Coordination Meeting the aviation authority shall verify the name of the Contractor's Competent Person on site at all times.

Regular or at least weekly inspections shall be carried out by aviation authority safety personnel on construction area in search of any deviations from actual safe work procedure. If anything found, NCR's to be draft and ensure corrective action implemented. NCR's and explanation copies to be filed for future clarification. The aviation authority shall periodically audit the Contractor's file for compliance with its Hazard Communication Standard. Also inspect Material Safety Data Sheets for materials to be used or incorporated into the work. In the event emergency response procedure of the contractor shall monitor, compliance in securing the work site in accordance with standard and the contractor's own emergency preparedness plan.

The aviation authorities shall create a project safety file to be maintained at their office. This file shall include, but not be limited to, the following:

- Contractor's safety plan/program
- Safety phasing plans
- Project safety non-compliance notices
- General safety correspondence to the Contractor
- Accident investigation and related project photographs
- Hot work permits
- Material safety data sheets (MSDS)
- Maintenance of traffic (MOT) plans
- Contractor's toolbox safety meetings reports
- Heavy equipment (i.e. cranes, backhoes, etc.) inspection reports
- Emergency preparedness plan
- Lost-time statistical data
- Contractor safety performance monitoring reports
- Safety correspondence to the PM
- Contractor self-inspection reports
- Trench safety reports.

The aviation safety authorities shall include in the agenda for the Weekly Construction Coordination Meeting for safety issues, including but not limited to:

- Observations of non-compliance with the Project safety plan/program or a job site accident or incident
- Updates on project safety
- Outstanding safety issues not resolved by the Contractor
- Report on the progress of the toolbox meetings and training programs
- New activity safety alerts
- Maintenance of traffic (MOT) plans
- Current emergency preparedness plan.

5.4.1 Site Selection

Desalination plant site selection is very vital for the design, financing construction and operation of desalination plants. The site, comprised of onshore and offshore parts,

a. Must be located in a place where access and interconnections to the power supply grid, (or independent power production) and to the water supply networks are technically and economically feasible.

b. The area extent and shape (size and geometry) must be the appropriate so that the marine intake head structures, the marine pipelines, the inland pit the seawater pumping station, the inland pipelines, the

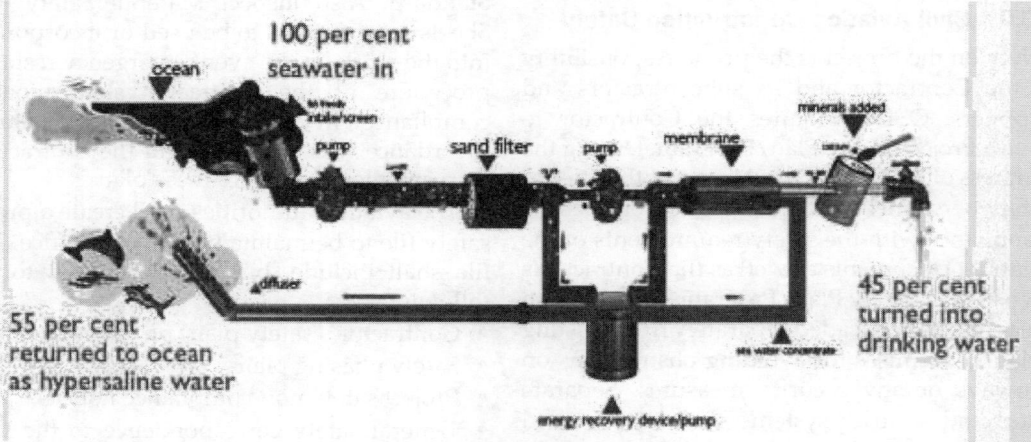

Fig. 5.2: Desalination process

5.4 DESALINATION PROJECT SAFETY

Desalination is the process converting seawater or backwater to mineral or drinking water. The balance waste water to be drain back to the main steam with out any environmental adverse impact. Seawater desalination requires the use of state of the art technology, a high proportion of exotic materials and innovative fabrication and construction techniques to counter the corrosive nature of seawater. Desalination also differs from conventional water treatments due to the basic requirement to extract salt from large volumes of pristine seawater and then returning the resultant hyposaline concentrate solution back to the ocean.

main facility structures, the post treatment system, the product delivery subsystem, and the power supply system are adequately accommodated and optimally located so that civil, electrical, piping interconnections are minimized.

c. Be suitably located in a marine environment where adequate quantity of feed water with a reasonable good, uniform and steady quality of feed seawater is abstracted at a reasonable level.

d. Be at a location where the brine, backwash wastewater and other wastes are disposed without environmental adverse effects.

e. Geologically and topographically are suitable for the construction and erection of the various structures.

f. Environmental, town planning and rural planning regulations, law requirements and restrictions are met.

g. The desalination plant shall have the social acceptance of the neighboring communities and other authorities.

5.4.2 Operation Activities

During the operation phase, the approach to plant operation includes the following objectives:

- Establish the plant process and equipment development goals.
- Accomplish all production performance targets and requirements, such as water quality and production goals.
- Reliable operations are maintained on a continuous basis,
- Equipment life is maximized,
- The safety of all personnel and equipment is assured,
- Full compliance is achieved for all applicable environmental regulations.

5.4.3 Project Planning

Planning is the important part for developing the Project Quality, Execution and HSE Management Procedure. Each new project must also have a Project HSE Management Plan to ensure that all material HSE considerations are addressed. HIRA and Safe work procedure shall be made for all ongoing activities.

The plan must also include

- Legal and risk identification, assessment and mitigation requirements relating to HSE and its safe work procedure
- The approach to managing HSE during each project phase
- The project HSE organization structure
- The health, safety and environment strategy for contractors, including how contractors HSE management plans will be integrated into the overall Plan

- A Regulatory Approval of Plan
- The Standards and Procedures shall be reviewed annually to ensure the plant operate under a leading practice framework
- HSE audits shall be conducted at regular intervals for checking work procedure and for ensuring operating sites undertake self-assessments against the Standards.

5.4.4 Desalination Plant Tests

The general 3-step approach overview to start-up includes checkout tests, prestart-up tests, and start-up and operating adjustments. These tests shall be conducted while in the time of starting a new plant, on the time after a repair or maintenance or any shutdown.

1. Checkout tests

Is to assure that the systems will be able to operate in a safe and efficient manner, without harm to operating personnel or damage to equipment. Initial start-up will consist of a dry run, followed by a water test run of the pretreatment train individual components. If deficiencies are found in any pretreatment train components, modification will be made before further tasks are accomplished.

2. Prestart-up tests

The hydraulics of the system will be evaluated, including pumps and valves. The sensors and applicable control loops will be evaluated and initial membrane performance (flux rate and product quality) will be measured and documented. Any membrane element defects or installation deficiencies will be replaced and/or corrected before further testing is accomplished. After this test flushing and cleaning systems will be evaluated.

During the checkout testing

- Charging the plant with chemicals, lubricating the equipment, and providing all the other needed operating and process supplies and materials
- Activating all plant systems, subsystems, and equipment
- Testing pumps, seal systems, and flushing piping systems

- Performing field trials and updating of opera-ting procedures.

3. Start-up and operating adjustments

The systems will be brought into operation and the adjustments needed to achieve design performance will be made. The start-up and adjustment period will continue until the systems reach their steady state, which is determined by observing the key process variables for a period of time. The baseline establishment test run will be made using a fixed set of conditions. These fixed conditions will be used in subsequent tests for comparison and evaluation of performance.

5.4.5 HSE Element Concern Mainly in

In line with HSE Standards and Procedures sites are required to have systems in place to effectively respond to crises and emergencies and re-establish full functioning operations as swiftly and smoothly as possible. Safety personnel shall be experienced to find and eliminate all potential hazards. Training's shall be conducted among the employees according to the safe work procedure.

Requirements include:

- Identifying potential emergency situations and their impacts
- Defining response plans, roles and responsibilities
- Identifying, maintaining and testing resources to ensure their availability
- Training employees, contractors, visitors and external stakeholders
- Identifying, documenting, sharing and following-up on learning's from emergency response drills.

A crisis or emergency may be an extreme climatic event, disease outbreak, security issue or any other event that poses a significant threat to the safety or health of employees, contractors, customers or the public, or that can cause significant damage to the environment.

5.4.6 Training Activities

The main training program will focus on the operation and maintenance of the plant, including the intake system, pretreatment system, desalination system, plant control system, and plant effluent. Additional training responsibilities that should be fulfilled include safety training. Safety will be an integral part of the main training program as well as a special topic covered at initial and scheduled safety lectures.

The training program will be divided into three categories:

- Initial training of the start-up crew includes shift leaders, technicians, and operators
- On-the-job training of the start-up crew
- Training of new hires subsequent to plant start-up.

5.4.7 Major Project Challenges

1. Health concerns

The water fed into a desalination system may introduce biological and chemical contaminants that are hazardous to human health. Biological contaminants include viruses, protozoa, and bacteria. Assuring public health and environmental protection requires monitoring and appropriate regulation of all desalination facilities. Health checks among the employees and neighbors shall be conducted on regular basis.

2. Environmental considerations

Adequate and safe disposal of the concentrated brine produced by the plant presents a significant environmental challenge. Typical brines contain twice as much salt as the feed water and have a higher density. In addition to high salt levels, brine from seawater desalination facilities can contain concentrations of constituents typically found in seawater, such as manganese, lead, and iodine, as well as chemicals.

Desalination produces highly concentrated salt brines that may also contain other chemical pollutants. Safe disposal of this effluent is a challenge

- More comprehensive studies are needed to adequately identify all contaminants in desalination brines and to mitigate the impacts of brine discharge.
- Water managers should carefully monitor, report, and minimize the concentrations of chemicals in brine discharges.

- Federal or state regulators should evaluate whether new water-quality regulations are needed to protect local environments or human health.
- Under all circumstances, water managers must minimize brine disposal in close proximity to sensitive habitats, such as wetlands.
- Disposal of brine in underground aquifers should be prohibited unless comprehensive and competent groundwater surveys are done and there is no reasonable risk of brine plumes appearing in freshwater wells. Impingement and entrainment of marine organisms are among the most significant environmental threats associated with sea-water desalination.
- The effects of impingement and entrainment require detailed baseline ecological assessments, impact studies, and careful monitoring.
- Intake pipes should be located outside of areas with high biological productivity and designed to minimize impingement and entrainment.

5.4.8 Desalination Environment Management Protection Plan

- Establish, implement, maintain and improve the Project Environmental Management Plan
- Implement the Performance Requirements for the Project.
- Engage an independent environmental repre-sentative.
- Carry-out environmental audits against compliance.
- Liaise with and coordinate relevant agencies for the smooth and efficient delivery of the Project.
- Obtain relevant approvals, licenses and permits before starting site works.
- Ensure that prior to commencement of work, top management have complied with relevant performance requirements such as preparing and implementing an environmental management plan.

- Review top management, line management and technicians performances against the performance requirements and take corrective action as necessary.

5.5 OIL REFINERY AND PETROCHEMICAL SAFETY

Petroleum is a complex mixture of organic liquids called crude oil and natural gas, which occurs naturally in the ground and was formed millions of years ago. Crude oil varies from oilfield to oilfield in color and composition, from a pale yellow low viscosity liquid to heavy black 'treacle' consistencies.

Crude oil and natural gas are extracted from the ground, on land or under the oceans, by sinking oil well and are then transported by pipeline or ship to refineries where their components are processed into refined products. An oil refinery is an organized and coordinated arrangement of manufacturing processes designed to produce physical and chemical changes in crude oil to convert it into everyday products like petrol, diesel, lubricating oil, fuel oil and bitumen.

As crude oil comes from the well it contains a mixture of hydrocarbon compounds and relatively small quantities of other materials such as oxygen, nitrogen, sulphur, salt and water. In the refinery, most of these non-hydrocarbon substances are removed and the oil is broken down into its various components and blended into useful products.

The refining process

Every refinery begins with the separation of crude oil into different fractions by distillation. The fractions are further treated to convert them into mixtures of more useful saleable products by various methods such as cracking, reforming, alkylation, polymerisation and isomerisation. These mixtures of new compounds are then separated using methods such as fractionation and solvent extraction. Impurities are removed by various methods, e.g. dehydration, desalting, sulphur removal and hydrotreating.

The petrochemical process

Petrochemical industry manage the operations and processes that transform crude oil and other raw materials into chemical products. There are two major categories of petrochemical companies:

- Companies that refine oil and other raw materials and convert them into basic chemicals such as ethylene,
- Companies that use these basic chemicals to create materials that can then be used in other industries to produce finished products, such as plastics, detergents, paints, fertilizers, polyester, etc.

Major products

Petroleum products are usually grouped into three categories: light distillates (LPG, gasoline, naphtha), middle distillates (kerosene, diesel), heavy distillates and residuum (heavy fuel oil, lubricating oils, wax, asphalt). This classification is based on the way crude oil is distilled and separated into fractions (called distillates and residuum). The out put products are:

- Liquid petroleum gas (LPG)
- Gasoline (also known as petrol)
- Naphtha
- Kerosene and related jet aircraft fuels
- Diesel fuel
- Fuel oils
- Lubricating oils
- Paraffin wax
- Asphalt and tar
- Petroleum coke.

The refinery and petrochemical safety concerns

Refining and Gas Processing group looks after the refinery operations, gas processing and operations of petrochemical plant located within the premises of refinery and gas processing plants. Refining and transforming chemical elements into finished products are both complex processes that involve the following procedures for the refining and petrochemical manufacturing industry:

- Creating and maintaining safe work procedure to measure and combine ingredients, and to avoid dangerous chemical reactions
- Building-up of efficient safety teams for taking safety regulations into account
- Hazardous area classification (HAC)
- Toxic chemical safety
- Project health, safety and environment
- Fire risk assessment and procedure
- Permit to work systems
- Comprehensive safety audit
- Emergency warning system and procedures.

5.5.1 Hazards in Hydrocarbon Industry

The Hydrocarbon Industry is subjected to high safety risk on account of potential hazard from fire and explosion due to handling, processing and storage of highly flammable liquid with low flash point, as well as gas and vapour and operation of the facilities at elevated temperature and pressure. Crude oil and gas are handled during oil exploration and there is potential hazard on account of fire and explosions from leaks or blowouts. Similarly at Petroleum refining, gas processing and petrochemical installations, there are emerging hazards on accounts of process upsets, extreme physical conditions in addition to potential fire and explosion from accidental release of flammable hydrocarbon.

In recent past, capacity built up in petroleum refining, coupled with induction of newer technologies like catalytic hydrotreating, isomerisation for product quality improvements and value addition through technologies like hydrocracking, fluidized catalytic cracking, delayed coking to extract maximum from bottom of the barrel and diversification towards petrochemicals, the complexity of operation has increasing so is our responsibility in operation of the assets in a safe and efficient way without disturbing the people and environment.

The plant and equipment of refineries are generally modern, and the processes are largely automatic and totally enclosed. Routine operations of the refining processes generally present

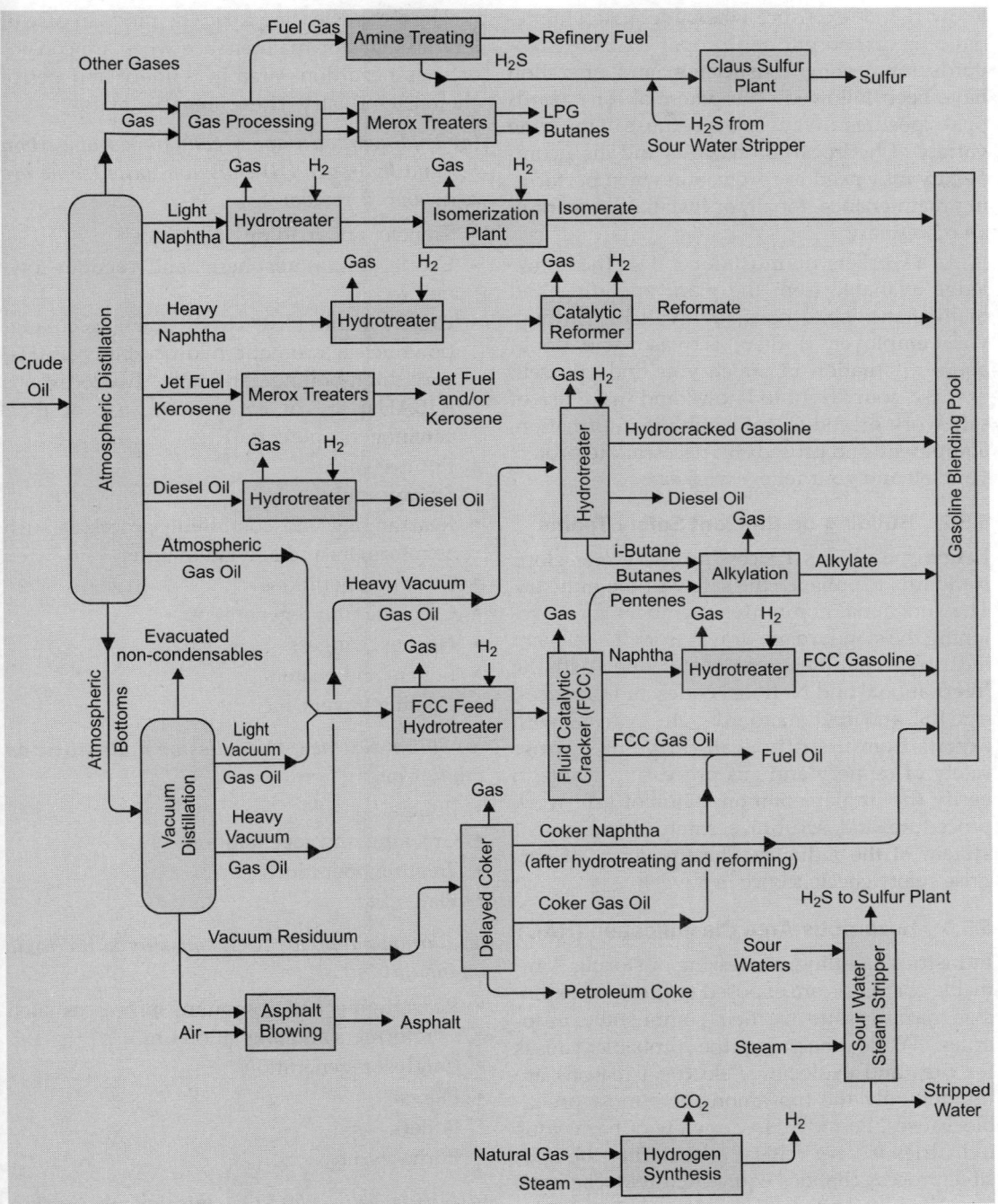

Fig. 5.3: Refinery flowchart

a low risk of exposure when adequate maintenance is carried out and proper industry standards for design, construction and operation have been followed. The potential for hazardous exposures always exists because of the wide variety of hydrocarbon hazards and their complexity may need for protection when performing maintenance, repair, or installation work in an oil refinery.

As a worker you must depend on the knowledge available from the plant operating and maintenance staff, normally available through your employer. If there is reasonable doubt about a situation in which you find yourself, exercise your "right to know" and make use of safe work procedure to obtain the information, equipment, and procedures necessary to protect yourself and your fellow workers.

5.5.2 Building-up Efficient Safety Teams

Refining and Gas Processing group develops standards to enhance the safety of the industry. The functional committee members for developing the standard are drawn from the relevant industry to put their experiences along with the International and National codes and practices, e.g. BSI and ISO standards, etc. A team with experts from industry shall carry-out surprise safety of refinery and gas processing plants to verify the implementation status of safe work procedure and scrutinize safety management system of the industry. The frequency of surprise safety check is once in a week.

5.5.3 Hazardous Area Classification (HAC)

Industries handling, processing or storing flammable chemicals are exposed to fires and explosion hazards due to their combustible properties. While analyzing the probable causes for fires and explosions, electrical reasons are undoubtedly the top among the 'most probable' causes. Periodic HAC review of hazardous industries is very critical considering the possible process changes equipment replacements, etc. All rules and regulations according to the local authorities and safe work procedure shall be implemented while in stacking, transporting and handling these chemicals.

The areas releasing toxic chemicals in case of any leakages or maintenance work are considered as hazardous area in refinery and petrochemical industry. These areas are mainly:

(a) Hydrocarbon vapors compounds of carbon (C) and hydrogen (H) emission units (some are toxic and corrosive)

- Transfer and loading operations
- Crude unit, atmospheric, and vacuum towers
- Cracking units ("cat", hydrocracking, coking-polynuclear aromatic hydrocarbons [PAHs] and high-boiling aromatic hydrocarbons [HBAHs] are of concern because of their carcinogenic potential)
- Pumps, valves
- Storage tanks
- Rearranging and combining processes such as reformers and alkylation units
- Treating operations
- Cracking unit regeneration
- Heat exchangers
- Boilers and heaters
- Cooling towers, etc.

(b) Sulfur dioxide (SO_2) emission units (toxic on inhalation)

- Boilers
- Cracking unit regeneration
- Treating operations
- Flares, etc.

(c) Carbon monoxide (CO) emission units (toxic on inhalation)

- Rearranging and combining processes such as reformers and alkylation units
- Catalyst regeneration
- Flares
- Boilers
- Furnaces, etc.

(d) Nitrogen dioxide (NO_2) emission unit (asphyxiate)

- Flares
- Boilers, etc.

(e) Hydrogen sulphide (H₂S) emission units (toxic on inhalation)

- Sour crudes
- Liquid wastes
- Pumps
- Crude tower
- Cracking operations
- Hydrogeneration
- Rearranging and combining processes such as reformers and alkylation units, etc.

(f) Catalyst dust emission units (some are toxic and corrosive, may damage eyes and respiratory system)

- Cracking units
- Catalyst regeneration
- Rearranging and combining processes such as reformers and alkylation units.

(h) Chlorine (Cl or Cl₂) emission unit (corrosive to skin and tissue): Caustic unit.

(i) Ammonia (NH₃) emission unit (toxic on inhalation): Compressors

The principal exposures to hazardous substances occur during emergency, shutdown or maintenance work, since these are a deviation from routine operations. Plant turnarounds require careful planning, scheduling and step-by-step procedures to make sure that unanticipated exposures do not occur. Any plant shutdown requires a complete plan in writing to cover all activities, the impact on other operations and emergency planning. Plans are normally formulated by plant personnel in conjunction with contractors. The safety plan shall be implemented on work face and employees shall enforced to wear all required PPE's.

5.5.4 Toxic Chemical Safety

In a refinery and petrochemical industry hazardous chemicals can come from many sources and in many forms. In crude oil, there are not only the components sought for processing, but impurities such as sulfur, vanadium and arsenic compounds. The oil is split into many component streams that are further altered and refined to produce the final product range. Most, if not all, of these component stream chemicals are inherently hazardous to humans, as are the other chemicals added during processing. Hazards include fire, explosion, toxicity, corrosiveness, and asphyxiation. Information on hazardous materials manufactured or stored in a refinery should be supplied by the client's representative when a work permit is issued.

Releases of hydrofluoric acid, carbon monoxide, methanol and hydrogen sulfide may present occupational exposure hazards. Hydrogen sulfide leakage may occur from amine regeneration in amine treatment units and sulfur recovery units. Carbon monoxide leakage may occur from fluid and residue catalytic cracking units and from the syngas production section of the hydrogen plant. Carbon monoxide and air mixtures are explosive, and spontaneous explosive re-ignition may occur. Hydrogen sulfide poses an immediate fire hazard when mixed with air. Workers may be exposed to potential inhalation hazards (e.g. Hydrogen sulfide, carbon monoxide, poly cyclic aromatic hydrocarbons (PAHs) during routine plant operations. Dermal hazards may include contact with acids, steam, and hot surfaces. Chemical hazards should be managed based on the results of a job safety analysis and industrial hygiene survey and according to the occupational health and safety guidance provided in the safe work procedure. Protection measures include worker training, work permit systems, use of personal protective equipment (PPE), and toxic gas detection systems with alarms.

The following precautions shall be taken

- Highlight major potential air contaminants which can escape from a typical refinery and petrochemical operation and their major sources. Safety precautions taken against these potential hazards shall be emphasized. Training shall be conducted for the work force according the safety precautions.
- Reviews common hazardous chemicals and chemical groups typically present and their most significant hazards to workers.
- Care should be exercised at all times to avoid inhaling solvent vapors, toxic gases, and other

respiratory contaminants. Because of the many hazards from toxicity, corrosiveness, burns and skin contact, most plants require to wear all required PPE's.

Oxygen-deficient atmosphere

The potential release and accumulation of nitrogen gas into work areas may result in the creation of asphyxiating conditions due to the displacement of oxygen. Prevention and control measures to reduce risks of asphyxiated gas release include:

- Design and placement of nitrogen venting systems according to industry standards,
- Installation of an automatic Emergency Shutdown System that can detect and warn of the uncontrolled release of nitrogen (including the presence of oxygen deficient atmospheres in working areas), initiate forced ventilation, and minimize the duration of releases,
- Implementation of confined space entry procedures as described in the safe work procedure with consideration of facility-specific hazards.

5.5.5 Project Health, Safety and Environment

The overall objective of the safety department is to assure the Company that HSE - sensitive areas have been identified in a systematic way and that the major projects, engineering and operational systems have been or will be developed to control and manage the identified risks. The assurance is provided by reviewing proposals at various key stages in their development.

The safety management in hydrocarbon industry is a multidisciplinary function and every person needs to keep vigil against potential cause of fire and accidents and strive to eliminate them. An incident is a failure in the control of any hazard that results in unplanned event like fire, explosion, run away reaction, release of toxic or flammable material, injury or fatality. A workplace incident is an indication that prevention was ineffective and that prompt changes need to be made. Everyone manning work stations should draw lessons from past industry incidents to prevent recurrence.

Process safety

Process safety programs should be implemented, due to industry-specific characteristics, including complex chemical reactions, use of hazardous materials (e.g. toxic, reactive, flammable or explosive compounds), and multistep reactions.

Process safety management includes the following actions

- Physical hazard testing of materials and reactions
- Hazard analysis studies to review the process chemistry and engineering practices, including thermodynamics and kinetics
- Examination of preventive maintenance and mechanical integrity of the process equipment and Utilities
- Worker training
- Development of operating instructions and emergency response procedures.

Safe work practices and procedures

Personnel

- Hearing protection and safety glasses must be worn in all operating areas or as posted
- Respiratory protection or equipment must be fitted. Facial hair is unacceptable where the mask must make an airtight seal against the face
- Shirts must be long-sleeved and worn with full-length pants or coveralls
- Clothing must not be of a flammable type such as nylon, dacron, acrylic or blends. Fire-resistant types include cotton, nomex, and proban shall be wear
- Other PPE required may include acid hood, impervious outerwear, rubber boots, face shields, rubber gloves, disposable coveralls, monogoggles and fall-arrest equipment
- Smoking is allowed only in designated areas.

Vehicles

- Vehicle entry is by permit only and keys are to be left in parked vehicles
- Vehicles must be shut down at the sound of any emergency alarm

- Vehicles must be equipped with ground straps or cables
- Vehicles shall fit with spark proof protection units
- MML display units to be work in order.

5.5.6 Fire Risk Assessment and Fire Protection Systems

Objectives

Fire Risk Assessments are conducted with the following objectives:
- To carry-out a systematic and critical evaluation of Fire Safety of the occupancy
- To suggest recommendations to improve the fire safety standards.

Scope of the study

A broad outline of the scope is given below:
- Identifying potential fire, explosion hazards and risks in the process and storage areas of the premises and suggesting appropriate preventive measures
- Reviewing the existing fire protection systems and suggest modifications wherever necessary as per applicable national and international standards
- Identify deviations with respect to fire safety procedures and suggest action plan to correct deviations
- Advising further scope on the compliance with statutory requirements related to fire safety and Explosions
- Providing guidelines for preparation of fire emergency and formation of team for fire fighting, first aid, rescue teams and allocating specific responsibilities. Reviewing the existing onsite emergency plans with respect to Fire risk wherever such plans are available.

The principal hazards at refineries and petrochemical plant are fire and explosion. Refineries process a multitude of products with low flash points. Although systems and operating practices are designed to prevent such catastrophes, they can occur. Constant monitoring is therefore required. Safeguards include warning systems, emergency procedures, and permit systems for any kind of hot or other potentially dangerous work. These requirements must be understood and followed by all workers. The use of matches, lighters, cigarettes, and other smoking material is generally banned in the plant except in specially designated areas.

Fire protection systems

Objectives

1. Carry-out a fire risk assessment to identify critical areas in the plant based on the fire load calculations.
2. Review the existing fire prevention and protection system against national and international Standards.
3. Carry-out a gap analysis of the existing maintenance practices pertaining to fire protection system.
4. Evaluate the preparedness (knowledge/training) of the personnel to handle emergency

Scope shall focus on

- Identification of fire hazards in Plant premises,
- Mapping of critical areas (with respect to fire hazard) in the plant based on the fire load,
- To evaluate the existing fire protection system based on relevant following standards.

Fire and explosion hazards generated by process operations include the accidental release of syngas (containing carbon monoxide and hydrogen), oxygen, methanol, and refinery gases. Refinery gas releases may cause 'jet fires', if ignited in the release section, or give rise to a vapor cloud explosion (VCE), fireball or flash fire, depending on the quantity of flammable material involved and the degree of confinement of the cloud. Methane, hydrogen, carbon monoxide, and hydrogen sulfide may ignite even in the absence of ignition sources, if their temperature is higher than their auto ignition temperatures of 580 C, 500 C, 609 C, and 260 C, respectively. Flammable liquid spills present in petroleum refining facilities may cause pool fires. Explosive hazards may also be associated with accumulation of vapors in storage tanks (e.g. sulfuric acid and bitumen).

Recommended measures to prevent and control fire and explosion risks from process operations include the following:

- Designing, constructing, and operating petroleum refineries and petrochemicals according to international standards for the prevention and control of fire and explosion hazards, including provisions for segregation of process, storage, utility, and safe areas. Safety distances can be derived from specific safety analyses for the facility, and through application of internationally recognized fire safety standards
- Providing early release detection, such as pressure monitoring of gas and liquid conveyance systems, in addition to smoke and heat detection for fires
- Evaluation of potential for vapor accumulation in storage tanks and implementation of prevention and control techniques (e.g. nitrogen blanketing for sulfuric acid and bitumen storage)
- Avoiding potential sources of ignition (e.g. by configuring the layout of piping to avoid spills over high temperature piping, equipment or rotating machines)
- Providing passive fire protection measures within the modeled fire zone that are capable of withstanding the fire temperature for a time sufficient to allow the operator to implement the appropriate fire mitigation strategy
- Limiting the areas that may be potentially affected by accidental releases by:
  a. Defining fire zones and equipping them with a drainage system to collect and convey accidental releases of flammable liquids to a safe containment area, including secondary containment of storage tanks
  b. Installing fire and blast partition walls in areas where appropriate separation distances cannot be achieved
  c. Designing the oily sewage system to avoid propagation of fire.

Fire and Accidents Investigation

This group carries out investigation of fires and accidents in refinery and gas processing plants and brings out the salient features (direct, indirect and root causes) responsible for the occurrence of such incidents and suggest remedial measures to prevent re-occurrences. The analysis and the incidents shared with the industry members during committee meeting.

5.5.7 Comprehensive Safety Audit and Inspections

External Safety Audits of the Refineries and Gas Processing plants by a specialized group of highly experienced executives from the industry as well as other statutory bodies to strengthen safety measures and also emphasis the implementations of safety rules and regulations of various statutory bodies in the Refineries for enhancing safety. External safety audits are carried out at a frequency of at least half a years or as per the directives of client.

The focus areas in such audits are

1. Design and layout of the installation
2. Operation safety and plant operating practices
3. Maintenance and inspection aspects
4. Safety management and fire protection systems
5. Environment compliance
6. Training and development activities.

A team with experts from industry carry-out surprise safety of refinery and gas processing plants to verify the implementation status of safe work procedure and scrutinize safety management system of the industry. The frequency of surprise safety check is once in a week.

5.5.8 Permit to Work Systems

No work takes place in a refinery without a safe work permit. A safe work permit is a document issued by an authorized representative of the client permitting specific work for a specific time in a specific area. Safe work permits are valid only for a limited time and must be renewed following expiry or normally after any one-hour stoppage, after an emergency warning on the site or for other safety reasons. After such an event, any required gas testing or other testing must be repeated to ensure a safe return to the work.

The types of safe work permits required typically include the following. Specific categories may vary from site to site.

- **Hot work**—covers any work that involves heat or an ignition source, including welding, grinding, and the use of any kind of motor. In high-risk areas, a spark watch may be required.
- **X-ray and radiation**
- **Benzene**—required when a benzene exposure hazard exists.
- **Confined space entry work**—involving potential ignition hazards.
- **Confined space entry cold work**—involving work that will not produce a spark.
- **Hoisting**—permit.
- **Electrical**—for other than routine work.
- **Camera**—typically requires a hot work permit when lighting is required.
- **Asbestos**—required whenever an asbestos exposure hazard exists.
- **Vehicle movement.**
- **Hydrant**—permits the use of plant fire hydrants.

5.5.9 Emergency Warning System and Procedures

In oil refineries there will be both plant alarms or whistles and individual unit alarms. All workers must receive training in recognizing and responding to these alarms. Verbal messages usually accompany the alarms. There will be different alarms for a fire or toxic emergencies. The purpose of preparing a well defined off site emergency plan is to systematically document and define various types of catastrophic situations where it will be necessary for the company to safeguard its own people and to assist district administration along with other statutory authorities to take control of the situation, rescue and evacuation of people living around the industry. The infrastructure set up required at the time of emergency and the envisaged action by District Administration, Fire Services, Police Dept., Medical Services, Transport Dept. and other voluntary organizations, till normalcy is achieved after the occurrence of the event. This plan takes in to account all emergency scenarios of the various industries of the areas to publish the risk contours in the various industrial zones to prepare industry to meet all potential emergencies.

Procedures while hearing an alarm

- When an alarm sounds, secure all equipment and shut down all vehicles.
- Note the wind direction (wind socks) and proceed to the appropriate assembly area.
- Do a head count to make sure all personnel are accounted for and report the result to a client contact person.
- Know the local designated safety areas or assembly point and emergency phone number(s). If you are the one who is first aware of an emergency, then call the emergency number by:
  1. Reporting your name
  2. Describing the emergency
  3. Identifying its location
  4. Indicating whether anyone is injured
  5. Proceed to the assembly area.

Electrical precautions

- Electrical tagging and lockout procedures must be understood and followed by all workers.
- All electric tools, cords, and equipment must be grounded or double-insulated.
- Use explosion-proof fixtures where required.
- All electric safety procedures shall be followed.

Part II

SAFETY DOCUMENTATION TRAINING PROGRAMME

Assignment of Duties

1

Some of Assignment of Duties are defined:

a. Assignment of Project Manager
b. Assignment of Site Engineer
c. Assignment of Safety Manager
d. Assignment of Safety Engineer
e. Assignment of Safety Officer
f. Assignment of Site Supervisor

Markings

All top management and line management employee (excluding the low management employee) should be assigned their duties, defined, signed and documented by their responsible head of department personnel, as per OSHA:18001; 4.4.1 Specification.

What is the Infrastructure of Assignment of Duties?

When a person is recruited for job position according to his professional talent, the company has to define to the recruited person about, what company needs from him and what he must do exactly for the company. For this course assignment of duties is held. This document should be defined, discussed and signed by both the senior management person of the company and by the recruited person.

Does Assignment of Duties Change from Industry to Industry?

Yes its contents change according to the type of Industry, activities performed, atmospheric condition, hazards and risks involved and the company policy. But the bottom-line concept of assignment of duties program never changes.

What is the Legal Eligibility of Assignment of Duties?

According to the company point of view, it is a part of company policy and if the recruited person fails to perform his duties as signed and agreed the company can discipline him or terminate him from his position. But if a serious accident occurs and it occurs in case of incorrectly performing his duties, he is to be considered responsible to the accident and legal department can take action against him.

How is Assignment of Duties Defined SDTP Document Filling?

See the attached documents:

SDTP	1.1 Assignment of Duties Project Manager	Document References	
		SDTP. ASS. 001	
		Rev. No	Doc. Date
		0	1/1/2009
		Pages	
		1 of 2	

As per OSHA 18001 Specification 4.4.1/Local Rule No.

This company has the responsibility to provide adequate preventive measures to the project workers against the dangers of occupational accidents and occupational diseases that may occur during the work phase, and also against the fire hazards and other hazards that may cause from the use of machines and other equipment. He/she shall also adopt all other preventive measures enforced by the Ministry of Labor and Social Affairs. He/she shall also be responsible not only for his own safety but also for all employee's working under him.

In terms of the provision of the said Law: I	(General Manager Name)	(Representative of Employer)
Designated	(Recruited Project Manager Name)	Employee as	Project Manager

To assist in ensuring compliance with the provision of the said Law, at the following premises/site:

(Company Name and Industry/Site Name and No.)	As from Date	(Date of Document Signing)	For the Duration of Project

Your assigned duties include, but are not limited to the listed duties; you are to familiarize yourself with the contents of the said Law.

Duties and Responsibilities

1.	Conduct or carry-out hazard identification and risk analysis (HIRA) for the tasks at hand and to ensure that appropriate and correct controls are in place and are communicated to all relevant personnel. Take the necessary corrective action where required.
2.	Where it is not possible to remove any remaining hazards you are to inform employees thereof and what precautionary action is to be taken.
3.	Ensure that any employee you appoint/designate to assist you is suitably competent and or qualified to perform their duties.
4.	Ensure that those employees who are required to operate machinery are competent and/or qualified.
5.	Ensure that all sub contractor(s) are aware on any danger/hazard, potential danger/hazard on the premises.
6.	Ensure that only authorized persons are allowed to enter onto and or work on the premises.
7.	Ensure all excavations are inspected and results recorded daily.
8.	Ensure all plant and equipment is inspected daily and the findings recorded on the register provided.
9.	Ensure that all HSE Operational Controls, Method Statements and Work Instructions are communicated to all employees and adhered to at all times.

10.	Ensure all work at heights is done under strict supervision at all times.
11.	Assist with the compiling of Method statement and the development of Safe Working Procedures.
12.	Ensure that Weekly HSE Inspections and Monthly HSE Audits are conducted.
13.	To assist the employer to comply with the legislation as defined.
14.	Ensure that HSE Registers and Inspections are conducted and completed.
15.	Endorse Safety Officer Reports/minutes of the meeting.
16.	Report and investigate any incident and accident as per policy and Legal requirements.
17.	Ensure the subcontractors/suppliers comply with all the requirements set out in the site safety plan and legal requirements.

AUTHORIZED/DESIGNATED BY		
(Sign of General Manager)	General Manager	(Date of Document Signing)
Signature	Designation	Date
ACCEPTANCE		
I	(Recruited Project Manager Name)	Hereby acknowledge receipt of
Accept and understand the requirements of this appointment.		
(Sign of Recruited Project Manager)	Project Manager	(Date of Document Signing)
Signature	Designation	Date

SDTP	**1.2 Assignment of Duties Site Engineer**	Document References

Document References
SDTP. ASS. 002

Rev. No	Doc. Date
0	1/1/2009

Pages
1 of 2

As per OSHA 18001 Specification 4.4.1/Local Rule No.

This company has the responsibility to provide adequate preventive measures to the project workers against the dangers of occupational accidents and occupational diseases that may occur during the work phase, and also against the fire hazards and other hazards that may cause from the use of machines and other equipment. He/she shall also adopt all other preventive measures enforced by the Ministry of Labor and Social Affairs. He/she shall also be responsible not only for his own safety but also for all employee's working under him.

In terms of the provision of the said Law: I	(Project Manager Name)	(Representative of Employer)
Designated	(Recruited Site Engineer Name)	Employee as	Site Manager

To assist in ensuring compliance with the provision of the said Law, at the following premises/site:

(Company Name and Industry/Site Name and No.)	As from Date	(Date of Document Signing)	For the Duration of Project.

Your assigned duties include, but are not limited to the listed duties; you are to familiarise yourself with the contents of the said Law.

Duties and Responsibilities	
1.	Conduct or carry-out hazard identification and risk analysis (HIRA) for the tasks at hand and to ensure that appropriate and correct controls are in place and are communicated to all relevant personnel. Take the necessary corrective action where required.
2.	To continuously think of and review the consequences and risks associated with the execution of respective activities, whilst ensuring that the appropriate actions and controls (i.e. Excavation Permit) are in place before any tasks are carried out.
3.	Ensure that risk assessments conducted, are reviewed periodically and are upto date.
4.	To ensure that no unsafe acts to be undertaken by yourself or any of your subordinates, or any unsafe conditions exist that could cause harm/damage or endanger any persons.
5.	Where it is not possible to remove any remaining hazard(s) you are to inform employees thereof and what precautionary action is to be taken?
6.	Ensure that only authorized persons are allowed to enter onto and or work on the premises.
7.	To assist the employer to comply with the legislation as defined.
8.	Ensure that any employee you appoint/designate to assist you is suitably competent and or qualified to perform their duties.
9.	Ensure all work at heights is done under strict supervision at all times.

10.	Ensure that those employees who are required to operate machinery are competent and or qualified.
11.	Ensure that all subcontractor(s) are aware on any danger/hazard, potential danger/hazard on the premises.
12.	Ensure that all Permits to work systems are obtained prior to proceeding with any excavation, hot works, confined space entry, etc and inspections and results recorded daily.
13.	Ensure all plant and equipment is inspected daily and the findings recorded on the register provided.
14.	Ensure that all HSE Operational Controls, Method Statements and Work Instructions are communicated to all employees and adhered to at all times.
15.	Ensure that HSE Registers and Inspections are conducted and completed.
16.	Assist with the compiling of Method statement and the development of Safe Working Procedures.
17.	Ensure the subcontractors comply with all the requirements set out in the site safety plan and legal requirements.
18.	Ensure that employees/subcontractors have the correct PPE and are applying/use PPE at all times.
19.	Report any incidents/accidents immediately to your immediate line manager and the Construction/Contracts/Sr. Project Manager as well as the Safety Manager/Officer.
20.	Receive and sign for any safety inspection findings, NCR's raised or any safety improvement notices issued.

AUTHORIZED/DESIGNATED BY		
(Sign of Project Manager)	Project Manager	(Date of Document Signing)
Signature	Designation	Date
ACCEPTANCE		
I	(Recruited Project Manager Name)	Hereby acknowledge receipt of
Accept and understand the requirements of this appointment.		
(Sign of Recruited Site Engineer)	Site Engineer	(Date of Document Signing)
Signature	Designation	Date

SDTP	**1.3 Assignment of Duties Safety Manager**	Document References		
		SDTP. ASS. 003		
		Rev. No	Doc. Date	
		0	1/1/2009	
		Pages		
		1 of 2		

As per OSHA 18001 Specification 4.4.1/Local Rule No.

This company has the responsibility to provide adequate preventive measures to the project workers against the dangers of occupational accidents and occupational diseases that may occur during the work phase, and also against the fire hazards and other hazards that may cause from the use of machines and other equipment. He/she shall also adopt all other preventive measures enforced by the Ministry of Labor and Social Affairs. He/she shall also be responsible not only for his own safety but also for all employee's working under him.

In terms of the provision of the said Law: I	(General Manager Name)	(Representative of Employer)
Designated	(Recruited Safety Manager Name)	Employee as	Safety Manager

To assist in ensuring compliance with the provision of the said Law, at the following premises/site:

(Company Name and Industry/Site Name and No.)	As from Date	(Date of Document Signing)	For the Duration of Project

Your assigned duties include, but are not limited to the listed duties; you are to familiarize yourself with the contents of the said Law.

	Duties and Responsibilities
1.	Overall responsibility for ensuring SHE Management System is implemented and maintained in accordance with OHSAS 18001 and ISO 9001:2000, and for reporting directly to the Managing Director for the management of the System
2.	Collect and present to the Management Review team data on Customer complaints, corrective actions and supplier problems and any other measurements of the effectiveness of the System. Provide schedule, agendas and Minutes for Management Review Meetings.
3.	Develop and control documentation relating to the System.
4.	Ensure that the System, including the policies, is reviewed on a regular basis or in the event of any organizational restructuring.
5.	Ensure that the policies are reviewed and communicated to all management team members and staff. Evaluate awareness of the same.
6.	Provide or arrange training in Safety, Health, Quality and Environmental matters to staff, including auditor training for internal auditors.
7.	Verify whether activities and related results comply with planned arrangements by regular auditing.
8.	Deal with suggestions for improvements and proposed amendments made by staff, and make appropriate recommendations to the management.
9.	

10.	
11.	
12.	

AUTHORIZED/DESIGNATED BY		
(Sign of Project Manager)	General Manager	(Date of Document Signing)
Signature	Designation	Date
ACCEPTANCE		
I	(Recruited Safety Manager Name)	Hereby acknowledge receipt of
Accept and understand the requirements of this appointment.		
(Sign of Recruited Safety Manager)	Safety Manager	(Date of Document Signing)
Signature	Designation	Date

SDTP	1.4 Assignment of Duties Safety Engineer	Document References

		Document References
		SDTP. ASS. 004
		Rev. No / Doc. Date
		0 / 1/1/2009
		Pages
		1 of 2

As per OSHA 18001 Specification 4.4.1/Local Rule No.

This company has the responsibility to provide adequate preventive measures to the project workers against the dangers of occupational accidents and occupational diseases that may occur during the work phase, and also against the fire hazards and other hazards that may cause from the use of machines and other equipment.

A Full-time Safety Engineer requires the following qualifications:
- Degree/Diploma in any Engineering or branch of engineering for works undertaken.
- Diploma in Safety: Membership of the Occupational Safety and Health Institute.
- Including relevant experience on construction sites.

In terms of the provision of the said Law: I		(Safety Manager Name)	(Representative of Employer)
Designated	(Recruited Safety Engineer Name)	Employee as	Safety Engineer

To assist in ensuring compliance with the provision of the said Law, at the following Premises/Site:

(Company Name and Industry/Site Name and No.)	As from Date	(Date of Document Signing)	For the Duration of Project

Your assigned duties include, but are not limited to the listed duties; you are to familiarize yourself with the contents of the said Law.

Duties and Responsibilities	
1.	To ensure that management (project supervisory and management staff) are advised with written compliance with the law, local regulations and client's requirements.
2.	Supervise on-site safety and promote the safe conduct of site work.
3.	Responsible for overseeing the implementation of the project, site or facility safety plan and keeping of up to date records for the safety file to enable auditing to be carried out.
4.	Conduct Employee Safety Inductions and Day Visitor/Supplier Safety Inductions and/or toolbox talks and keep records of attendees.
5.	Conduct Employee safety toolbox talks as well as keeping records of training and attendees.
6.	To stop unsafe acts being undertaken by anyone on the site or work to commence/continue in unsafe conditions.
7.	Ensure workers use correct PPE for the type of work being carried out.
8.	Report any incidents, accidents or near miss immediately, as per the Company procedure, (this may include liaison with police and/or authorities) and prevent that the area be tampered with.

9.	Ensure injured parties are cared for and the Administration Department, Line Management and the Safety Manager are notified of any accidents.
10.	Inform the Construction/Contracts/Snr. Projects/Plant/HR Manager and Safety Manager of any areas where realistic improvements can be made for site working conditions.
11.	Maintain communication by attending meetings (close action points from minutes) and close any violation or discrepancies from Consultant or Client.
12.	Ensure site welfare facilities (rest areas, water, toilets, etc.), sufficient for the number of personnel, are being provided by the site supervisory staff.
13.	Ensure that subcontractors comply with the company safety rules and regulations as Site Safety Plan.
14.	Be proactive in addressing operational safety issues to ensure that the risk to workers and others are managed, minimized and prevented.
15.	Promote a positive safety attitude to personnel, subcontractors, consultants and clients.
16.	Investigate incidents, accidents and near miss events to determine the root, basic and immediate causes and ensure that corrective/preventative actions are implemented for all safety related events.
17.	Ensure and facilitating communication and coordination between the company and subcontractors.

AUTHORIZED/DESIGNATED BY		
(Sign of Project Manager)	General Manager	(Date of Document Signing)
Signature	Designation	Date
ACCEPTANCE		
I	(Recruited Safety Engineer Name)	Hereby acknowledge receipt of
Accept and understand the requirements of this appointment.		
(Sign of Recruited Safety Manager)	Safety Engineer	(Date of Document Signing)
Signature	Designation	Date

SDTP	1.5 Assignment of Duties Safety Officer	Document References

		Document References
		SDTP. ASS. 005
		Rev. No	Doc. Date
		0	1/1/2009
		Pages
		1 of 2

As per OSHA 18001 Specification 4.4.1/Local Rule No.

This company has the responsibility to provide adequate preventive measures to the project workers against the dangers of occupational accidents and occupational diseases that may occur during the work phase, and also against the fire hazards and other hazards that may cause from the use of machines and other equipment.

Has a responsibility to provide for any workplace with over 150 workers, at any given time, to have a fulltime Safety Officer appointed. If less than 150 workers, then a "part time" Safety Officer is acceptable providing he is given enough time to perform his duties effectively as Safety Officer.

In terms of the provision of the said Law: I		(Safety Manager Name)	(Representative of Employer)
Designated	(Recruited Safety Officer Name)	Employee as	Safety Officer

To assist in ensuring compliance with the provision of the said Law, at the following Premises/Site:			
(Company Name and Industry/Site Name and No.)	As from Date	(Date of Document Signing)	For the Duration of Project

Your assigned duties include, but are not limited to the listed duties; you are to familiarize yourself with the contents of the said Law.

	Duties and Responsibilities
1.	To ensure that management (project supervisory and management staff) are advised with written compliance with the law, local regulations and client's requirements.
2.	Supervise on-site safety and promote the safe conduct of site work.
3.	Responsible for overseeing the implementation of the project, site or facility safety plan and keeping of up to date records for the safety file to enable auditing to be carried out.
4.	Conduct Employee Safety Inductions and Day Visitor/Supplier Safety Inductions and/or toolbox talks and keep records of attendees.
5.	Conduct Employee safety toolbox talks as well as keeping records of training and attendees.
6.	To stop unsafe acts being undertaken by anyone on the site or work to commence/continue in unsafe conditions.
7.	Ensure workers use correct PPE for the type of work being carried out.
8.	Report any incidents, accidents or near miss immediately, as per the Company procedure, (this may include liaison with police and/or authorities) and prevent that the area be tampered with.

9.	Ensure injured parties are cared for and the Administration Department, Line Management and the Safety Manager are notified of any accidents.
10.	Inform the Construction/Contracts/Snr. Projects/Plant/HR Manager and Safety Manager of any areas where realistic improvements can be made for site working conditions.
11.	Maintain communication by attending meetings (close action points from minutes) and close any violation or discrepancies from Consultant or Client.
12.	Ensure site welfare facilities (rest areas, water, toilets, etc.), sufficient for the number of personnel, are being provided by the site supervisory staff.
13.	Ensure that subcontractors comply with the company safety rules and regulations as Site Safety Plan.
14.	Be proactive in addressing operational safety issues to ensure that the risk to workers and others are managed, minimized and prevented.
15.	Promote a positive safety attitude to personnel, subcontractors, consultants and clients.
16.	Investigate incidents, accidents and near miss events to determine the root, basic and immediate causes and ensure that corrective/preventative actions are implemented for all safety related events.
17.	

AUTHORIZED/DESIGNATED BY		
(Sign of Safety Manager)	Safety Manager	(Date of Document Signing)
Signature	Designation	Date

ACCEPTANCE		
I	(Recruited Safety Officer Name)	Hereby acknowledge receipt of
Accept and understand the requirements of this appointment.		
(Sign of Recruited Safety Officer)	Safety Officer	(Date of Document Signing)
Signature	Designation	Date

SDTP	**1.6 Assignment of Duties Site Supervisor**	Document References

Document References	
SDTP. ASS. 006	
Rev. No	Doc. Date
0	1/1/2009
Pages	
1 of 2	

As per OSHA 18001 Specification 4.4.1/Local Rule No.

Has a responsibility to provide adequate preventative equipment to protect workers against the dangers of employment accidents and occupational diseases that may occur during the work, and also against the fire hazards and other hazards that may result from the use of machines and other equipment. He shall also adopt all other preventative methods ordered by the Ministry of Labor and Social Affairs.

Further, anyone who is placed in control of other people is not only responsible for his own safety, but that of the people he is overseeing/supervising.

In terms of the provision of the said Law: I	(Project Manager Name)	(Representative of Employer)
Designated	(Recruited Site Supervisor Name)	Employee as	Site Supervisor

To assist in ensuring compliance with the provision of the said Law, at the following Premises/Site:

(Company Name and Industry/Site Name and No.)	As from Date	(Date of Document Signing)	For the Duration of Project

Your assigned duties include, but are not limited to the listed duties; you are to familiarize yourself with the contents of the said Law.

Duties and Responsibilities	
1.	To assist the employer to comply with the legislation as defined.
2.	Endorse Safety Officer Reports/Minutes of the meeting (where applicable).
3.	Report any incident or accident immediately to the Safety Representative and direct line management.
4.	Report and assist in investigating any incident with investigating team.
5.	Ensure that subcontractors comply with all the requirements set out in the site safety plan and legal requirements.
6.	Ensure Risk assessments are conducted within your area of responsibility and upto date.
7.	Assist with the compiling of Method statement and the development of Safe Working Procedures.
8.	Ensure all excavations are inspected and results recorded daily.
9.	Ensure all work at heights is done under strict supervision at all times
10.	Ensure all plant and equipment is inspected daily and the findings recorded on the register provided.

11.	Ensure that employees and subcontractors use the correct tools and equipment at all times.
12.	Conduct Pre-Start Review Meeting with teams/gangs prior to commencing daily activities whereby orientation of the tasks at hand is being carried out.
13.	Ensure that the right gang is utilized for the right job while ensuring that the tasks are executed in a controlled and safe manner.
14.	Ensure that all personnel under your supervision/control is trained and competent for the various tasks/types of work they are expected to perform.
15.	Conduct Man Job Tasks Analysis/Observations on a frequent and regular basis as well as when starting a new job/activity to determine the levels of risk and exposure.
16.	Ensure regular checks/verification of relevant employee licenses/certificates, training and experience required to perform tasks.
17.	Ensure that no unsafe acts or conditions are undertaken or allowed either by yourself, your subordinates or any other persons performing tasks on the premises under your control.
18.	Ensure that Forman/Chargehands and laborers carry-out and perform housekeeping activities for all areas under your control.
19.	Ensure that employees/subcontractors utilize the appropriate PPE for the relevant job being executed whilst ensuring that the mandatory PPE (safety footwear, face protections, hard hat and coverall). Specialised/other PPE as required to execute a specific job/task.
20.	Notify/inform the Safety Department of any areas of the operation where achievable and realistic improvements can be implemented to improve the working conditions and reducing/minimizing employee risk and exposure.
21.	Accept, sign and perform immediate corrective/preventative actions with regards to any safety notices issued to your gangs/area of responsibility.

AUTHORIZED/DESIGNATED BY		
(Sign of Project Manager)	Project Manager	(Date of Document Signing)
Signature	Designation	Date

ACCEPTANCE		
I	(Recruited Site Supervisor Name)	hereby acknowledge receipt of
Accept and understand the requirements of this appointment.		
(Sign of Recruited Site Supervisor)	Site Supervisor	(Date of Document Signing)
Signature	Designation	Date

Training Program

2

Some of Training Program are Defined

a. Excavation work.
b. Eye protection.
c. Housekeeping.
d. Working at heights.
e. Working on scaffold

Markings

All working environment should be discussed, trained and documented before committing any job activities as per OSHA:18001 specification.

What is the Infrastructure of Training Program?

It is a criteria of making every group of workers of a company to perform their job activities as per safe work procedure. So prior to job activities the employees need to be well trained in their area of activity for developing their expertness, professionalism and safe work practice according to OSHA standard. After training procedure, the training documents should be signed both by the trainer and workers trained and document filled for father enquiry.

Does Training Program change from Industry to Industry?

Yes its contents change according to the type of Industry, activities performed, atmospheric condition, hazards and risks involved and company policy. But the bottom-line concept of training program never changes.

What is the Importance of Training Program?

According to the company's point of view, it is a part of company policy to train their employees prior to committing job activities for performance of safe work procedure according to the job activity hazard identification and risk assessment practice . From the part of employees, after training they will get clear information about the job activity they need to perform, risk and hazards they need to avoid and safe work practice they need to follow.

How is Training Program Defined SDTP Document Filling?

See the attached documents:

2.1 Training of Excavation Work

Underground Services Damage

Digging into underground services can lead to:

1. Fires and explosions from ruptured gas pipes
2. Contamination of fresh water supplies
3. Flooding from ruptured water and sewerage pipes
4. Soil contamination from burst fuel pipes
5. Loss of communication and electric power services
6. Electrocution.

Preventing Damage to Cables and Pipes

1. Dig trial holes to confirm the position of cables and pipes
2. Dig along-side the line of the cable or pipe, not directly above
3. Have an observer to guide the operator
4. Use shovels and spades to make the final exposure.

Avoid using

1. Picks or forks
2. Jack hammers near plastic pipes and electric cables
3. Explosives within 30 meters of a gas pipe.

Ground Support Methods

You must not work in excavations deeper than 1.5 meters which have not been

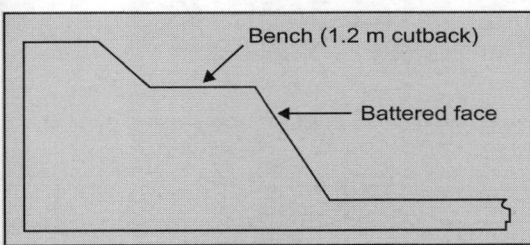

Bench (1.2 m cutback)

Battered face

Battering: Involves digging the excavation face so it is sloped rather than vertical. Benching: is used for excavations which are deeper than three (3) meters.

Shoring: Should be used when benching and battering are impractical.

Cracked formed by drying out.
Water penetration softness clay which can slide or fall into the trench due to lack of cohesion

Cracks formed

Unstable lumps

Clay swells inward

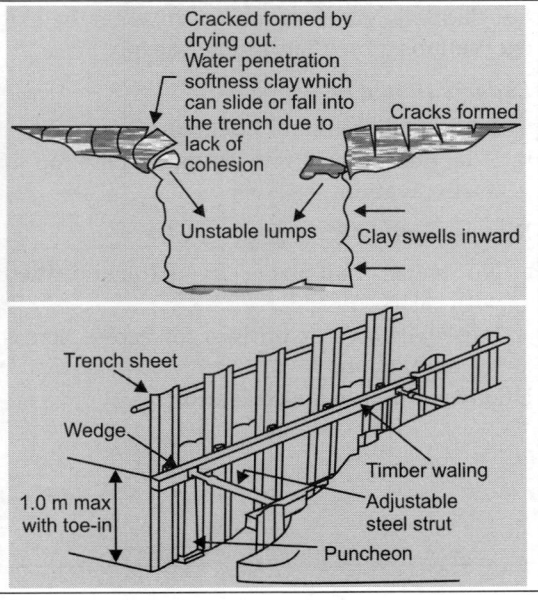

Trench sheet

Wedge

1.0 m max with toe-in

Timber waling

Adjustable steel strut

Puncheon

Preventing Collapses

Things you can do to prevent the excavation from collapsing include:

a. Store excavated rock and soil at least 500 mm from the edges

b. Keep vehicles and heavy machinery away from the edges

c. Pump ground water out of the excavation

d. Use ditches or windows to divert surface water away from the excavation.

Ventilation and Lighting

Install artificial lighting where:

- Daylight does not provide enough lighting.
- **Work needs to be carried out at night**
  - Excavations must be kept free of toxic and explosive gases.
  - Do not use portable petrol or diesel engines inside excavations.
  - Store compressed gas cylinders outside the excavation.

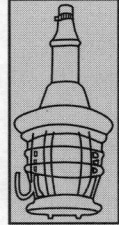 

Excavation Access

Use ladders, stairs or ramps to enter or exit excavations more than 1 meter deep.

Ladders should be

a. Set at no more than 75 degrees;

b. Extended at least one (1) meter past top of the excavation.

Ramps should be

c. No steeper than one (1) in six (6); and fitted with cleats.

Use walkways or bridges for access across the excavation.

Do not jump into an excavation.

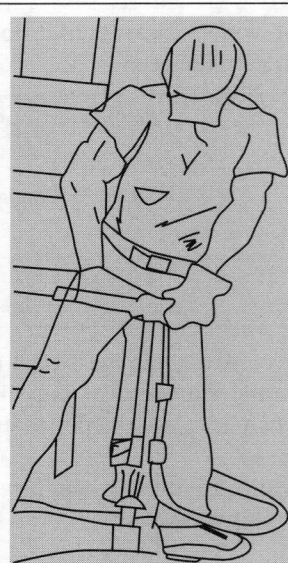

Barricading and Signs

1. Set up barriers or barricades around the edges of the excavation.
2. Place warning signs to warn people of the excavation.
3. Divert traffic away from the excavation area.
4. Install flashing amber lights for night-time warning.

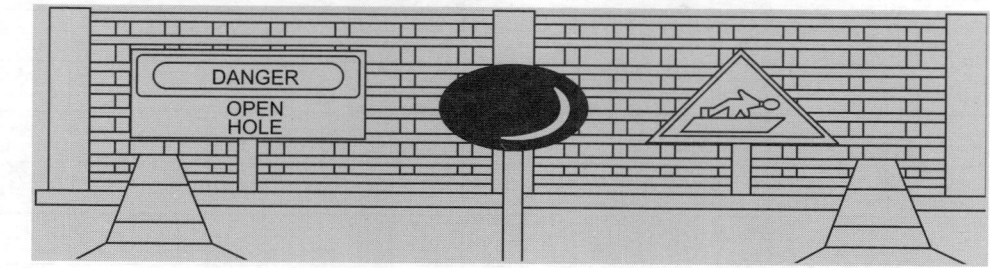

Name of Attendees	Designation	Signature	Date

Name of Attendees	Designation	Signature	Date

2.2 Training of Eye Protection

Need for Protection

Flying debris

Dust and dirt

Welding flashes

Fragments of metal, glass

Chemical splashes

Fumes from chemicals

Types of Eye Protection

Safety glasses

Goggles

Splash shields

Face shields

Selecting Eye Protectors

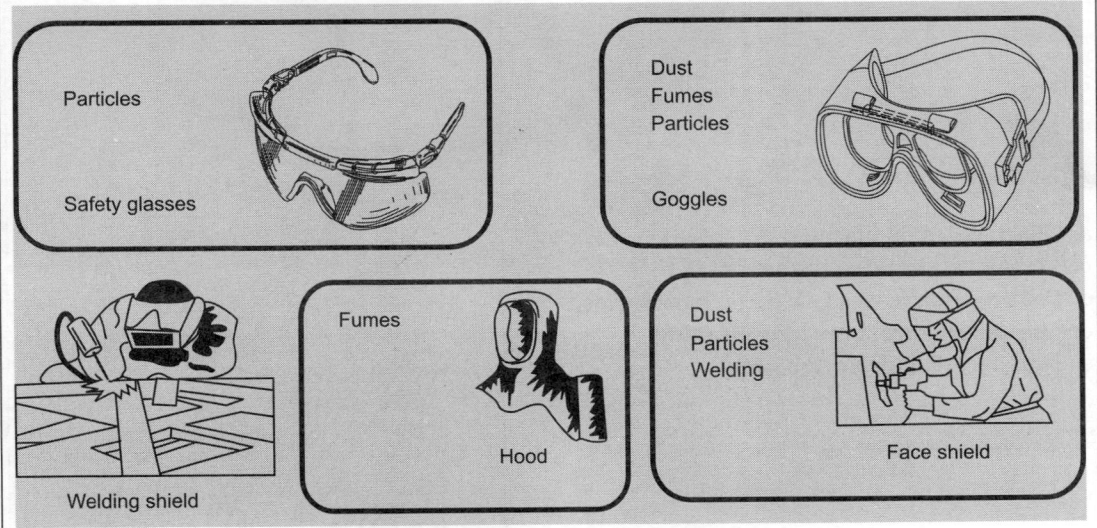

Particles

Safety glasses

Dust
Fumes
Particles

Goggles

Welding shield

Fumes

Hood

Dust
Particles
Welding

Face shield

Wearing Eye Protectors Correctly

Nearly all eye protectors are adjustable. The main ones are:

Safety glasses have adjustable arms	
Goggles have elastic straps which go around the head and can be altered for a snug fit.	
Face shields have an adjustable headband.	

Eye Protection Maintenance

a. Clean lenses regularly b. Make sure elastic straps are not perished c. Put protector lenses face up on hard surfaces d. Use antifogging spray after cleaning e. Store face shields upright or.	

Not Wearing Eye Protection

People choose not to wear eye protection for the following reasons:

- They are not used to them and feel uncomfortable
- The wearer feels hot and sweaty
- The protectors are not trendy; and
- They feel foolish if no-one else is wearing them.

When to wear Eye Protection

You should wear eye protection when:

- There are signs telling you to do so

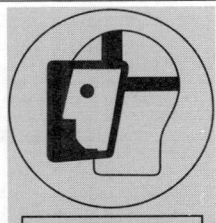

- You are working with flying particles

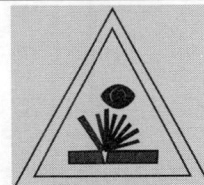

- You are welding

- You are working with chemicals.

Name of Trainer	Designation	Signature	Date

Name of Trainer	Designation	Signature	Date

2.3 Training of Housekeeping

Cost of Poor Housekeeping

Poor housekeeping can lead to:
1. Splinters, cuts and eye injuries
2. Cuts to hands and fingers
3. Slip and trip accidents
4. Crush injuries
5. Wasted time
6. Fires
7. Serious injury or death
8. Poor quality work.

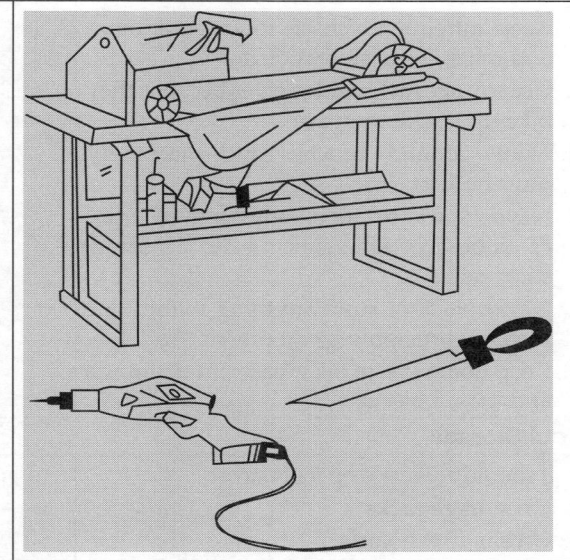

Benefits of Good Housekeeping

Effective housekeeping requires:
1. Organizing your workplace
2. Keeping work areas uncluttered
3. Storing everything in its proper place.

The benefits of good housekeeping are:
1. Less risk of accidents, injuries and fires
2. A more pleasant workplace
3. Less time wasted trying to find—tools, equipment and materials.

Floors and Access-ways

a. Only keep frequently used tools in your work area.
b. Keep infrequently used tools, equipment and materials in store rooms
c. Use storage racks and shelves because they take up less floor space
d. Floors around benches and machinery must be kept clear
e. Never stack or store anything in front of doorways emergency exits or safety showers
f. Never stack or store anything within one meter of fire fighting equipment
g. Keep floors free of oil, grease, mud, liquids and loose objects.

Tool Storage

You should always keep tools stored:
1. On storage racks
2. On shadow boards
3. In Toolboxes.

Never leave tools
1. Sitting on top of machines
2. Overhanging the edges of bench tops
3. Resting on top of vehicles
4. Lying on the floor.

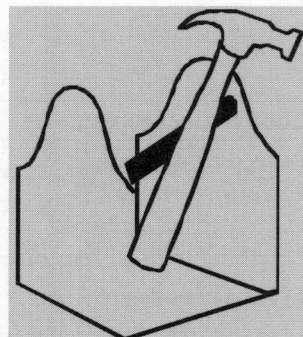

Materials Storage

a. Store sheet material in vertical or horizontal racks
b. Store round bar, pipe and conduit in racks which prevent rolling or movement
c. Store small items on shelves and in trays

Avoid storing materials:
a. On top of shelf units
b. Underneath workbenches
c. On window sills
d. On top of wall beams.

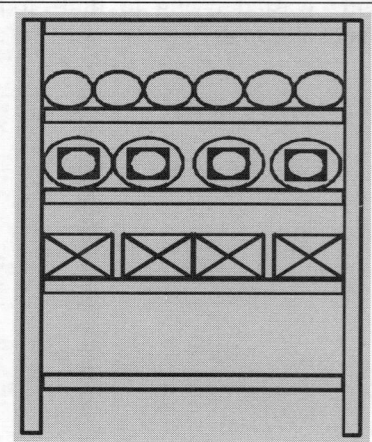

Storing Flammable Liquids Containers

Store small containers of flammable liquids in metal flame—proof cabinets.

These include

a. Tins of paint
b. Plastic bottles of oil
c. Aerosol packs of paint and degreaser.

Do not store flammable liquid containers on top of benches, on the edge of shelves, or on wall purling where they can be punctured or exposed to heat.

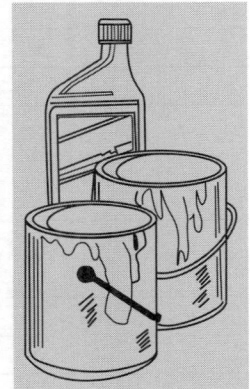

Rubbish and Waste Hazards

- Metal filings—splinters, cuts and eye injuries
- Sharp metal scrap—cuts to hand and finger
- Scrap material on the floor—slips and falls
- Paper, cardboard and wood shavings fire
- Food scraps and wrappers—vermin and disease
- Oily rags—fire.

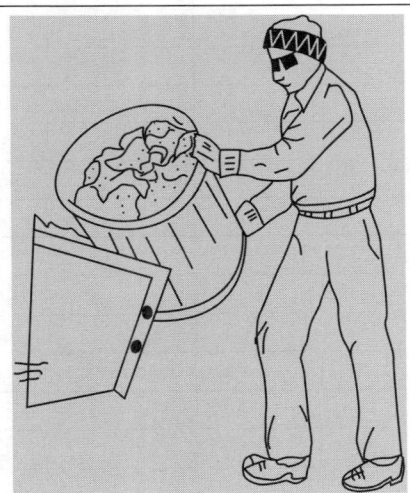

Waste and Rubbish Disposal

- Clean bench tops and equipment to remove metal filings and scrap
- Dispose of metal waste into scrap metal bins
- Immediately dispose of paper and cardboard
- Put food scraps into a bin and keep the lid on to avoid attracting flies and vermin
- Oily rags and waste must be disposed of separately into metal bins with close fitting lids because of the fire risk.

Clean up as you go rather than waiting until the end of the day!

Name of Attendees	Designation	Signature	Date

Name of Attendees	Designation	Signature	Date

2.4 Training of Work at Heights

Using Ladders Safely

When using a ladder you should:
1. Always face towards the ladder while climbing up or down it
2. Only move up or down the ladder one rung at a time
3. Keep a three point of contact at all times
4. Keep your body centered within the ladder stiles
5. Climb down from the ladder if you need to reposition it.

Scaffold Construction

- For scaffolds, 2 meters or more above the work plane, the following construction is mandatory
- Work platform must be at least 450 mm wide
- Scaffold planks must be securely lashed at both ends.

Access ladded secured and lashed

Rigid guard rails 900 mm to 1100 mm high

Toe board 200 mm high

Mid rail or mesh no less than 450 mm above platform

The Scafftag

a. Any scaffold higher than 4 meters must only be erected, modified, repaired and inspected by a certified scaffolds.
b. Never alter or modify any scaffolding unless you are a certified scaffolds.
c. A scaffold must be fitted with a scaffold tag at every access point.
d. Never work on any scaffold, 4 meters or higher, which is not fitted with a valid scaffold tag.

Scaffold Safety

- Keep walkways free of obstacles, tools and equipment
- Keep platforms free of grease and mud
- Climb from one level to another using the ladders provided
- Use a crane, hoist or winch to carry materials up to and down from the scaffold.

You should never

- Exceed the safe working load of the scaffold
- Stand on the hand-rails
- Work from a defective scaffold.

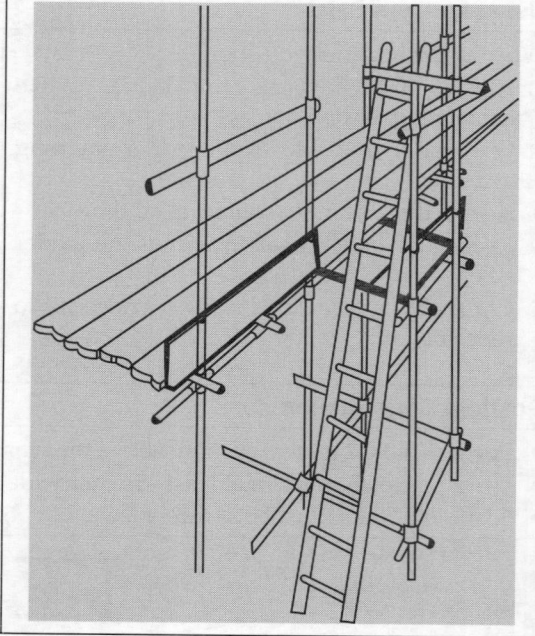

Mobile Tower Scaffolds

When working from a mobile tower scaffold make sure:

- It is no higher than three times the smallest base width
- Outriggers are fitted
- Caster brakes are locked on.

You must not be on the scaffold while it is being moved.

Only use this type of scaffold on level, surfaced areas.

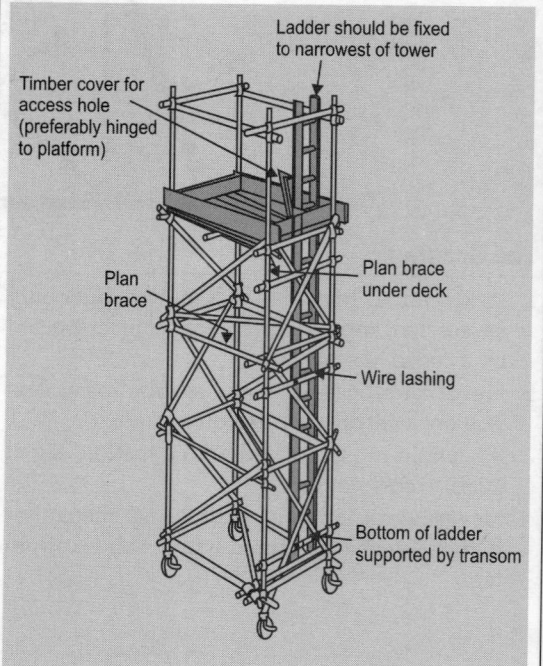

Ladder should be fixed to narrowest of tower

Timber cover for access hole (preferably hinged to platform)

Plan brace

Plan brace under deck

Wire lashing

Bottom of ladder supported by transom

Elevating Work Platforms

- Barricade under and around the elevating work platform
- Place all tools and equipment in bags or buckets
- Only enter and exit the basket via the gate.
- Keep the basket gate closed and locked while working inside the basket
- Wear a full body harness attached to an anchor point inside the basket
- Keep the front of the basket within 300 mm of the work area.

Power Line and Stability Hazards

Maintain a distance of:
- Two (2) meters from any electrical distribution wires on poles
- Six (6) meters from any electrical transmission wires on towers.

Never exceed the Safe Working Load of the basket.
1. 200 kg for fiberglass basket
2. 250 kg for steel basket.

Do not position too close to excavations.

Set back at least one meter from the edge of the excavation for each meter of excavation depth.

Forklift Cage

Make sure the fork lift cage is:

a. Correctly positioned onto the forklift tines
b. Locked into position so it cannot slide off the tines
c. Always wear a fully body harness attached to an anchor point located inside the cage
d. Never stand on the forklift tines
e. Never use a forklift pallet as a work platform.

Safety Harnesses

A safety harness:

- Stops the fall and spreads the impact shock a large area of the body
- Allows a fall to be stopped or arrested without causing bodily injury

   The maximum free-fall distance allowed is two meters.

The safety harness must be attached to:

- A lanyard with a built in shock absorber
- A self-retracting lanyard.

   The other end of the lanyard must be attached to a static line or anchor.

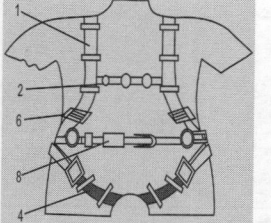
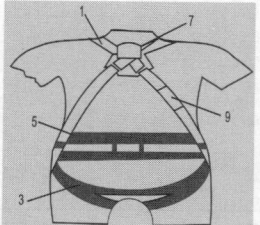

1. Shoulder strap
2. Secondary strap
3. Sit strap (primary strap)
4. Thigh strap
5. Back support for work positioning
6. Adjustment element
7. Fall arrest attachment element
8. Buckle
9. Marking.

Using Fall Protection Equipment

- Check the harness for any signs of wear or damage
- Check the lanyard is firmly attached to the lanyard
- Make sure the anchor point is positioned above where you are working
- If you are using a strap or belt type lanyard, ensure there is a minimum of slack so that you can not fall any further than two meters
- Report any faults immediately
- Never use faulty or damaged equipment.

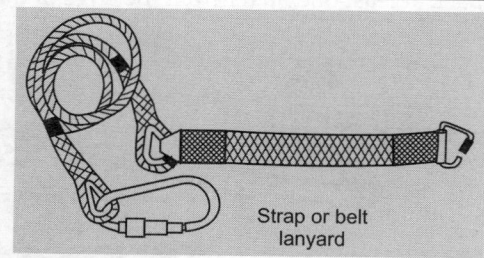

Strap or belt lanyard

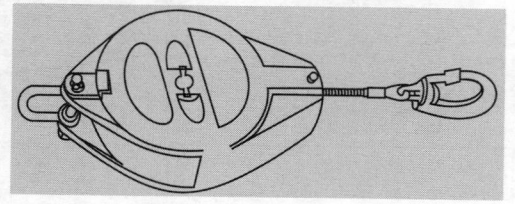

Self-retracting lanyard

Name of Attendees	Designation	Signature	Date

Name of Attendees	Designation	Signature	Date

2.5 Training on Scaffold Erection and Working on Scaffolds

What is a Scaffold?

An elevated, temporary work platform
Three basic types:

- Supported scaffolds—platforms supported by rigid, load bearing members, such as poles, legs, frames, and outriggers
- Suspended scaffolds—platforms suspended by ropes or other non-rigid, overhead support
- Aerial Lifts—such as "cherry pickers" or "boom trucks".

Hazards

Employees working on scaffolds are exposed to these hazards:

- Falls from elevation—caused by slipping, unsafe access, and the lack of fall protection
- Struck by falling tools/debris
- Electrocution—from overhead power lines
- Scaffold collapse—caused by instability or overloading
- Bad planking giving way.

Fall Hazards

Falls may occur
- While climbing on or off the scaffold.
- Working on unguarded scaffold platforms
- When scaffold platforms or planks fail.

Protecting Workers from Falls

If a worker on a scaffold can fall more than 10 feet, protect them by:
- Guardrails
- Personal Fall Arrest Systems (PFAS).

Guardrails

- Install along open sides and ends
- Front edge of platforms not more than14 inches from the work, unless using guardrails and/or PFAS
- Top rails—38 to 45 inches tall
- Midribs halfway between top rail and platform.

Personal Fall Arrest Systems (PFAS)

- You must be trained how to properly use PFAS
- PFAS include anchorage, lifeline and body harness.

Fall Protection Requirements

- Can use PFAS instead of guardrails on some scaffolds
- Use PFAS and guardrails on suspension scaffolds
- Use PFAS on erectors and dismantlers where feasible.

The ends of this scaffold are not properly guarded

Falling Object Protection

- Wear hard hats
- Barricade area below scaffold to forbid entry into that area
- Use panels or screens if material is stacked higher than the toe board
- Build a catch platform or erect a net below the scaffold that will contain or deflect falling objects.

Overhead Power Lines

- The possibility of electrocution is a serious consideration when working near overhead power lines
- Check the clearance distances listed in the standard.

Scaffold Support Examples

Base plate

Mud sills

Good support

Inadequate support—in danger of collapse

Essential Elements of Safe Scaffold Construction

- Use appropriate scaffold construction methods
- Proper scaffold access
- Properly use a competent person.

Scaffold Platform Construction

Platforms must:
- Be fully planked or decked with no more than 1 inch gaps
- Be able to support its weight and 4 times maximum load
- Be at least 18 inches wide.

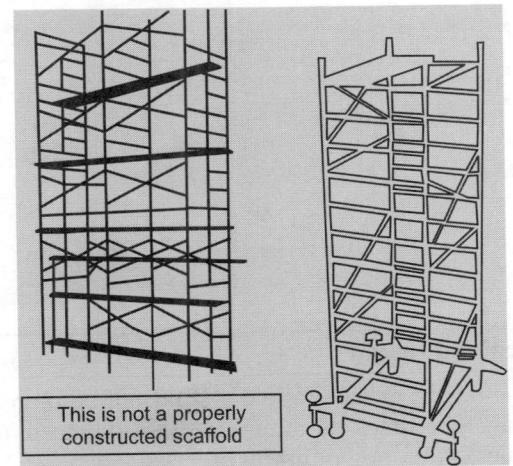

This is not a properly constructed scaffold

Scaffold Platform Construction

- No large gaps in front edge of platforms
- Each abutted end of plank must rest on a separate support surface
- Overlap platforms at least 12 inches over supports, unless.

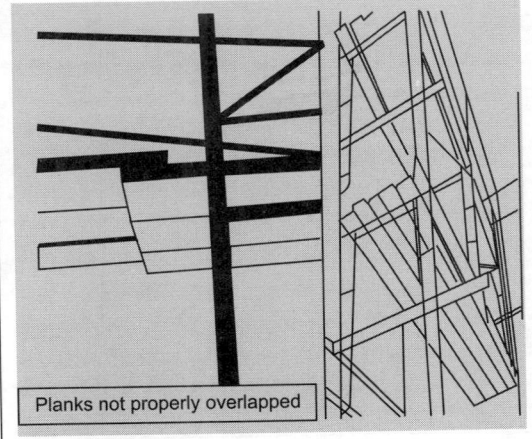

Planks not properly overlapped

Scaffold Platform Construction

- No paint on wood platforms
- Use scaffold grade wood
- Fully planked between front upright and guardrail support
- Component pieces used must match and be of the same type
- Erect on stable and level ground
- Lock wheels and braces.

Scaffold Height

The height of the scaffold should not be more than four times its minimum base dimension unless guys, ties, or braces are used.

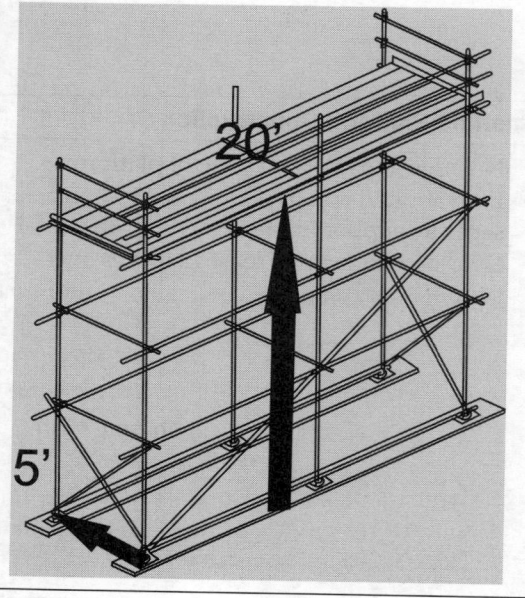

Platform Ends

Each end of a platform, unless clearance or otherwise restrained by hooks, must extend over its support by at least 6 inches.

NO CLEARANCE

Supported Scaffolds

Platforms supported by legs, outrigger beams, brackets, poles, uprights, posts, and frames
- Restrain from tipping by guys, ties, or braces
- Scaffold poles, legs, posts, frames, and uprights must be on base plates and mud sills or other firm foundation.

This support is not adequate!

Proper Scaffold Access

Provide access when scaffold platforms are more than 2 feet above or below a point of access

Permitted types of access

- Ladders, such as portable, hook-on, attachable, stairway type, and bulletins
- Stair towers
- Ramps and walkways
- May use building stairs and come out window.

Scaffold Access

- No access by cross braces
- When using ladders, bottom rung not more than 24 inches high.
- Can use some end frames.
- Can access from another scaffold, structure or hoist.

Do not access by cross braces

End Frame

Suspension Scaffolds

- Train employees to recognize hazards
- Secure/tie to prevent swaying
- Support devices must rest on surfaces that can support four times the load.

Competent person

- Evaluate connections to ensure the supporting surfaces can support load
- Inspect ropes for defects before shift PFAS must have anchors independent of the scaffold support system

Platforms suspended by ropes or wires. Rope must be capable of supporting 4 times the load.

Moving Scaffolds

Employees cannot be on a moving scaffold unless:

- Surface is level
- Height to base ratio is 2 to 1
- Outriggers are installed on both sides of scaffolds.

Employees cannot be on scaffold part beyond the wheels.

Competent person must be on site to supervise.

Fatal Fact—Moving a Lift

- Employee was operating an aerial lift, with an extendable boom rotating work platform
- The boom was fully extended and the machine apparently ran over some bricks, causing the boom to flex or spring, throwing the employee from the basket
- The employee fell 37 feet to a concrete surface.

Do not use Shore or Lean-to Scaffolds

- Shore scaffold supported scaffold which is placed against a building or structure and held in place with raps
- Lean-to scaffold supported scaffold which is kept erect by tilting it toward and resting it against a building or structure.

Using Scaffolds

- Do not work on snow or ice covered platforms or during storms or high winds
- Use tag lines on swinging loads
- Protect suspension ropes from heat and acid.

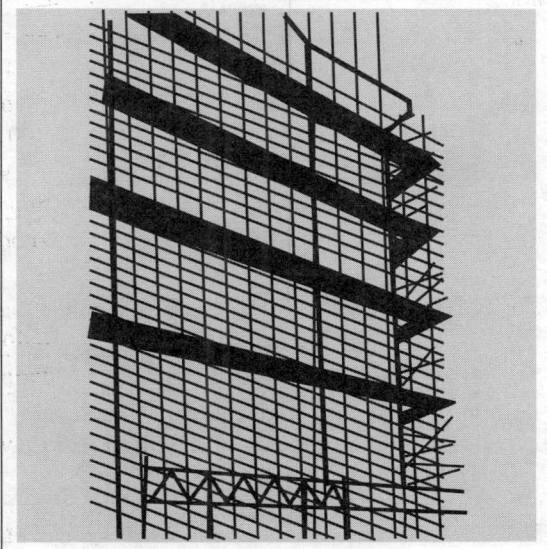

A covered scaffold has special wind load considerations

Fatal Fact—Ice and No Guardrails

- Laborer was working on the third level of a tubular welded frame scaffold which was covered with ice and snow
- The scaffold was not fully decked, there was no guardrail and no access ladder
- The worker slipped and fell head first 20 feet to the pavement below.

Overhand Bricklaying from Supported Scaffolds

A guardrail or personal fall arrest system is required on all sides except the side where the work is being done

Competent Person

- Person capable of identifying and promptly correcting hazards
- Determines if it is safe to work on a I scaffold during storms or high winds
- Trains workers to recognize hazards
- Selects qualified workers to conduct.

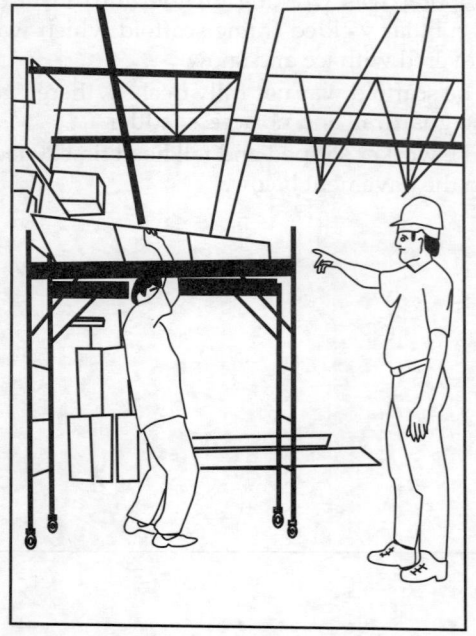

Scaffold Inspection

- Competent person or certified Scaffold Inspector, needs to inspect scaffolds for visible defects before each shift and after any alterations
- Defective parts must be immediately repaired.

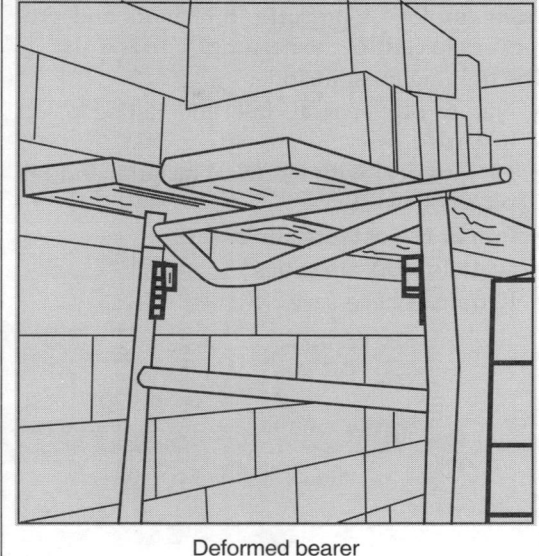

Deformed bearer

Scaffold Erection

- Scaffolds can only be erected, moved, dismantled or altered under the supervision of a competent person and by trained and certified Scaffolders.
- Competent person directs these workers and determines the feasibility of fall protection

Training Requirements

Train employees on scaffold hazards and procedures to control the hazards.

The training must include:

- Nature of electrical, fall, and falling object hazards
- How to deal with electrical hazards and fall protection systems
- Proper use of the scaffold
- Scaffold load capacities
- Retrain as necessary.

Avoid the Main Hazards of Scaffolds

- Getting struck by falling tools or debris
- Electrocution
- Falls from elevation
- Bad planking
- Scaffold collapse.

Name of Attendees	Designation	Signature	Date

Name of Attendees	Designation	Signature	Date

Induction and Special Training

3

Induction and Training are Defined

a. Training for newcomers.
b. Special training in case of any fire accident.

Markings

Every newcomers and visitors should be inducted before entering a new site because they never new about the hazards and risk involving in site activities.

What is the Infrastructure of Induction Training Program?

Induction training is a criterion of making each and every new employee and visitor to introduce or make aware about site premises including site traffic and walkways, prohibition acts, warnings, mandatory requirements/site activities including hazard and risks and their safe working procedures/ awareness of welfare activities provided according to OSHA standard.

After training procedure, the training documents should be signed both by the trainer and newcomer or visitor and the document is to be filed for further enquiry.

What is the Purpose of Special Training Program?

Special Training Program is conducted for mainly five major occasions:

a. If new Plant, machinery and any type of new tool is introduced to industry.
b. After an accident, incident, near miss and stoppage of work activity or site.
c. On introducing new rules or regulation to site or industry.
d. Refresher training after a period of time.
e. Emergency evacuation drills and Fire equipment training.

What is the Importance of Induction Training Program?

New employee and visitors are always unaware about the site premises and activities. According to the companies point of view, it is a part of company policy to train and make aware these employees, prior to entering the site straight. If they are unaware of the rules, regulations and safe practices their attitude and approach will turn into accident naturally.

The new employees and visitors, after induction training will get clear information about the site premises, job activities, risk and hazards they need to avoid and safe practice they need to follow.

How is Induction Training Program Defined in SDTP Document ?

See the attached documents to know how to do Safety Induction Training:

3.1 Site Induction Programe

Site Rules

Management, Supervisors, Operatives and Subcontractors

- Safety Helmets, High Visibility Waistcoat and Safety Footwear—mandatory. Operatives and Supervisors—overalls
- Work Activity Briefing/Toolbox Talk—before starting work or new work activity. This talk must be provided by the site Supervisor/Safety Officer.
- Wear the personal protective equipment (PPE) specified on the method statement/risk assessment/task sheet.
- Alcohol and drugs—not permitted on site.
- Any drugs prescribed by your doctor—inform your supervisor/first aider
- Any health problems (diabetes, epilepsy, etc.) identified to the first aider/medic.
- Eat, drink or smoke in designated areas smoking elsewhere on site—(Rs. 200/- fine or permanent exclusion from site)

- Drink plenty of water, make sure that you do not expose your skin in excessive amount of sunlight.
- Look after your own health and safety, do not put others at risk.

Work Areas

- Do not enter restricted/unauthorized areas (e.g. chemical and inflammable liquids storage area/other section or other work areas) without permission. You may need a permit to work (e.g. confined space areas/hot work areas and other hazardous areas)
- Keep your work area tidy. Clear lightweight-rubbish regularly to reduce Foreign Object Debris (FOD) damage to respiratory system, eye and skin.
- Keep access/egress routes (especially those required in an emergency) clear of debris, materials, cables and parked vehicles/plant, etc.

- Edge protection must be provided to prevent falls from working platforms, decking, form-work, slab edges.
- Guard rails and toe-boards are provided to prevent you from falling.
- Unauthorized removal/alteration is prohibited.
- Holes in floors, etc. doors and opening in vertical risers and shafts must be protected. Covers or suitable edge protection must be provided.
- Waste must not be thrown off the building or to a floor below. Use Rubbish skips provided at site.
- Hazardous materials:
  1. Read the product safety in formation sheet MSDS/COSHH assessment
  2. Read understand and implement the requirements of the method statement and risk assessment.
  3. Implement the control measures.
  4. Wear required and correct PPE'S

Plant and Transport

- Only trained and authorized plant department operatives may operate or drive plant and transport.
- Vehicle may only be driven on authorized routes.
- Obey the road/safety signs and maximum speed limit of 25 km/h.
- Never carry or lift unauthorized passengers. Passengers must sit in/on properly designed and fixed seat.
- Do not use Tale handler forks/JCB 3 CX buckets or forks to provide a working platform.
- Private cars are not allowed on site or in the store/workshops compound.
- Only trained and authorized slingers and signalers wearing red overalls may sling loads and give signals to crane operators.
- Only trained and authorized Traffic Marshal/Banks-men may give signal to traffic and plant.
- Reversing vehicle should be accompanied by banksmen.

We will not be able to achieve this result without your co-operation and active contributions to a positive health and safety culture.

Site layout

- Parking
- Access
- Entry–pass system
  - Visitors pass
  - Landside pass
  - Materials pass
  - Vehicle pass
  - Camera pass.
- Keep vehicles and machines separate from people

Major Risks

- Exposure limit of vapors and gas in case of leakage.
- Hot work and stacking of compressed gas cylinders.
- Working near heavy inflammable oil stacking capacity tanks
- Fire hazards.
- Vibration and noise hazards.
- Holes—never interfere with edge protection
- Working close to people/plant and equipment
- Mechanical handling/fork lift trucks
- Electrical Hazards—hand tools/cables
- Heat, ergonomics and occupational diseases.

Project Specific Risks "Your Health"

Conditions and substances that can affect your health:

- Manual handling
- Dust/fumes
- Toxic and inflammable vapors
- Noise and vibration
- Flying metal debris

"Wearing the correct personal protective equipments and personal hygiene are essential to avoid health problems."

Precaution and Control Measures

- Site investigations
- Barriers and signage's
- Hazard identification and risk assessments
- Safe work procedures
- Training and safety talks
- Permit to work that apply on the site
- Personal Protective Equipment (PPE)
- Health checks
  Safe system of work = Safe place of work

Major Causes of Accidents

- Slips, trips and falls
- Stepping on/against objects
- Falling/flying objects
- Negligence towards safety
- Over confidence
- Manual handling injuries—account for more days' sickness than any other single cause.
- Not assessing Permit-to-Work systems.
- Unauthorized entry to assess denied areas.

Health and Safety targets

- We want this to be the safest sit the country and certainly to have disabling accidents.

SAFETY FIRST

Our target is zero accidents

Responsibilities

Our Responsibilities—Legal duty

- To make sure that we provide a safe place and system of work for people working for us.

- Assessing the risks and making a plan (method statement) to remove or control the risks so that you can do the work safely.
- Explaining the safe work procedure to you before you start work.
- Emergency/evacuation procedure drills to conducted periodically.

Your Responsibilities—legal Duty

- To follow the safe systems of work and site rules and regulations.
- To work safely—both for yourself and other people.
- To assess and follow Permit-to-work systems.
- To wear the correct and adequate PPEs.
- Not to interfere with safety equipment.
- Not to obstruct any emergency escape roots.
- Not to enter in any assess restricted areas without entry permit.
- Always co-operate with site Emergency/ Evacuation procedure.
- Always follow signage's and safety posters.

Personal Protective Equipment (P.P.E.)

PPE—What is it

- Personal Protective Equipment (PPE) must be provided when necessary by reason of hazards encountered that are capable of causing injury or impairment
- FPE is not a substitute for engineering, work practice, and/or administrative controls
- PPE is last line of defense against Hazards
- Use of PPE does not eliminate the hazard so if the equipment fails then exposure occurs
- Must be worn to provide protection

Minimum Requirements

1. Safety Helmet
2. Safety Footwear with Steel Topes and Mid-sole Insert.
3. Coverall, Safety goggles and face mask.
4. Wear all required PPE as per safe work task.

All PPEs must fit correctly and be kept clean by you.
Do not use DAMAGED PPE.

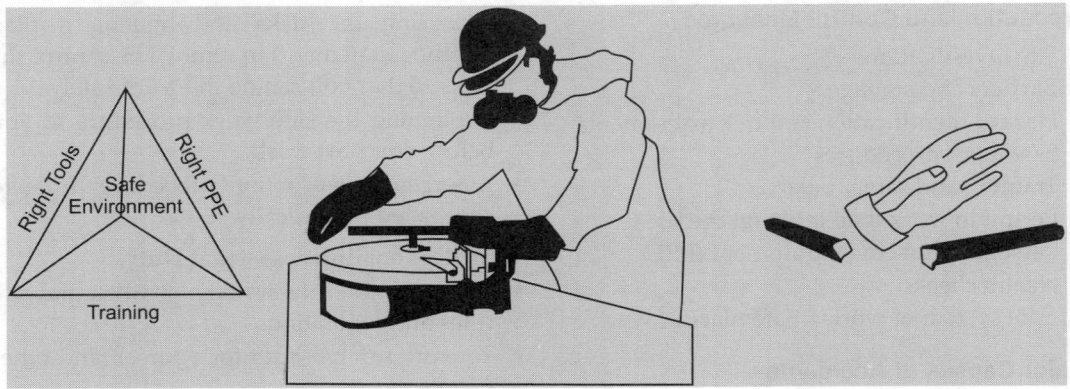

Use it at all times

Common PPE to be Worn

- Eye protection—if there is a danger of eyes being damaged.
- Gloves to protect hands.
- Safety Harness—if there is a danger of falling.
- Ear protection or Ear muff—when working in noisy areas.
- Dustmask—if the environment is dusty.
- Breathing Apparatus—if working in confined space and gas, vapor atmosphere.

Eye Protection

- Spectacles
- Goggles
- Face shields
- Welding Goggles/helmets

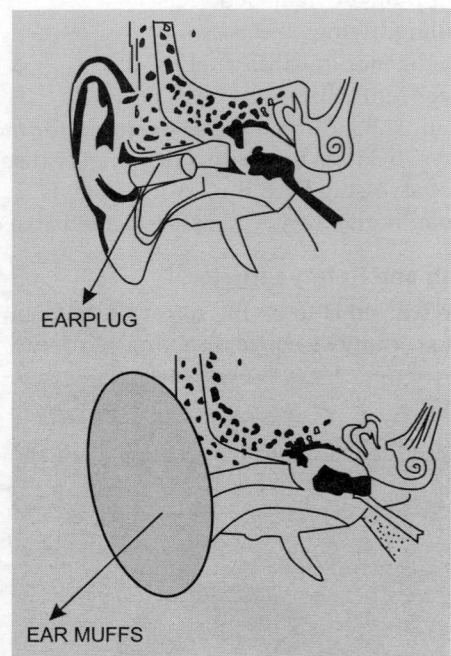

EARPLUG

EAR MUFFS

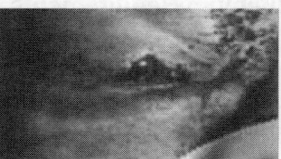

Hearing Protection

- Earplugs (fit inside the ear)
- Ear muffs (cover the ears)

Material	Good for
Leather	Abrasion protection, heat resistance
PVC	Abrasion protection, water limited chemical resistance
Rubber	Degreasing, paint spraying
Cloth/nylon, latex coated	Hand grip
Latex	Electrical insulation work
Chain mail	Cut protection

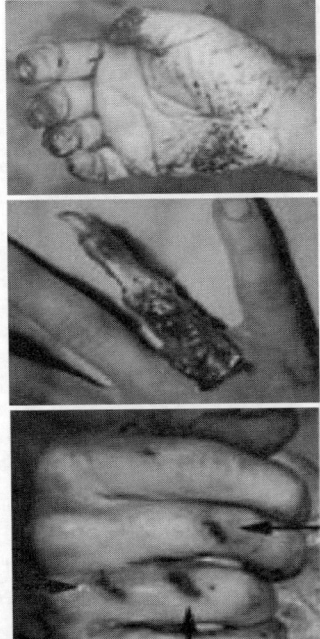

SAFETY FIRST ALL TIMES

Hand and Skin Protection

- Required when employees are in areas where their hands and body are exposed to skin absorption of harmful substances, severs cuts or lacerations, chemical or thermal bums, etc.
- Protection must be compatible with hazard

While coming from and going into your work location ensure you have the proper and adequate clothing and PPE.

Care for PPE

1. Always check PPE for damage before and after you use it
2. Clean PPE before storing
3. Dispose of and replace damaged PPE
4. Properly store PPE and avoid conditions that could damage it, such as heat light moisture, etc.

Scaffolding and Ladders

- To be erected, maintained, altered, and dismantled only by authorized scaffolder.
- Working platform to befitted with guardrails (top and intermediate) and toe-boards.
- Where material is stacked above toe-board height, brick guard or suitable debris netting must be provided and maintained.
- Access ladders must be securely lashed/ clamped at both stiles from inside. Never monkey-climb a scaffold, always use the ladder.
- Ladders/Scaffoldings must be inspected at the beginning of every month. Color code as schedule.
- Never interfere with or make alterations to scaff-tags. A red scaff-tag means that the scaffold is unsafe and must not be used.
- If you think the scaffold is unsafe, immediately report it

Ladders/Working at Heights

Use full body safety harness when working at elevation

Ladders

Before Climbing a Ladder Make Sure that it is:

1. Clean mud or grease from boots
2. Take precautions to prevent vehicles or people knocking against the ladder
3. Tools can not fall from your pockets
4. Ladders must be erected at an angle of 75 (1 out 4 up)
5. Positioned on firm, even ground.

Full Body Harness

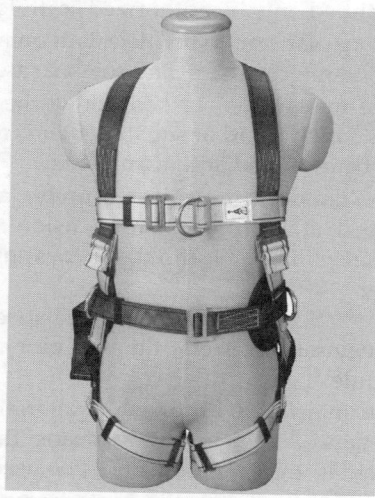

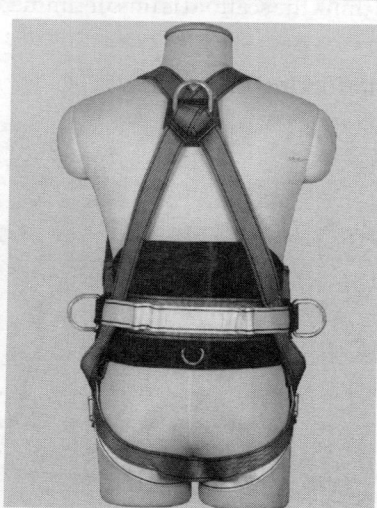

Use full-body harness while working at height beyond 1.8 meters. Safety belts are not permitted

Electrical

Portable Tools and Appliances

- All portable electrical tools above 110 V must be sent to engineer for dispensation
- Choose or provide with the correct tools/appliance to do the job.
- Make sure that the leads, hoses and the tool/appliance is in good order. Inform your supervisor if maintenance/repair are needed.
- Never remove the guards, which are provided for your safety.
- You are not authorized to carry-out repairs (electricians/fitters).

Heat Stress

- Environmental Factors
  - Age
  - Temperature
  - Humidity
  - Radiant heat
  - Air velocity
- Personal Factors
  - Weight
  - Fitness
  - Acclimatization

Heat Stress and Body's Response

- Blood Circulation
  - Circulated to Skin
  - Increases Skin Temperature.

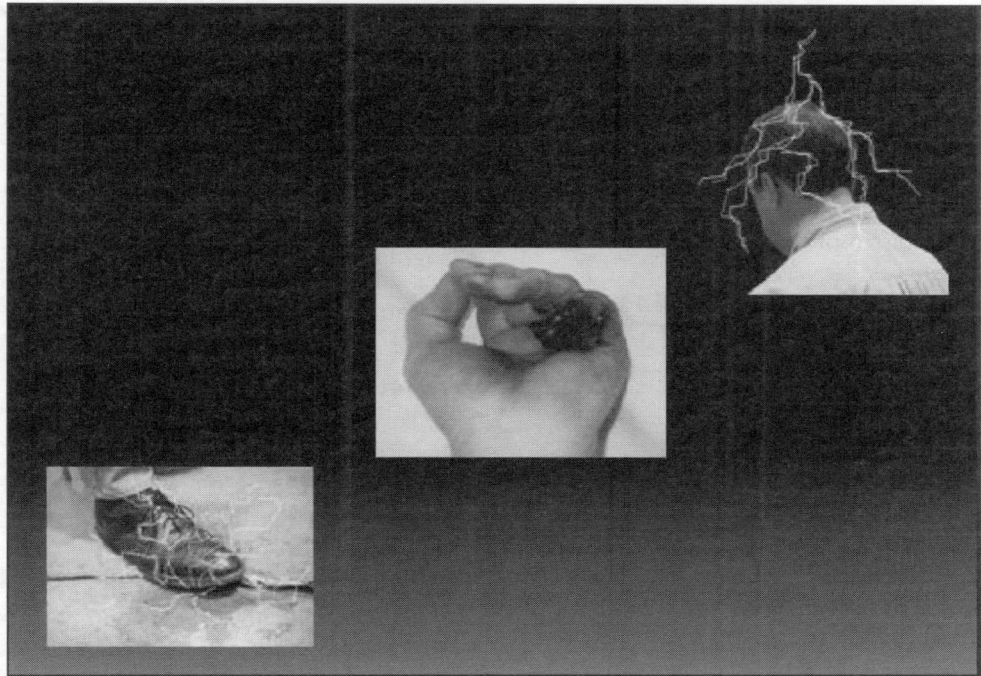

- Excess Heat Exits
  Via Skin.
- More Activity Less
  Blood.
- Sweating
  - Evaporation Cools the Skin.
  - Humidity
    Dependent.

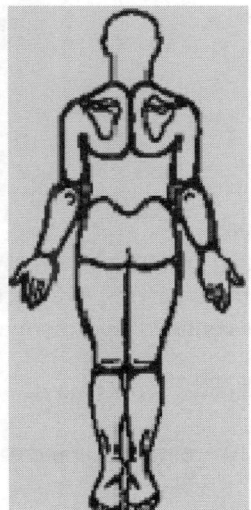

Heat Stress: Heat Disorders

- Heat Rash
- Heat Cramps
- Heat Exhaustion
- Heatstroke

Prevention Methods

- Acclimatization.
- Spend as little time as possible in direct sunlight.
- Schedule heaviest outside work during the cooler times of the day.
- Drink plenty of cool water ore electrolyte replacement fluids even if not thirsty.
- Recognize early symptoms and take appropriate action to prevent serious heat disorders.

Pressure

1. Unexpected release of pressure gas or liquid
2. Explosion or implosion of vessels

Compressed air is a hazardous power supply if used incorrectly

a. Never play with compressed air and hydraulic tools
b. Turn off the supply before changing of tools or adding hoses
c. Chicago fitting must be wire tied to prevent accidental disconnection.

Permit to Work

You may need a "Permit to Work" when carrying out the following work activities:

1. Entry into confined spaces (manholes, pits, tanks, etc.).
2. Working airside works/access to restricted areas.
3. Hot works, Working in Gas and vapor presence area.
4. Deep excavations and working near chemical and compressed gas storage area.
5. Work near live services.
6. Work on electricity system.
7. Pressurized system (pneumatic/hydraulic).

You are not Permitted (Unless Authorized)

- Drive vehicles
- Operate plant
- Erect/dismantle false work/scaffolds/ladders
- Carry-out electrical/mechanical repairs.
- Dig deep excavations
- Light fires

Accident Reporting and First Aid

- If you have an accident or near miss you must report it to your supervisor/safety officer.
- Make sure a record is kept in the accident book to prove any claim you make.
- The names and locations of First Aiders are listed on the information board.

Fire and Emergency Procedures

"If you hear a fire alarm"

- Make your work safe and then go to your nearest assembly point.
- Do not return to work until are told it is safe to do so.

If you See a Fire

- Only attempt to put it out if it is safe to do so.
- Shout fire to raise the alarm and/or set off the alarm.
- Make your way to the assembly point.
- Tell other people to go to their assembly point.

Fire Protection

Find out in Advance:

- Where is the nearest fire extinguisher?
- Where is the nearest fire alarm?
- Where is the nearest telephone?
- What is the emergency call number for the fire
- Department?

MANDATORY

WARNING

SAFE CONDITION SIGN

PROHIBITION

ROAD TRAFFIC

Safety Signs

"REMEMBER"

If you try to put out a fire
- Always keep yourself between the fire and your escape route.
- Know how to use a fire extinguisher.

Welfare Facilities
- Canteen
- Toilets
- Site office
- First-aid facilities
- Firstaider

  Site   Moe. No   Site Office

Recall
- Attend briefings/Toolbox talks. Obey safety signs and instructions.

- Inform your supervisor or the safety officer if you think something is unsafe.
- The working environment is important. Reduce exposure to:
  1. Dust (damp down/clear up regularly)
  2. Noise (do not remove mufflers, etc.)
  3. Chemicals (do not pour oils/chemicals down the drain)
  4. Fumes (adequate ventilation)
  5. Vibration (work rotation/wear gloves)
- Wear the correct PPE for the job.
- Learn about emergency equipment, assembly points and emergency exits
- Heat stress while working at hot sun
- Never put the safety of others at risk.
  1. Your work mates/management visitors and the public

- You must not enter the airport "airside" without the correct authorization.

Your contribution is extremely valuable
We must develop a positive health and safety culture

Please help us to maintain a safe place of work

Any Questions ????????????!!!!!!!!!!!
Thank you

Name of Attendees	Designation	Signature	Date

Name of Attendees	Designation	Signature	Date

3.2 Fire Safety at Work

Aims of the Course

To give an understanding of
- The nature of fire
- Fire prevention at work
- Local fire procedures
- How to select and use a portable fire extinguisher safely.

Objectives of the Course

- At the end of the course you will
- Be able to reduce fire risk at work
- Know what to do when the fire alarm sounds
- Know what to do if you find a fire
- Be able to correctly and safely select and use a fire extinguisher.

Why do we do this Training

- Because a fire accident occurs and led to the stoppage of activities for four hours
- Because management care about your safety
- Because many people are unnecessarily injured by fire at work each year
- Because the law requires employers to train its staff in fire safety
- Because fire risk assessment flagged a need to train staff in Fire Safety.

Lesson Plan

- Precourse assessment
- Disaster—a case study
- The Nature of Fire

- Fire Procedures
- Fire Extinguishers.

To Begin : Precourse Knowledge

- What three things are needed for fire to exist
- What do you think is the most common cause of fire in industrial premises?
- What would you think is the average maximum time allowed for fire evacuation?
- What color is a modern CO_2 extinguisher?
- What extinguisher would you use if your computer caught fire?

Fire at Bradford

11th May 1985
- Time : First flame—well alight?
- How many died?
- Why?
- Two to three minutes
- 56 people lost their lives
- Very poor standards of fire safety.

What is Fire?

The rapid oxidation of a fuel evolving heat, particulates, gases and non-ionizing radiation

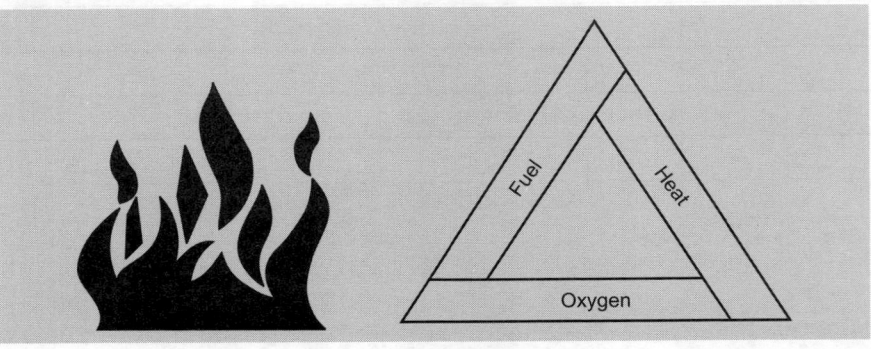

Fire Prevention

Fire involves three elements, this is called "**The Fire Pyramid**"

1. Oxygen
2. Fuel
3. Heat

The Fire Pyramid shows the three elements of fire. If any element is removed the fire will go out.

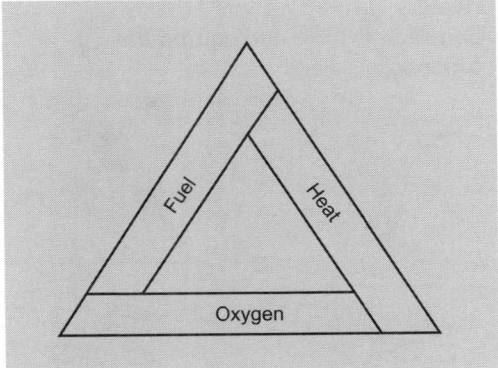

Removing one of these constituent elements will collapse the reaction and the combustion process will cease or be extinguished. This is commonly achieved by:

1. Starvation—removing or limiting the fuel
2. Smothering—removing or limiting the oxygen
3. Cooling—removing or reducing the heat.

Sources of Ignition
- Smoking
- Electrical equipment
- Heaters
- Contractors tools and equipment
- Arson

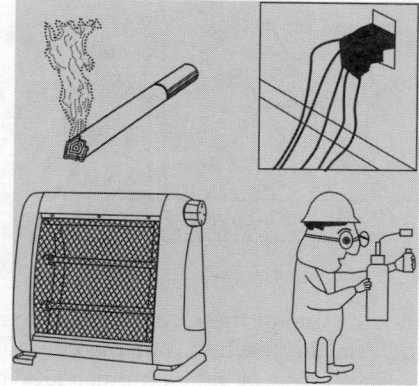

Fuels
- Paper and boxes, etc.
- Plastics
- Solvents
- Carpets
- Furniture
- Waste materials.

Classes of Fire
- A—free burning materials/paper, wood/plastics, etc.
- B—flammable liquids, petrol, meths, solvents, etc.
- C—flammable gases, methane, hydrogen, etc.
- D—metals, potassium, sodium, magnesium, etc.
- F—cooking fats
- Electricity can be involved in any class of fire.

Fire Prevention
- Be mindful of fire safety
- Do not block fire exits, call points or extinguishers
- No smoking policy
- Take care when cooking
- Observe good security
- Do not wedge fire doors open

Fire Procedure—Fire Alarm
- Leave the building immediately
- Use the nearest exit

- Walk quickly but do not run closing doors behind you
- Do not delay your exit to collect your belongings
- Attend the fire assembly point and report to the Fire Marshal
- Do not return until told to do so.

Action on Discovering a Fire
- Shout fire and raise the alarm by breaking a fire call point (red square with glass panel).
- Call the Fire Service by dialing and remember these no. correctly. (Civil Defense Fire or other locally available fire defense services.)
- Fight the fire if you are competent and you consider it safe to do so.
- Follow the evacuation procedure.

To Operate Fire Extinguisher
Pass
- Pull the Pin
- Aim the nozzle on the base of fire
- Squeeze the lever
- Sweep from side-to-side.

Fire Prevention

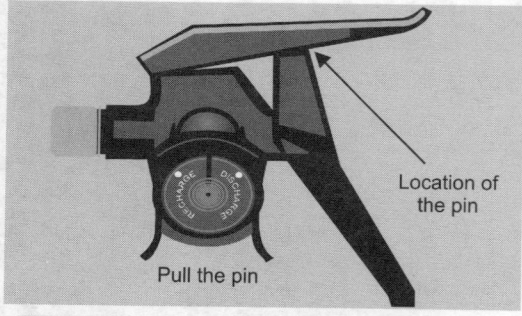

Location of the pin

Pull the pin

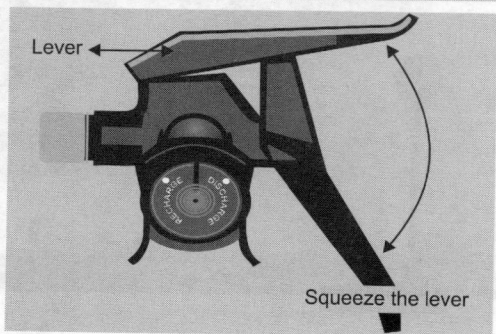

Lever

Squeeze the lever

Proper use of Fire Extinguisher

Aim Low

Sweep from side-to-side

Be Aware

- CO_2 reduces the amount of oxygen in a room
- Always know were the EXIT is
- Never turn your back on a fire

Fighting a Fire—do not Fight the Fire

- It is bigger than a waste paper bin
- One extinguisher is not enough
- Smoke is affecting your breathing
- You cannot see the way out
- Gas cylinders or chemicals are involved
- Your efforts are not reducing the size of the fire.

Extinguishers—international Changes

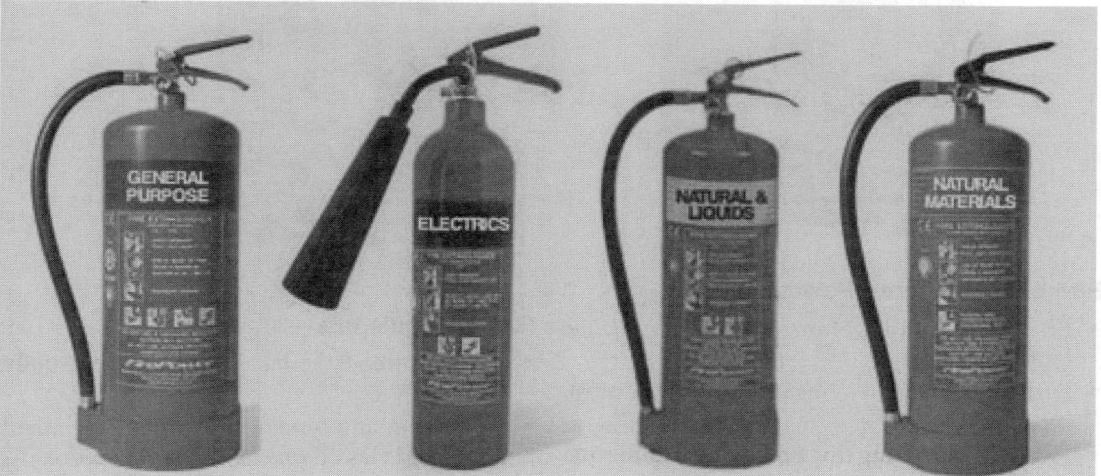

Fire Extinguishers—Water

- Red body
- Suitable for use on class A fires, wood and paper, etc.
- Not suitable for combustible liquids, cooking fats, etc.
- Not safe to use on fires involving electricity
- Extinguishes by cooling.

Fire Extinguishers—Foam

- Cream body (old type) or red body with cream label
- Suitable for class A and B fires
- Not suitable for use on fires involving electricity
- Extinguishes by cooling and sealing the surface of a burning liquid.

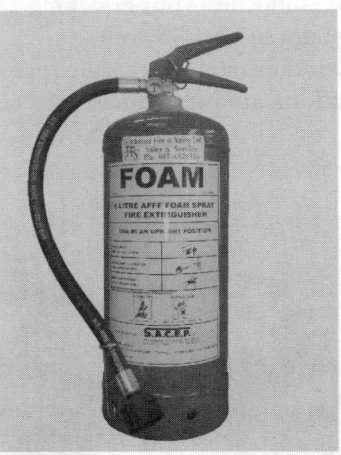

Fire Extinguishers—Carbon Dioxide

- Black body (old type) or red body with black label (new type)
- Best on class B and C fires but safe to use on any type of fire
- Safe to use on fires involving electricity
- Extinguishes by reducing oxygen levels and cooling.

Fire Extinguishers—Powder

- Blue body (old type) or red body with blue lable
- Best on class B fires but safe to use on most types of fire
- Works by coating the fuel and smothering the fire.

Fire Extinguishers—Blanket

- Any color body or label but they are usually red or white
- For use on any type of fire but best on small contained class B fires and people on fire
- Extinguishes by asphysiating.

Fire Safety at Home

- Fit and maintain a smoke detector
- Check round at night
- Close doors at night
- Do not smoke in bed
- Prepare an escape plan.

Post Course Knowledge

Answers to questions

- Fuel, heat and oxygen
- Arson !
- Two minutes
- Red
- Carbon dioxide (CO_2)

Name of Attendees	Designation	Signature	Date

Name of Attendees	Designation	Signature	Date

Toolbox Talks 4

Some of Toolbox Talks are Defined

a. Excavation.

b. Scaffolding.

c. Risk work.

d. Fire prevention.

e. Safe manual lifting.

Markings

For all site work activities, Toolbox talks must be conducted prior to start/shift work by site supervisor to the workers in the presence of site safety personnel for remembering employees about the hazards and risk involved and precaution to be taken in their work activities.

What is the Infrastructure of Toolbox talk Program?

It is a criterion of giving a safety talk prior to start or shift of work about the work activities, hazards and risk involved in their work, plant and machinery to be used and the remembrance of their safety measure to be taken, safe working procedure and method to be followed and PPE to be worn, to every group of workers by their current supervisors in the presence of safety personnel, for performing their job activities as per safe work procedure. So prior to job activities, the employees will be well trained in there area of activity for developing their expertise, professionalism and safe work practice according to OSHA standard.

After training procedure, the training documents should signed both by the trainer and workers trained and the document filled for further enquiry.

Does Training Program Change from Industry to Industry?

Yes its contents change according to the type of industry, activities performed, atmospheric condition, hazards and risks involved and company policy. But the bottom-line concept of Toolbox talk training program never changes.

How is Training Program Defined in SDTP Document?

See the attached documents:

SDTP	4.1 Toolbox Talk	Document References

<table>
<tr><td rowspan="4">SDTP</td><td rowspan="4" align="center">4.1 Toolbox Talk</td><td colspan="2">Document References</td></tr>
<tr><td colspan="2">SDTP. TBT. 001</td></tr>
<tr><td>Rev.No</td><td>Doc. Date</td></tr>
<tr><td>0</td><td>1/1/2009</td></tr>
<tr><td></td><td></td><td colspan="2" align="center">Pages
1 of 1</td></tr>
</table>

<table>
<tr><td colspan="2" align="center">1. Excavation</td></tr>
<tr><td>Talk Presented by :</td><td>Site :</td></tr>
<tr><td>Signature :</td><td>Date :</td></tr>
</table>

Talk Content/Discussion

How do I Project Excavations?

1. Solid barricading 1.5 meter away from excavation edges.	5. Flashing orange lights in the night.
2. Follow excavation permit to work systems.	6. Remove soil heaps, materials, tools from edges.
3. Shoring, stepping, Sloping (were applicable) to protect sides.	7. Prohibit unauthorized personnel entry to all employees and adhered to at all times.
4. Toe boards fixed to protect materials, tools falling in.	8. Standard visible warning signs.

What is the Safe Access to Excavation?
Ladder, stairway access.

Points to remember:

1. Induct workers in safe working method.

2. Needs to appoint a competent person to inspect excavation every day.

3. No machineries and equipment should work on edges while workers busy with excavation.

4. Use life lines and fall arresters wherever necessary.

5. Keep always access free from grease, oil and dirt.

6. Place a watchman for high risk jobs.

7. Be vigilant, when high rainfall would destabilize the excavation wall sides.

Attendance			
Name	**Sign**	**Name**	**Sign**

SDTP	4.2 Toolbox Talk	Document References

Document References
SDTP. TBT. 002

Rev.No	Doc. Date
0	1/1/2009

Pages
1 of 1

2. Scaffolding

Talk Presented by :		Site :	
Signature :		Date :	

Talk Content/Discussion

What is a Safe Scaffold?

1. Ensure scaffold made with standard (ISO marked) materials.	5. Ensure all metal parts are free of cracks and bends.
2. Ensure scaffold made by authorized and certified scaffolders.	6. Ensure that the scaffold is in level from bottom.
3. Scaffold is signed and "safe for use", green scaff-tag placed.	7. Scaffold fixed with enough supports, braces and ties.
4. Ensure guardrails and toe boards in place on all open sides.	8. Access of the scaffold are free from obstructions.
What is the Safe Access to go Top on Scaffolds?	Ladder, stairway access.

Points to remember:

1. Has the safety briefing made to all people working on/for scaffolding.

2. Needs to appoint a scaffold inspector to inspect scaffolds everyday.

3. Scaffold being tested to hold four times its maximum intended load.

4. Is the front of the scaffold holds distance from 14 inches of the work.

5. Is scaffold meets electrical safety clearance distances.

6. Have all planks being properly secured to the scaffold structure to prevent slipping and falling hazards.

7. Are all the people working on scaffold medically fit.

8. Are all protruding parts are coped.

Attendance			
Name	**Sign**	**Name**	**Sign**

SDTP	4.3 Toolbox Talk	Document References

		Document References
		SDTP. TBT. 003
		Rev.No / Doc. Date
		0 / 1/1/2009
		Pages
		1 of 1

3. Risk Work

Talk Presented by :		Site :	
Signature :		Date :	

Talk Content/Discussion

What is the Harm and Consequence Need to Face while Committing Risk Work?

1. Fatality	5. Amputation
2. Permanent disability	6. Laceration
3. Lost time incident	7. Strain/Sprain
4. Multiple fracture	8. Severe Cut

How can you Avoid Risk Work?	Think safely, avoid shortcuts.

Points to remember:

1. A safe system of the work to be cultured.

2. Wear all required personal protective equipment.

3. Needs to fix all edge protection and toe-boards as per standard.

4. Needs to fix and secure firmly all ladder and staircase access.

5. Housekeeping to be done regularly.

6. Permit to work system to be followed for all hazard jobs.

7. Needs to provide strict supervision and appoint watcher for high risk works.

Attendance			
Name	**Sign**	**Name**	**Sign**

SDTP	4.4 Toolbox Talk	Document References	
		SDTP. TBT. 004	
		Rev.No	Doc. Date
		0	1/1/2009
		Pages	
		1 of 1	

4. Fire Prevention

Talk Content/Discussion

How can a Fire Occur on Site?

1. Combustible material accumulation.	5. Hot works without permit.
2. Bad housekeeping.	6. Poor quality electric installation.
3. Smoking on site.	7. Poor maintenance of machineries and equipment.
4. Bad stacking practice.	8. Lack and poor efficiency of fire fighting equipment.

What is the Attitude Needed to Prevent Fire?	Avoid negligence and overconfidence, be always cautious about fire safety awareness.

Points to remember:

1. Induct workers in safe working method and enforce to do housekeeping on regular basis.

2. Keep all heat sources away from combustible material, gas, liquid stacking area. Store in ventilated and cool place.

3. Do not remove or tamper any charged and positioned fire extinguishers.

4. Keep always clear and free from obstructions on the fire fighting equipment access.

5. Always post fire watchers and place suitable fire extinguishers on hot work area. Follow permit to work systems.

6. Prohibit smoking on working area. Only allow smoking on smoking zone or area.

7. Do not use naked lights on working area, especially on hazardous and danger areas.

Attendance			
Name	**Sign**	**Name**	**Sign**

SDTP	4.5 Toolbox Talk	Document References

Document References
SDTP. TBT. 005

Rev.No	Doc. Date
0	1/1/2009

Pages
1 of 1

5. Safe Manual Lifting

Talk Presented by :		Site :	
Signature :		Date :	

Safe Manual Lifting Demonstration

Safe Team Lifts

Talk Content/Discussion

What are the Thinking to be Made before Manual Lifting is Planning?

1. Can the lift be avoided?	5. Can a mechanical aid be used?
2. What has to be moved?	6. Can one person manage?
3. From where and where to go?	7. Has training been given?
4. Is the route to be taken clear of obstructions and tripping hazards?	8. PPE's available are suitable and withstanding?

Which Part of Body the Harm more Effected?	Back Injury

Points to remember:

1. Induct workers in safe working method and enforce to wear all PPEs.
2. Needs to check your route of travel.
3. Needs to get a good grip to the load.
4. Needs to lift with your legs, not with your back.
5. Always carry lifted load close to the body.
6. Ensure that you can see over the object you are carrying and Watch where you are going.
7. Housekeeping to carry-out frequently to clear travel route free from grease, oil and dirt for avoiding slip and trip hazards.

Attendance			
Name	**Sign**	**Name**	**Sign**

Material Safety Data Sheet (MSDS)

5

Some of MSDS Program is Defined

a. Nitoflor FC 145.
b. Nitoflor TF 100000.

Markings

Material safety data sheet of every hazardous chemicals and substances to be received while these are hand over to the company by the manufacture because the precaution against the risk and hazard involved in handling those materials are only mention in it.

What is the Importance of Material Data Safety Sheet Environment?

While receiving each and every hazardous chemicals and substance, its material safety data sheet also needs to receive from the concern chemical, petroleum and compressed gas manufactures to know the risk and hazard involved in handling, stacking, fire prevention, first aid treatment to be given in case of accident and the usage of that particular product.

MSDS sheets copies to be file in safety office, first aid room and one copy should be displayed on the storage area as per OSHA standards to give information about risk and hazard including fire prevention and first aid treatment to be given in handling these hazardous substances, to the employees, nurse as well as visitors also.

How Material Data Safety Sheet Described?

See the attached sheets and find some chemical MSDS are exhibited:

SDTP	5.1 Material Safety Data Sheet	Document References

Document References
SDTP. MSDS. 001

Rev. No	Doc. Date
0	1/1/2009

Pages
1 of 4

MSDS of Nitoflor FC 145

1: Identification of the Substance/Preparation of the Company/Undertaking

Product Name:	NITOFLOR FC 145 BASE
Application:	Base component of two-part solvent based epoxy coating

2: Composition/Information on Ingredients

Composition:	Epoxy resin, xylene, 1-methoxy-2-propanol, di-isobutyl ketone, inert fillers, pigment.

Hazardous ingredient (s)	Symbol	Risk Phrases	Other Information	%
Epoxy resin (number average molecular wt<700)	XI	R 36/38, 43	Case No: 25068-38-6	>30<60
Xylene, mixture of isomers	XII	R 10, 20/21, 38	Case No: 1330-20-7	>10<25
Di-isobutyl ketone	XI	R 10, 37	Case No: 108-83-8	>2.5<10
1-methoxy-2-propanol	none	R10	Case No: 107-98-2	>2.5<10

All constituents of this product are listed in EINECS (European Inventory of Existing Commercial Chemical Substances) or ELINCS (European List of New Chemical Substances) or are exempt. Refer to section 8 for Occupation Exposure Limits.

3: Hazards Identification

Flammable: Harmful by inhalation and in contact with skin
Swallowed: May cause sensitization by skin contact.

Irritant   Flammable

4: First Aid Measures

Eyes:	Irrigate immediately with copious quantities of water for several minutes. Obtain medical attention urgently.
Skin:	Wash immediately with soap and water or suitable skin cleanser as soon as possible. Obtain medical advice if skin disorders develop.
Inhalation:	Remove from exposure, rest and keep warm and obtain medical attention urgently.

Ingestion	Obtain medical attention urgently. Do not induce vomiting. Beware of aspiration if vomiting occurs.

5: Fire Fighting Measures

Suitable Extinguishing Media:	Carbon dioxide, powder, foam or water fog. Do not use water jets.
Special Exposure Hazards:	Toxic fumes.
Special Protective Equipment:	Self-contained breathing apparatus.

6: Accident Release Measures

Personal Precautions:	Immediately issue NO SMOKING and NO NAKED FLAMES warning. Wear suitable protective clothing, gloves and eye/face protection.
Environmental Precautions:	Prevent entry into drains, sewers and water courses.
Decontamination Procedures:	Soak up with inert absorbent or container and remove by best available means. Gather into containers. Dispose of as waste.

7: Handling and Storage

Handling:	Maintain good standards of personal hygiene. Avoid skin and eye contact. Do not eat, drink or smoke whilst using this product. Ensure adequate ventilation. In case of insufficient ventilation. Wear suitable respiratory equipment.
Storage:	Store in conformity with local regulations. Store away from sources of ignition. Store in a cool place away from sources of heat and out of direct sunlight to avoid pressuret build up.

8: Exposure Controls/Personal Protection

Occupational exposure Limits:

Substance	8 Hour TWA	STEL	Source/Other Information
2,4-dimethyl heptan-4-one (isobutylketone)	150 mg/m^3		EH 40
1-Methoxy-propane-2-of	360 mg/m^3	1080 mg/m^3	EH 40
xylene, mixed isomers	435 mg/m^3	650 mg/m^3	EH 40

Engineering control measures:	Use only in wall ventilated areas. Local exhaust ventilation is recommended.
Personal protective equipment:	Impervious gloves (e.g. PVC), goggles. In case of In-sufficient ventilation wear suitable respiratory equipment. Change contaminated clothing immediately and clean before re-use.

9: Physical and Chemical Properties	
Physical state:	Liquid
Color:	Pigmented.
Odour:	Aromatic.
Boiling Point/Range (°C):	135–150
Flash Point (Closed, °C):	23
Autoflammability (°C):	Auto ignition temperature: approx 300
Explosive Properties (%)	Lower explosive limit (LEL): 1.0%. Upper explosive limit (UEL): 11.5%
Oxidising Properties:	Not determined
Vapour pressure (kPa at 20°C):	approx 0.3
Relative density (at 20°C):	1.30–1.40
Water Solubility:	Not determined.

10: Stability and Reactivity	
Stability:	Contains volatile solvent stable: Reacts with compounds containing amino, hydroxyl or carboxyl groups to give stable, cross linked materials.
Conditions to Avoid:	Any source of Ignition.
Materials to Avoid:	Strong oxidizing agents, Strong acids, Strong alkalis.
Hazardous Decomposition Products:	Thermal decomposition—Oxides of carbon, Toxic fumes.

11: Toxicological Information

The following toxicological assessment is based on a knowledge of the toxicity of the product's components: Oral LD50, rat 2 g/kg. Classified as a skin sensitizer.

Health Effects	
On Eyes:	Irritating and may injure eye tissue if not removed promptly. Vapor may cause Irritation.
On Skin:	Irritation, May cause sensitization. Harmful by absorption through the skin. May defeat the skin.
By Inhalation:	Harmful by inhalation. Vapours may cause headaches, dizziness and central nervous system depression. High concentrations of vapor can cause narvous system depression. High concentrations of vapor can cause narcosis and asphyxiation.
By Ingestion:	May cause irritation of mouth, throat and digestive tract. Aspiration of liquid during ingestion or from vomiting may cause severe chemical pneumonitis.
Chronic:	Repeated or prolonged skin contact may lead to skin disorders.

12: Ecological Information

Environmental Assessment:	Contains volatile organic compounds.
Mobility:	Mobile liquid contains voltaic components, Insoluble in water.
Persistence and Degradability:	Expected to be not readily biodegradable.
Bioaccumulative Potential:	Not expected to be bioaccumulative.
Ecotoxicity:	Expected to be ecotoxic to fish/daphnia/algae.

13: Disposal Considerations

Disposal must be in accordance with local and national legislation.

Unused Product:	Dispose of through an authorized waste contractor to a licensed site.
Used/Contaminated Product:	As for unused product.
Packaging:	Must be disposed of through an authorized waste contractor.

14: Transport Information

UN/SI Number: 1263

IMO:	Class 3.3	Packing Group: III	Marine pollutant: No
IATA/ICAO:	Class: 3	Packing Group: III	
ADR:	Class: 3	Item: 31 (c)	

Transport Name: Paint Related Material (xylene mixture)

Hazchem/Kemler code: 3y/30

15: Regulatory Information

Hazard Label Data:	
Named Ingredients:	Epoxy resin (number average molecular weight < 700) Xylene (mixture of isomers)
UN Number:	1263
Symbol (s)	xn
Risk Phrases:	Flammable. Harmful by inhalation and in contact with skin. Irritating to eyes and skin. May cause sensitisation by skin contact.
Safety Phrases:	Keep away from sources of ignition—no smoking. In case of contact with eyes, rinse immediately with plenty of water and seek medical advice. Wear suitable protective clothing, gloves and eye/face protection. Use only in well-ventilated areas. Contains epoxy constituents. See information supplied by the manufacturer.

EC Directives:	Dangerous substance Directive 67/548/EEC and adaptations. Dangerous Preparations Directive, 88/379/EEC. Safety Data sheets Directive 91/155/EEC.
Statutory Instruments:	Chemicals (Hazard Information and packaging) Regs. 1993 (SI 1746). Control of Substances Hazardous to Health Regs 1988 (SI 1657). Health and safety at Work, etc. Act, 1974. Environmental Protection (Duty of care) Regs 1991 (SI 2839) Highly Flammable Liquids and Liquefied Petroleum Gases Regs. 1972 (SI 917).
Codes of Practice:	Waste Management. The Duty of Care.
Guidance Notes:	Occupational skin diseases: Health and safety precautions (EH 26). Occupational exposure limits (EH 40). The storage of highly flammable liquids (CS 2).

The above publications are available from HMSO.

16: Other Information

Also refer to the safety Datasheet (s) for information on other components of this product.

The data and advice given apply when the product is used for the stated applications. The product is not sold as suitable for any other application. Use of the product for applications other then as stated may give rise to risks not mentioned in this sheet. The product should not be used other than for the stated application or applications without seeking advice from AL Gurg Fosroc.

If this product has been purchased for supply to a third party for use at work, it is the purchaser's duty to take all necessary steps to secure that any person handing or using the product is provide with the information in this sheet.

If is the responsibility and duty of the employer to inform employees and others who may be affected of any hazards described in this sheet and of any precautions which should be taken.

This sheet does not constitute or substitute for the users own assessment of workplace risk, as required by other health and safety legislation.

SDTP	5.2 Material Safety Data Sheet	Document References

Document References
SDTP. MSDS. 001
Rev. No — 0
Doc. Date — 1/1/2009
Pages — 1 of 4

MSDS of Nitoflor TF 10000

Heavy duty epoxy resin floor screed

Uses	Properties
Nitoflor TF 10000 provides an extremely durable, high strength, chemical, impact and abrasion resistant floor and is suitable for the following.	The values given below are typical figures achieved in laboratory tests. Actual values obtained on-site may show minor variations from those quoted.

Uses	Properties		
• Ideally suitable for use in heavy engineering plants. Process areas oil refineries, workshops and other industrial plants where a floor topping of up to 10 mm thick is required.	Mechanical characteristics @ 23°C Cura or 7 days Compressive strength (ASTM C-579)	:	102 N/mm^2
• For use in areas of lighter duty where above average durability and low maintenance costs are required.	Flexural strength (ASTM C-580)	:	25 N/mm^2
• For use in overcoating and sealing with one of the Nitoflor FC apoxy resin floor coating range where high degrees of cleanliness and chemical resistance are required.	Tensile strength (ASTM C-307)	:	18 N/mm^2
	Bond strength to concrete	:	Cohesive strength of the concrete
• Trafficable–can be open to traffic after 24 hours.	Indentation Modified US MIL- D-3134	:	No indentation from a height of 2.4 m
	Shore D Hardness (ASTM D-2240)	:	70–75

Advantage

Advantage	Curing characteristics			
• Durable–capable of withstanding heavy loads. Highly resistant to abrasion	Curing characteristics			
• Chemical resistant–withstands a wide range of chemicals.			25°C	30°C
• Low odor–no harmful vapours present.	Pot life	:	45 mins	30 mins
• High strength-strengths in excess of the concrete to which material is applied.	Initial hardness	:	16 hours	8 hours
• Locally produced–formulated for use in Middle East conditions.	Full cure	:	7 days	4 days
	Trafficable time	:	24 hours	16 hours

Description	
Nitoflor TF 10000 is a four component solvent free morlar consisting of an epoxy resin base, hardener, a specially selected and graded aggregate or high crushing strength and a Color pack.	Chemical resistance
It is trowel applied at a thickness between 5 and 10 mm for the heaviest industrial applications.	Nitoflor TF 10000 is resistant to the following chemicals at 30°C
Nitoflor TF1000 screed provides an impervious durable. Chemical resistant topping with a slightly granular texture to give non-skid finished surface.	Acids (m/v) Acetic acid 5% : Resistant Hydrochloric acid 18% : Resistant Sulfuric acid 20% : Resistant Phosphoric acid 25% : Resistant

Solvents

Isopropyl Alcohol   :   Resistant
Ethyl Alcohol   :   Resistant

These tests have been conducted in accordance with ASTM D1308 for spot tests and ASTM D543 for immersion tests as appropriate. Please contact your local Fosroc office for the complete list of chemical tested.

All the above properties have been determined by laboratory controlled tests and are in excess of those expected in practice.

Nevertheless, success in use will be determined by the implementation of good house-keeping practices.

Instructions for use

Surface preparation

A good floor topping is only good as the preparation. It is essential that Nitoflor TF 10000 is applied to sound. Clean and dry surfaces in order that maximum bond strength is achieved between the substrate and the flooring system.

New concrete floors

Should be at least 28 days (at 35°C). Laitance on new concrete floors is best removed by light grit blasting, mechanical scrabbling. Acid etching is not recommended.

Old concrete floors

Again, mechanical cleaning methods are strongly recommended on old concrete floors particularly where heavy contamination by oil and grease has occurred or existing coatings are present. These may well have been absorbed several millimeters into the concrete. To ensure adhesion, all contamination should be removed. Proprietary chemical degreaser may be used on small areas of light contamination only.

All dust and debris should be removed prior to laying Nitoflor TF 1000, preferably by industrial vacuum cleaning.

Steel surfaces

Steel surfaces should be degreased and grit blasted to SA21/2 immediately prior to application.

Priming

All surfaces treated with Nitoflor TF 10000 should be primed with Nitocote Primer Sealer, a solvent free epoxy resin primer designed for maximum absorption and adhesion to the substrate.

Add the entire contents of the hardener tin to the base tin and mix thoroughly. Once mixed, immediately apply he primer in a thin continuous film to the clean prepared surfaces.

Work the primer into the surface and avoid over application and pudding. On porous floors, the primer will be absorbed very quickly leaving characteristic light Colored dry patches. It is recommended that a second priming coat is applied in these areas.

While still wet, dress the surface with $1/2$ kg/m^2 of Antislip Grain No 3 to provide a key for the application of Nitoflor TF 10000. Ensure that the primer is touch dry prior to commencing application.

Mixing

It is important that Nitoflor TF 10000 is mixed correctly.

Empty the Color pack into the base container and mix for 15-30 seconds, until homogeneous.

Add the entire content of the hardener into the mix and stir for further 30 seconds, until homogenous.

Gradually empty the filter (aggregate) into the mix of base. Color pack and hardener and continue mechanical mixing, using Fosroc MR4 mixing paddle fitted to a slow speed forced action mixture, for a further 2-3 minutes, until all components are thoroughly blended.

Application

Nitoflor TF 10000 is laid up to 10 mm thick in one application.

The mixed Nitoflor TF 10000 is raked to the required thickness and tamped with a wooden float, or leveled with a screeding bar, to ensure full compaction.

To provide as close a textured finish as possible, finish off the process using a steel trowel regularly cleaned. The use of Fosroc Trowel EZ*↑ on the trowel will aid application.

Expansion Joints

Expansion joints in the existing substrate must be maintained through the Nitoflor TF 10000 scared.

For suitable joint sealants, please contact your local Fosroc office.

Sealing

Incase where a high degree of cleanliness, or where a specific color, or improved chemical resistance is required the surface of Nitoflor

TF 10000 can be sealed. Recommended Fosroc products include Nitoflor F°C 145*↑, Nitoflor F°C 140*↑ and Nitoflor F°C 130*↑. For application detail and other information see the relevant technical data sheets.

Overcoating can commence when the Nitoflor TF 10000 has cured for 24 hours at 35°C.

Cleaning

Tools and equipment should be solvent cleaned with Fosroc solvent 102* immediately after use.

Limitation

- Nitoflor TF 1000 should not be applied on to surfaces which are know to, or likely to suffer from, rising dampness, potential osmosis problems or have a relative humidity greater than 75% as measured in accordance with BS 8203 appendix a or by a Hammond concrete/mortar moisture tester type COCO.
- Fosroc does not recommend acid etching as a method of floor preparation. However the use of such method should be approved by the project consultant.
- Nitoflor TF 10000 is not recommended for outdoor applications.
- If movement or cracking of the takes place after application, reflective cracking of the toping may occur. All known expansion joints should be continued in the topping.

Technical Support

Fosroc offers a comprehensive range of high performance, high quality flooring, jointing and repair products for both new and existing floor surfaces. In addition, the company offers a technical support package to specifies, end-users and contractors, as well as on site technical assistance in locations all over the world.

Estimating

Supply

Nitoflor TF10000	:	12 liter packs
Nitocote primer sealer	:	1 and 4 liter packs
Fosroc solvent 102	:	5 liter can
Antislip grin no. 3	:	20 kg bags

Coverage		
Nitoflor TF 10000	:	1.2 m² per pack @ 10 mm thickness
	:	4–5 m²/liter

Note: The above coverage rates are given for guidance only as actual quantities used will very depending upon nature of the substrate and condition on site. A heavily scrabbled or uneven floor will increase the consumption.

Storage

Shelf life

When stored in dry, cool conditions, in the original, unopened pack, Nitoflor TF 10000 will have a shelf life or 12 months

Precautions

Health and safety

Nitoflor TF 10000 and Fosroc solvent 102 should not come into contact with the skin and eyes, or be swallowed. Ensure adequate ventilation and avoid inhalation of vapors. Some people are sensitive to resins, hardeners and solvents.

Wear suitable protective clothing, gloves and eye protection. If working in confined areas, suitable respiratory protective equipment must be used.

The use of barrier creams provide additional protection. In case of contact with skin, rinse with plenty of clean water, then cleanse with soap and water. Do not use solvent.

In case of contact with eyes, rinse immediately with plenty of clean water and seek medical advice. If swallowed seek medical attention immediately, do not induce vomiting.

Fire

Fosroc solvent 102 is flammable, keep away from sources of ignition. No smoking, in the event of the extinguish with CO_2 or foam. Do not use a water jet.

Flash points

Fosroc solvent 102 : 33°C

Additional Information

Fosroc manufactures a wide range of complementary products which include:
- Waterproofing membranes and water stops
- Joint sealants and filer boards
- Cementitious and apoxy grouts
- Specialized flooring materials

Fosroc additionally offers a comprehensive package of products specially designed for the repair and refurbishment of damaged concrete. Fosroc's systematic approach to concrete repair features the following:
- Hand—placed repair mortars
- Spray grade repair mortars
- Fluid microconcretes
- Chemically resistant epoxy mortars
- Anti-carbonation/anti-chloride protective coatings
- Chemical and abrasion resistant coatings.

Signages and Posters

6

Some of International Signages and Safety Posters are defined

a. International Signages

b. Safety Posters

Markings

In every industry as well as constructional sites, signage's needs to be displayed in every section to inform the employees and visitors about the risk and hazards involved in these sections and safety rules to be followed as per OSHS Standards. Safety posters needs to display as a reminder or as a part of training program for the workers who engaged in work activities.

What is the Importance of Signages on Working Environment?

Signages are considered as the mirror of the risk and hazards shown to the employees as well as visitors involved in work place activities. Every persons responsibility is to obey the signages placed in different places of work activities, if not the harm and consequence will be very high because the unknown activities and procedures can take their life.

Mainly signages are demonstrated to read in three ways:

a. Symbolic Signages   b. Test Signages   c. Both Mixed

Signages are colored to exhibit mainly:

a. Warning Signages—yellow background color—giving warning

b. Prohibited Signages—red background color—never allowed (except fire equipment Signages)

c. Emergency Signages—green background color—for emergency purpose

d. Mandatory Signages—blue background color—to be worn all times

What is the Importance of Safety Posters on Working Environment?

Safety posters are designed to the work place to build up the awareness of the workforce. If the Safety Posters are exhibits on main places of work face the workers in their absconding mind should be always aware that if they do this activity in negative way, this tragedy will happen and hurt them. Much more than the unsafe acts and unsafe condition can be avoided to a great extend.

Safety posters are to be designed with real or comic pictures of harm and consequences happens and the text to be translate to their regional languages of the workforce because the illiterate employees also need to understand and aware contents written in it. Safety posters are to be designed and to be placed for all the work activities included in the company premises.

How will you Read Signages/Signages and Safety Posters Described?

See the attached sheets, how to read Signages and some Signages and Safety Posters are exhibited:

How will you Read Signages

Main Types of Safety Signs

Symbolic signs

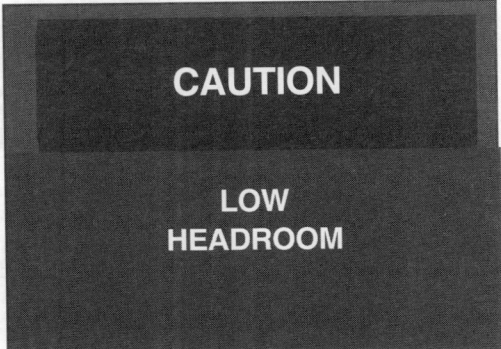

CAUTION

LOW
HEADROOM

Text signs

How well do you know Safety Signs?

1

2

3

4

The Meaning of the Shapes

Regulation  The message on the sign must always be obeyed.

Caution  Take proper precautions for the identified hazard.

Emergency Information  Indicates the type and location of emergency and fire fighting equipment.

Regulation Mandatory Signs

Wear mask

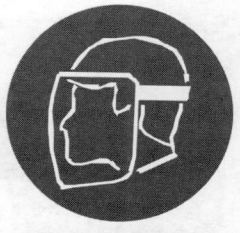

Wear face shield

Protective footwear must be worn in this area

Regulation Prohibition Signs

No digging   No open flame   No admittance

These actions are never allowed in the areas with these signs.

Regulation Danger Warning Signs

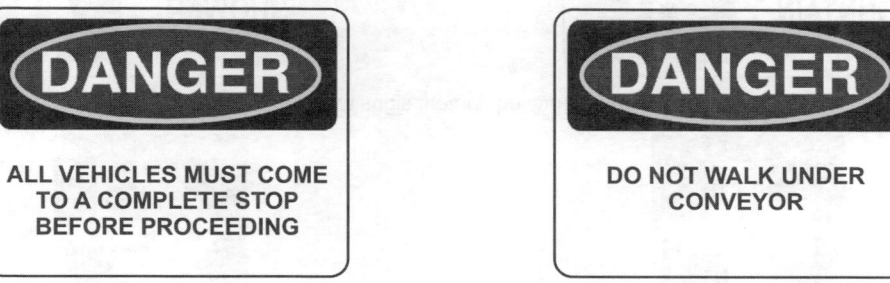

The instruction on the sign must always be obeyed.

Caution Signs

The symbol tells you the hazard that you must beware of.

Emergency Equipment Signs

 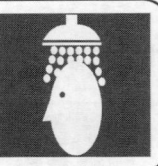

Symbolic emergency equipment signs with supporting text

First Aid

General indication of direction to emergency equipment

Fire Fighting Equipment Signs

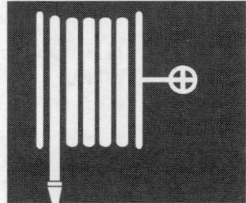 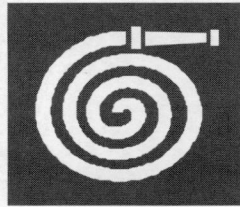 

Fire hose reel

Fire extinguisher

Fire hydrant hose

Use of Safety Signs

Safety signs are provided to help protect your safety and health so:

1. Always obey safety signs.
2. Never remove any safety sign unless you have permission from management.
3. Never deface or alter a sign.
4. Never obstruct visibility by stacking materials or equipment in front of signs.
5. Keep signs clean and legible.

SDTP	6.1 International Signages	Document References	
		SDTP. IS. 001	
		Rev.No	Doc. Date
		0	1/1/2009
		Pages	
		1 of 10	

Some Warning Signages	Color	Yellow

Danger
Buried cable

Mind your head

Mind the step

Danger
Live wires

Danger
Scaffolding incomplete

Explosive substance

Working progress

Danger
Deep excavation

High noise area

High radiation area

Danger
Construction work area

Caution
Slippery floor surface

Danger
Fragile roof

Caution
Out of order

Danger
Automatic machinery may start without warning

Danger
Moving machinery risk of trapped hands

Caution
Trip hazard

Danger
Men Working ahead

Danger
asbestos

Danger
Sharp edge

Danger
Overhead hazard

Harmful substance

Organic peroxide

Dangerous chemicals

Spontaneously combustible

Danger
Battery acid

Compressed gas

Danger
Overhead hazard

Risk of freezing

Toxic gas

Danger
Risk of falling

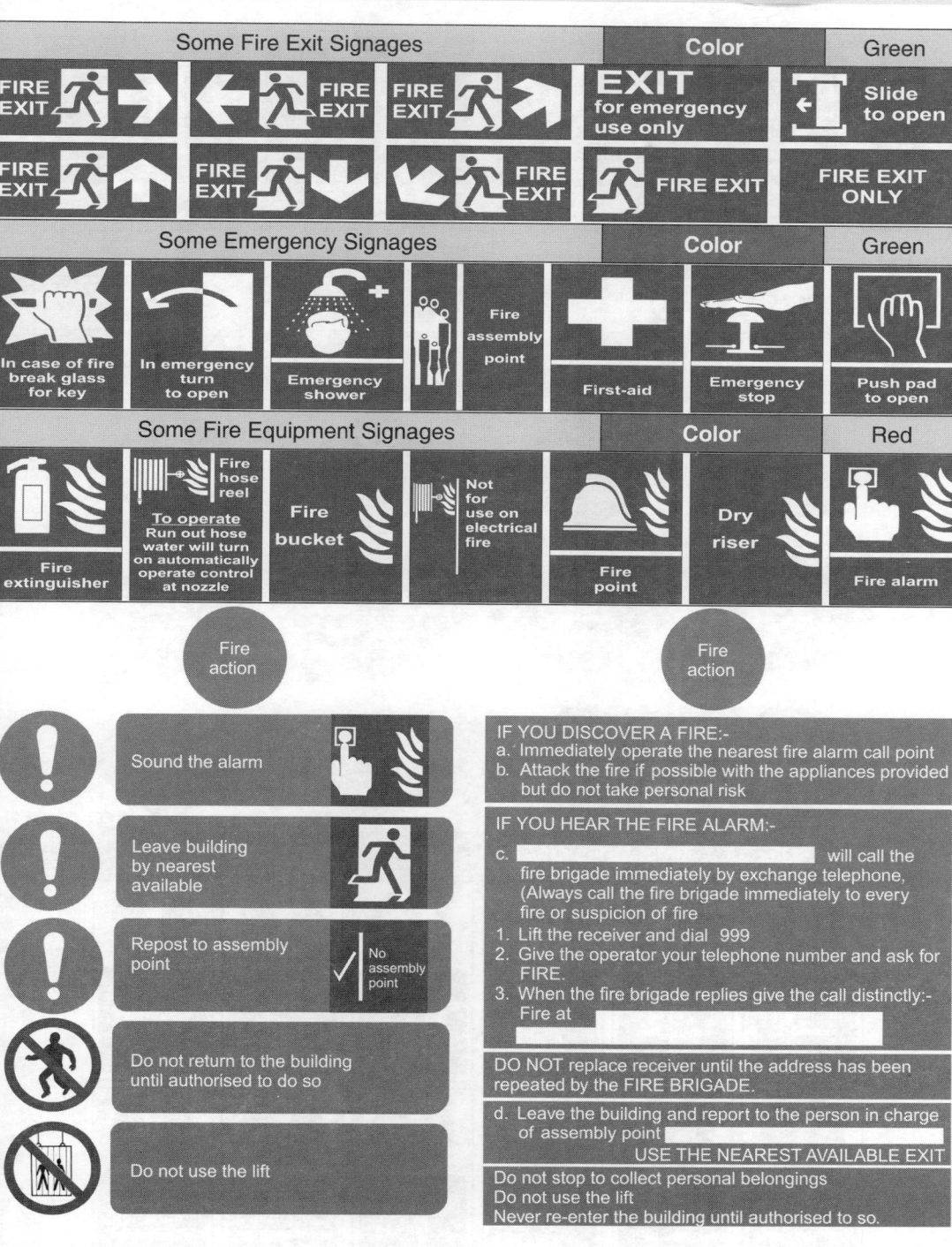

Some Fire Exit Signages — Color: Green

FIRE EXIT →

← FIRE EXIT

FIRE EXIT ↗

EXIT for emergency use only

Slide to open

FIRE EXIT ↑

FIRE EXIT ↓

FIRE EXIT

FIRE EXIT

FIRE EXIT ONLY

Some Emergency Signages — Color: Green

In case of fire break glass for key

In emergency turn to open

Emergency shower

Fire assembly point

First-aid

Emergency stop

Push pad to open

Some Fire Equipment Signages — Color: Red

Fire extinguisher

Fire hose reel — To operate Run out hose water will turn on automatically operate control at nozzle

Fire bucket

Not for use on electrical fire

Fire point

Dry riser

Fire alarm

Fire action

- Sound the alarm
- Leave building by nearest available
- Repost to assembly point — No assembly point
- Do not return to the building until authorised to do so
- Do not use the lift

Fire action

IF YOU DISCOVER A FIRE:-
a. Immediately operate the nearest fire alarm call point
b. Attack the fire if possible with the appliances provided but do not take personal risk

IF YOU HEAR THE FIRE ALARM:-
c. _____ will call the fire brigade immediately by exchange telephone, (Always call the fire brigade immediately to every fire or suspicion of fire
1. Lift the receiver and dial 999
2. Give the operator your telephone number and ask for FIRE.
3. When the fire brigade replies give the call distinctly:- Fire at _____

DO NOT replace receiver until the address has been repeated by the FIRE BRIGADE.

d. Leave the building and report to the person in charge of assembly point _____
USE THE NEAREST AVAILABLE EXIT

Do not stop to collect personal belongings
Do not use the lift
Never re-enter the building until authorised to so.

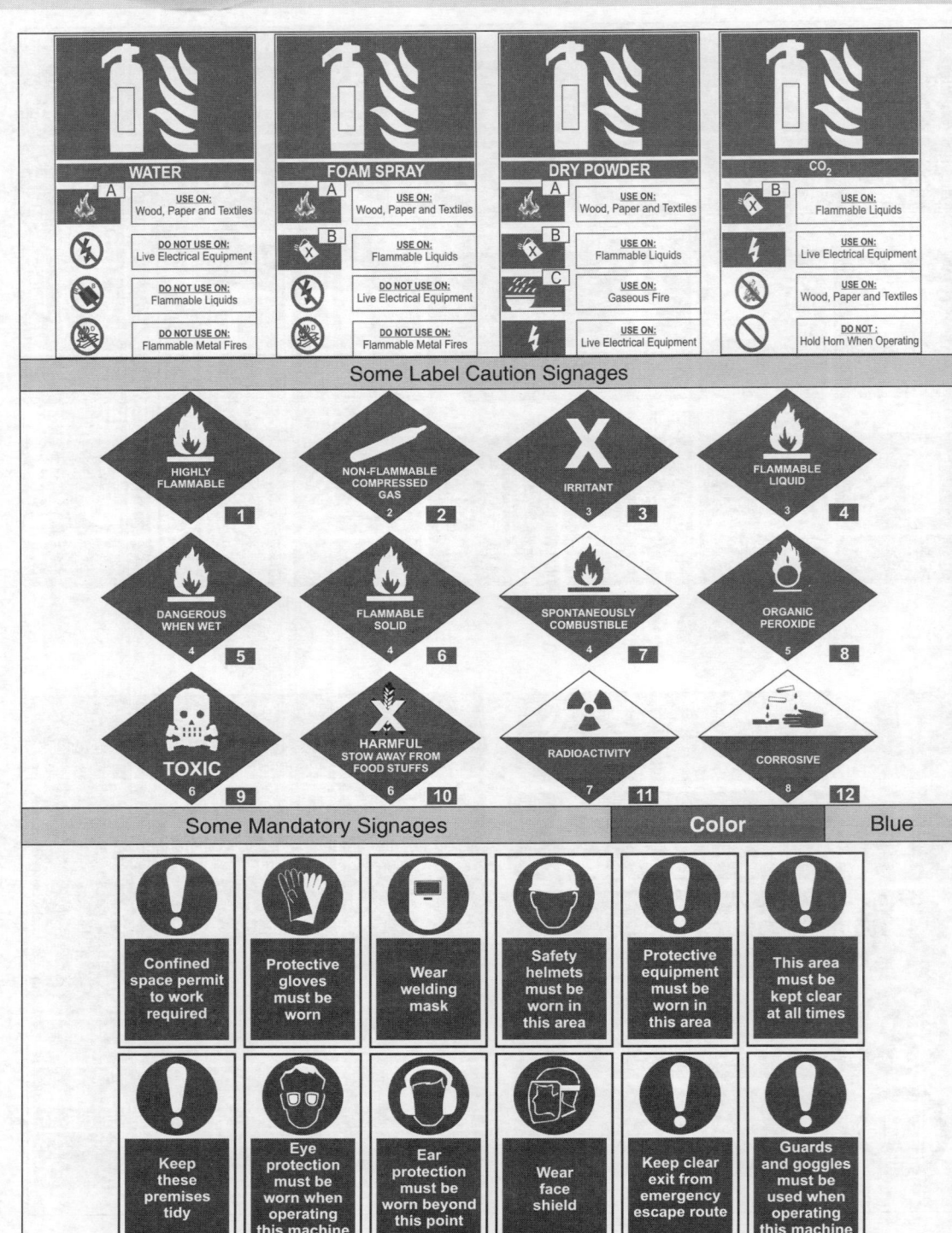

WATER

	USE ON: Wood, Paper and Textiles
A	DO NOT USE ON: Live Electrical Equipment
	DO NOT USE ON: Flammable Liquids
	DO NOT USE ON: Flammable Metal Fires

FOAM SPRAY

A	USE ON: Wood, Paper and Textiles
B	USE ON: Flammable Liquids
	DO NOT USE ON: Live Electrical Equipment
	DO NOT USE ON: Flammable Metal Fires

DRY POWDER

A	USE ON: Wood, Paper and Textiles
B	USE ON: Flammable Liquids
C	USE ON: Gaseous Fire
	USE ON: Live Electrical Equipment

CO$_2$

B	USE ON: Flammable Liquids
	USE ON: Live Electrical Equipment
	USE ON: Wood, Paper and Textiles
	DO NOT : Hold Horn When Operating

Some Label Caution Signages

HIGHLY FLAMMABLE	1	NON-FLAMMABLE COMPRESSED GAS	2 2	IRRITANT	3 3	FLAMMABLE LIQUID	3 4
DANGEROUS WHEN WET	4 5	FLAMMABLE SOLID	4 6	SPONTANEOUSLY COMBUSTIBLE	4 7	ORGANIC PEROXIDE	5 8
TOXIC	6 9	HARMFUL STOW AWAY FROM FOOD STUFFS	6 10	RADIOACTIVITY	7 11	CORROSIVE	8 12

Some Mandatory Signages · Color · Blue

| Confined space permit to work required | Protective gloves must be worn | Wear welding mask | Safety helmets must be worn in this area | Protective equipment must be worn in this area | This area must be kept clear at all times |
| Keep these premises tidy | Eye protection must be worn when operating this machine | Ear protection must be worn beyond this point | Wear face shield | Keep clear exit from emergency escape route | Guards and goggles must be used when operating this machine |

Some Multipurpose Signages

Danger
Highly Flammable

No Smoking

No Naked Flames

No Mobile Phones

Switch off engine

SAFETY INSTRUCTIONS

 Wear hard hat during all loading / unloading operations

 Wear safety shoes / coverall during workshop / yard dutles

 No unauthorized person permitied

 Smoking in designated ares only

 Speed limit 10 km/hr

 Weatherlord General Safety rules must be strictly followed

DANGER   KEEP OUT

Highly Flammable

No Smoking
No Naked Lights

SITE SAFETY

Under the health and safety at work act, 1974, all persons entering this site must comply with regulations under this act. All visitors must report to the site office and obtain permission to proceed on the site or any work area. Safety sings and procedures must be observed and personal protection and safety equipment must be used at all time

 Construction work in progress.
Parents are advised to warn children of the dangers of entering this site.

 Safety Helmets must be worn

 Unauthorized entry to this site is strictly forbidden.

Safety Instructions

 No unauthorised persons admitted.

 PPE is required beyond this print.

Protective clothing must be worn

Respirator must be worn

Face shield must be worn

Protective gloves must be worn

Protective footwear must be worn

Ear protection must be worn

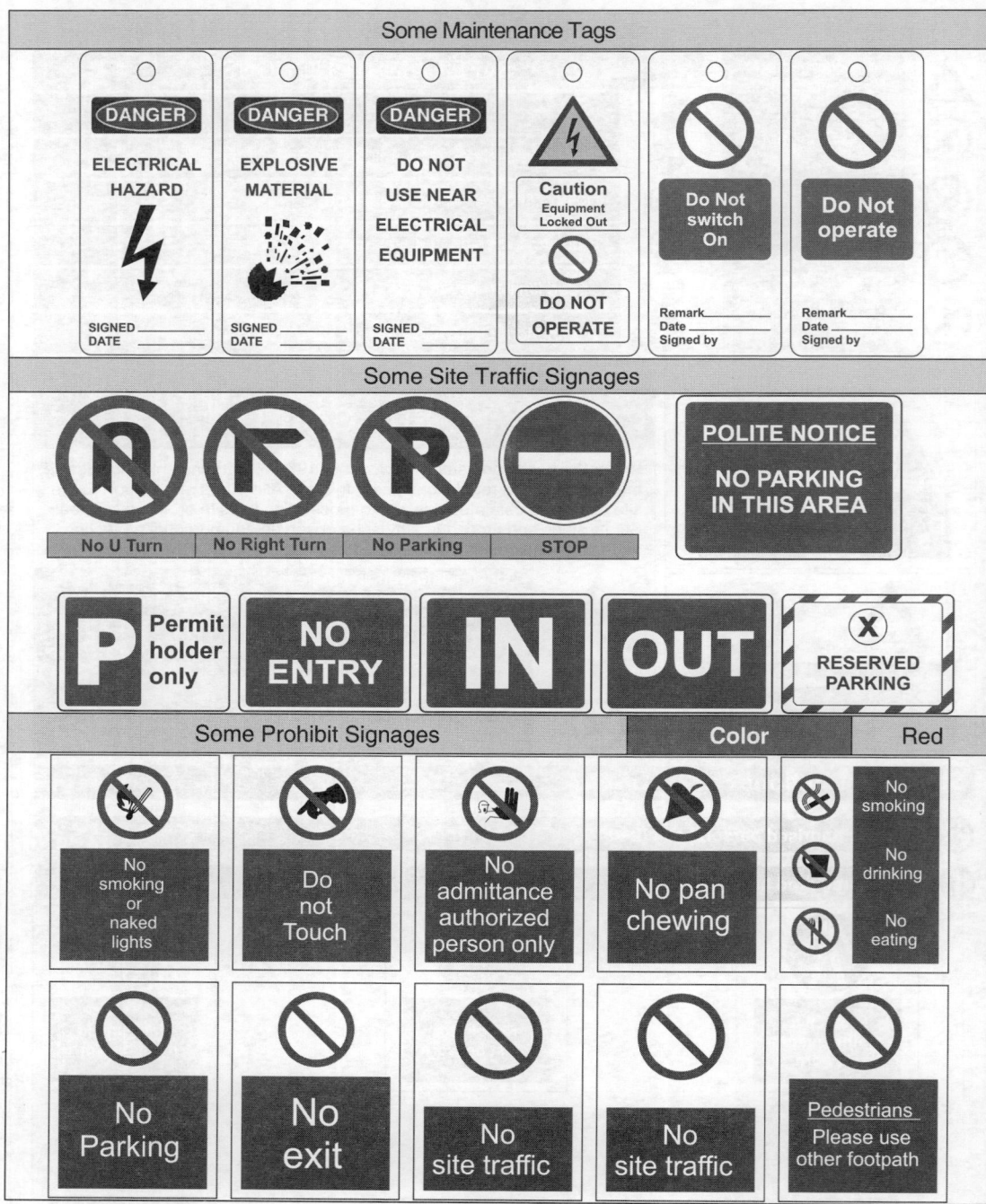

Some Maintenance Tags

DANGER — ELECTRICAL HAZARD — SIGNED _____ DATE _____

DANGER — EXPLOSIVE MATERIAL — SIGNED _____ DATE _____

DANGER — DO NOT USE NEAR ELECTRICAL EQUIPMENT — SIGNED _____ DATE _____

Caution — Equipment Locked Out — DO NOT OPERATE

Do Not switch On — Remark_____ Date _____ Signed by

Do Not operate — Remark_____ Date _____ Signed by

Some Site Traffic Signages

No U Turn — No Right Turn — No Parking — STOP

POLITE NOTICE — NO PARKING IN THIS AREA

P Permit holder only — NO ENTRY — IN — OUT — **X** RESERVED PARKING

Some Prohibit Signages

Color — Red

No smoking or naked lights — Do not Touch — No admittance authorized person only — No pan chewing — No smoking — No drinking — No eating

No Parking — No exit — No site traffic — No site traffic — Pedestrians Please use other footpath

SDTP	6.2 International Safety Posters	Document References	
		SDTP. SP. 001	
		Rev.No	Doc. Date
		0	1/1/2009
		Pages	
		1 of 1	

Accidents Start where Safety Ends!

The correct use of a cheater

- Extensions or cheaters must not be used on wrench handles until efforts to break out or make up the connection with the largest wrench available have failed.
- If an extension is used, place it on the largest readily available wrench.
- The extension must extend the full length of the handle so that it will not damage the wrench or slip off the handle.
- Never use an extension on a crescent-type wrench or an aluminum wrench.
- Aluminum extensions must be schedule 80 or greater
- Fiberglass extensions must not be used
- Avoid awkward body position
- Do not stand on cheater.

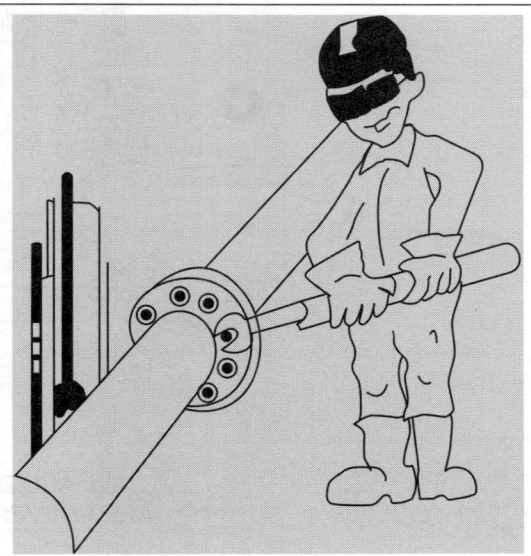

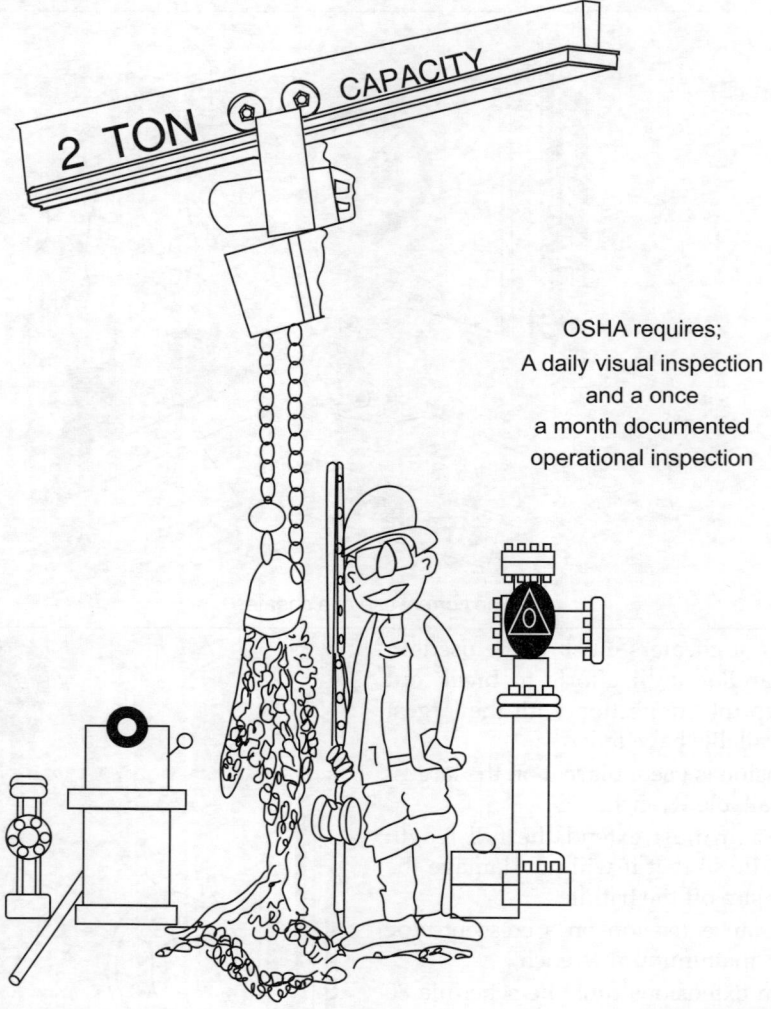

CRANE INSPECTION
AT START OF EACH SHIFT
TEST IT FIRST!

2 TON CAPACITY

OSHA requires;
A daily visual inspection
and a once
a month documented
operational inspection

ALL ACCIDENTS
ARE PREVENTABLE!

PARTNERS IN SAFETY
AND LOSS CONTROL

WE KNOW THAT
PREVENTION
IS THE CURE!

**KNOW WHERE
THE FIRE EXTINGUISHERS ARE LOCATED IN OUR AREA AND
MAKE SURE THEY ARE ACCESSIBLE**

PARTNERS IN SAFETY
AND LOSS CONTROL
KNOW MORE
LIVE LONGER.

FIRE EXTINGUISHERS ...

KNOW
WHERE
THE FIRE EXTINGUISHER
IS LOCATED IN YOUR AREA

IN CASE OF FIRE

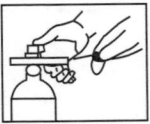

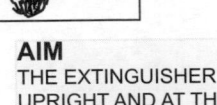

PULL
PULL THE
PIN

AIM
THE EXTINGUISHER
UPRIGHT AND AT THE
BASE OF THE FIRE

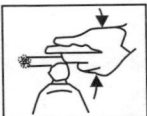

PARTNERS IN SAFETY
AND LOSS CONTROL
KNOW MORE
LIVE LONGER.

SQUEEZE
SQUEEZE
TRIGGER

SWEEP
EXTINGUISHER FROM
SIDE TO SIDE TO OVER
FIRE AREA

AND PUT IT OUT!

Heat Stress Management		
Severity	Environmental Heat (WGBT Index)	Awareness Issues/Actions
Level I	Less than 77°F.	No special precautions
Level II	77°F–79°F.	Replenish lost fluids as needed
Level III	80°F–82°F.	Replenish fluids frequently. Pace work a condignly. Break as needed in shade or an air-conditioned environment.
Level IV	83°F–86°F.	Replenish fluids frequently (every 15–20 minutes) Pace work a condignly. Break as needed in shade or air-conditioned environment. Be aware of early symptoms of hear illness: Fatigue, headache, muscle cramps. Limit tooks involving continuous mode rate to heavy work.
Level V	Greater than 86°F.	Replenish fluids frequently every (15–20 minutes) Pace work a condignly. Break as needed in shade or an air-conditioned environment. Be aware of early symptoms of heat illness: fatigue, headaches, muscle cramps. Severely limit tasks involving continuous moderate to heavy work. Make use of appropriate to external cooling devices (cool vests, vortex suits, etc.) Employ buddy system for high-risk tasks. Rotate workers to reduce heat exposure.

OF GOOD HOUSEKEEPING

ABC'S

STORE MATERIALS

Store all work materials, from paper products to flammable liquids, in approved clearly marked containers, kept in designated storage areas.

GOOD HOUSEKEEPING

Can help make any job you do:
Safer—Easier—more productive housekeeping should be a routine part of your job!

CLEAN AND STORE TOOLS PROPERLY

After you finish using them, make sure your tools are free of dirt, oil, etc. and in their proper storage area.

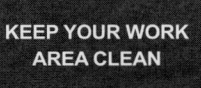
KEEP YOUR WORK AREA CLEAN

MAINTAIN LIGHTING

For maximum brightness, keep all lighting clean and unobscured by furniture, storage cabinets, etc.

CLEAN AND MAINTAIN MACHINES PROPERLY

Follow all routine cleaning and maintenance procedures, and report any problems immediately.

HOUSEKEEPING IN THE LAB

Clean up and dispose of spilled chemicals promptly and properly.

PARTNERS IN SAFETY AND LOSS CONTROL

A CLEAN WORK AREA
IS A SAFER WORK AREA

LADDER SAFETY
FOR BIG AND SMALL JOBS

COMMON SAFETY PRACTICES

Choose the right ladder

Check condition of ladder before use

Engage all ladder locks

Use 4-to-1 rule (for every 4 feet of ladder height, position the ladders base 1 foot away from the vertical support)

Secure (tie off) the ladder as close to the top as you can

Extend the ladder at least 3 feet above the supporting edge when climbing onto a roof or platform

Free hands when climbing. Carry tools, parts, etc. in an apron or belt or raise and lower them by a hand line. Hook equipment on a rung while you work.

Do not lean or stretch to reach your object work area, move the ladder.

PARTNERS IN SAFETY
AND LOSS CONTROL

STEP UP TO SAFETY...
USE A LADDER!

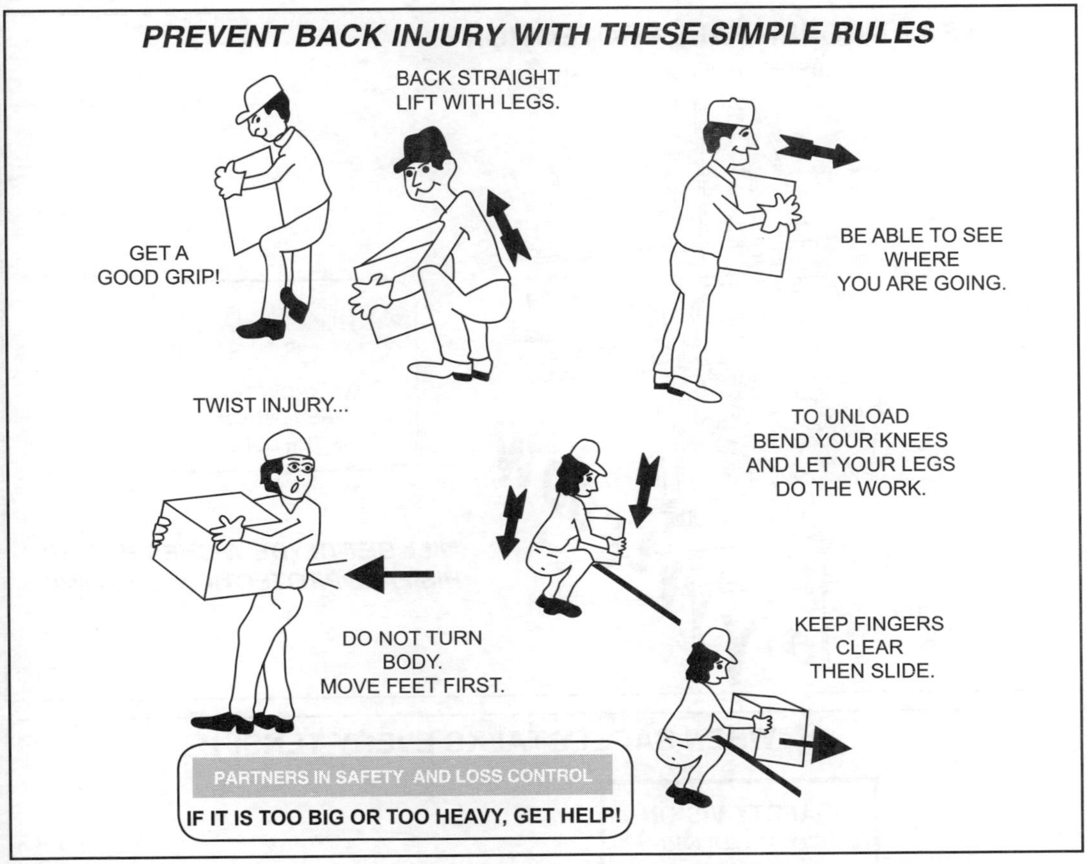

PREVENT BACK INJURY WITH THESE SIMPLE RULES

BACK STRAIGHT
LIFT WITH LEGS.

GET A
GOOD GRIP!

BE ABLE TO SEE
WHERE
YOU ARE GOING.

TWIST INJURY...

TO UNLOAD
BEND YOUR KNEES
AND LET YOUR LEGS
DO THE WORK.

DO NOT TURN
BODY.
MOVE FEET FIRST.

KEEP FINGERS
CLEAR
THEN SLIDE.

PARTNERS IN SAFETY AND LOSS CONTROL

IF IT IS TOO BIG OR TOO HEAVY, GET HELP!

PROPER LIFTING

GET A
GOOD GRIP

LIFT WITH
YOUR LEGS

DO NOT
STACK
HIGHER
THAN
EYE
LEVEL

GET HELP WITH
HEAVY LOADS

PARTNERS IN SAFETY AND LOSS CONTROL

IF IT IS TOO BIG, BULKY OR HEAVY GET HELP!

HARD HATS

A SAFE IDEA!

ARE YOU WEARING YOURS?

DISASTER WHILE RIDING SHOVEL NEAR EXCAVATION

DO NOT Drive

too close to the edge of the **EXCAVATION**

The **RESULT** can be a **DISASTER**

WHY SLEEPING ON SITE IS PROHIBITED

DO NOT REST NEAR MACHINES, STACKED BLOCKS AND IN EXCAVATION

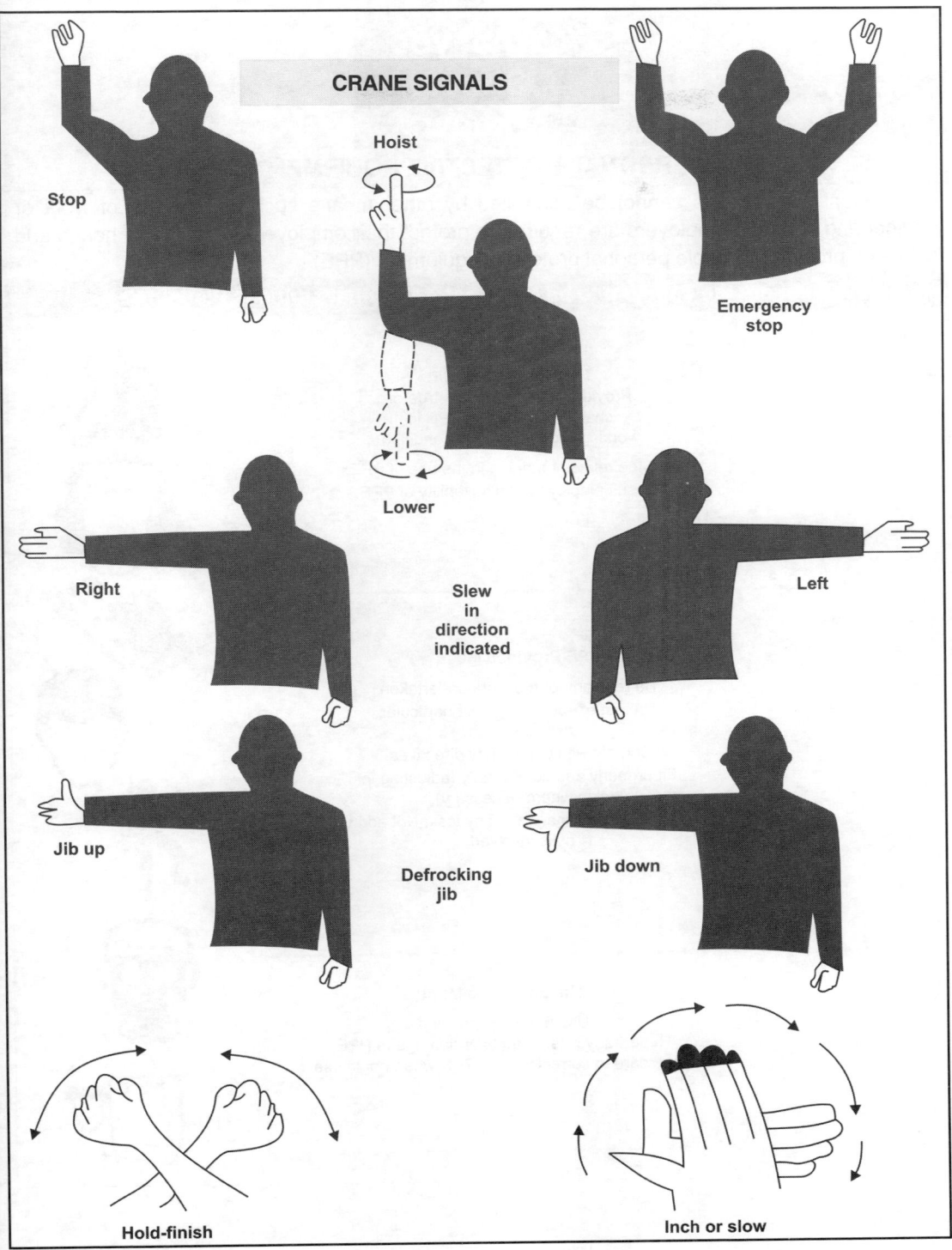

CRANE SIGNALS

What you Should know

PERSONAL PROTECTIVE EQUIPMENT

In situations where risks cannot be controlled by other means such as systems of work or engineering controls, employers are required to protect their employees from risk to health and safety by providing suitable personal protective equipment (PPE)

What are the Thinking to be Made before Manual Lifting is Planning?

The Employer Must:
Provide suitable PPE for free of charge maintaining PPE in working order and good condition

Provide relevant training in the use of PPE.
Consult employees on suitability of PPE.

PPE Provided Must:
Be relevant for the work undertaken
Protect effectively against particular risks involved.
Comply with community directives.
Fit properly and comfortably (adjusting in size where necessary).
Not hinder the performance of any task. Not add to the risks involved.

The Employee Must:
Use the PPE provided.
Report any loss, defects or damage to PPE.
Take care to correctly store PPE when not in use

What you Should know

BURNS AND SCALDS

1. Place the burnt area under cold running water immediately for at least 10 minutes. If it is a serious burn, ensure an ambulance is called.

2. If possible remove any item that may prevent swelling to burnt area, i.e. belt, boots, watches, or rings.

3. Place a clean starlike dressing over the burnt area.

4. Check that if required an ambulance has been reported to the correct individuals.

1. Do not apply any lotion, ointments or creams.

2. Do not attempt to remove any item of clothing that may be sticking to that area.

3. Do not touch or place anything other than a sterile dressing on a burn.

4. Do not burst any blisters that may form on or around the wound.

Your Prompt Action can prevent serious injury or even death

**Make Sure you Learn the Labels!
They are for your Protection.**

Flammable

Poison

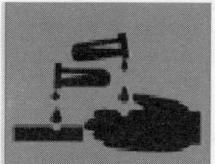

Corrosive

Explosive

Harmful

Organic peroxide

HOW TO AVOID SLIPS, TRIPS AND FALLS AT WORK

Watch out for...

- Hidden steps when stepping outside or turning a corner
- Smooth surfaces such as floors which have been waxed but not buffed
- Wet spots
- Oil and grease spots
- Carpets which are without rubber underlay or not tacked down
- Loose tiles or floorboards
- Electrical leads that are not secured to the floor or wall
- Open filling cabinet drawers
- Small movable objects on the floor such as pencils
- Furniture not in its proper place
- Loose or worn stair carpet
- Shoes with worn out soles or high heels
- Poor lighting conditions which may distort vision unsafe chairs
- Unsafe ladders and steps.

ONE MAN'S UNTIDINESS   IS ANOTHER MAN'S ACCIDENT!

Safety Committee Meeting

7

Safety Committee Meeting Minutes are mentioned

a. Safety Committee Meeting Minutes.
b. Meeting Attendance Register.

Markings

Safety Committee meeting should be conducted weekly by the assigned committee members as per OSHA standard for discussing, finding corrective action and setting target date for correcting procedures that may rise from audits and inspections reports or NNCR raised. It is also defined as a conference, to check the performance of last week activities and findings safe working methods for next week activities of the company.

What is the Infrastructure of Safety Committee Meeting Program?

It is a scenario where, the assign member who was selected by the top management to discuss, make decisions and to implement the decided decision about the companies work activities including its safety precaution aim the division of responsibilities, around a table on weekly or monthly basis is safety committee meeting. The minutes of meeting copies to be documented and officially circulated to all committee members.

The main contents of meeting minutes document are attendance register, points to discuss in meeting, action by whom, date when the actions to be completed and who's responsibility to complete that action.

The points of the safety committee meeting minutes mainly consider the matters of opening talk about the aim of company/company policy discussion and planning including its legal requirements and objectives/implementation and operation/checking and corrective actions including the deviation points discussion of audits and inspections/past and next week or month activity discussion and division of responsibilities/operational control documentation discussion/ and the next week or month safety committee meeting date announcement.

What is the Importance of Safety Committee Meeting Program?

It is the conference of assign and selected internal site members of a company to discuss and clear past defects happen in their work premises, and to discuss, plan and make decision for the future site work, safety and welfare activities. The decision made should be comply with company policy, HSE plan, safe work procedure and legal requirements.

The safety committee members should instruct and discuss in their sections, the new work, safety and welfare procedures that decided in the committee meetings to the line and low management.

So the importance of the safety committee meeting is that, every individual employee of the company site work premises will get a clear picture of what work procedure he need to follow, what safety precaution he needs to take and what is his responsibility.

How is Permit to Work Systems Defined in SDTP Document?

See the attached documents as an example of Safety Committee meeting points and attendance register, recorded as a document:

SDTP	7.1 Safety Committee Meeting Minutes	Document References

Document References | SDTP. SCM. 001

Rev. No	Doc. Date
0	1/1/2009

Pages
1 of 3

Site:				Meeting No:	
Date:		Time:		Venue:	

Present	Designation	Absent	Designation
See Attached Attendance Register	Chairperson		
	Safety Officer(s)		
	Project Manager		
	Site Agents		
	Engineers		
	Jr. Engineers		
	Secretary		
	Supervisor	Apologies	Designation
	General Foreman		
	Member(s)		
	Contractor(s)		
	Consultants Representative(s)		
	Subcontractor(s)		
	Hire representative(s)		

1.	CONFIRMATION—PREVIOUS MINUTES		
Meeting No.	Dated		Approved
1.1	The meeting started at............hrs.		
1.2	The chairman explained that the purpose of the Safety Committee meeting is to discuss company Management systems. The aim must be to comply with legal and company requirements and to follow OSHAS18001 standards. System failures need to be identified, dis.		
1.3	The following items will remain on the minutes as topics for discussion. Committee members are expected to report system failures and corrective measures (compliance) back to the meeting.		

2	Topic		Remarks	Action	Due by	Responsible
2.1	OHS Policy	Available	OHS Policy document should be prominently displayed and must be available to all interested parties.	Copies to be distributed and pasted on notice board. Training sessions to be held.	Date of policy made	General Manager and Safety Department

		Communi-cated	All employees should have a basic knowledge of what the purpose of the policy is.			
2.2	Planning	Site Safety Plan	Employes must be work accordance with the site safety plan. It is evident that unwanted incidents and unsafe condition are being condoned due to various reasons.	Signed copies to be issued to all committee members and members needs to communicate to below.	Date of Plan made	Project Manager
		Legal Require-ment	Comply with 1948 factories Act, ISO Standards, OSHA18001 Specifications etc.	Ensure compliance	Standing instruc-tion	All
		Objectives	• Comply with the Health and Safety Policy. • Everyone is responsible for organizing accident prevention at his or her own level on site. • Realizing the importance of Safety training. • Accident Prevention.	Ensure compli-ance through Zero Tolerance approach.	Standing instruc-tion	All
2.3	Implementation and Operation	Structure and Responsi-bility	All employees needs to Issued with a docu-ment defining their roles and responsi-bilities (Assignment of duties).	Project Manager	Date Issued	Safety Manager
		Training awareness and Competent	The following items need to be checked and compliance en-sured: * Induction training;	Safety officers to submit cur-rent programs to Project lead-ers for approval	Date sent	Safety De-partment
		Consul-tation and	Client needs and con-cerns to be addressed more effectively by	Visible felt lead-ership program.	Standing instruc-tion	Project Manager

		Commu-nication	discussing and enforc-ing the agreed site safety plan.			
		Data and Document control	All original, signed and approved documents should be filled on appropriate files under document controller.	All	Standing instruc-tion	All
		Operatio-nal Control	Registers to be kept by designated persons under control of project leadership. These registers must be updated and defects marked to be corrected on a ongoing basis.	Distributed copies to project leaders	Standing instruc-tion	General Manager
		Emergency prepared-ness and Response	Ensure escape routes and assembly points are clearly marked and signs in place	Drills to be scheduled	Date of drill planned	Safety Department
3	**Topic**	**General Discussion/ New Matters**		**Action**	**Due by**	**Responsible**
3.1	Checking and corrective Action	Audits	All audits reports to be distributed to top and line management, NCR and deviations to be discussed and corrective action should be taken and completed within target date. 1. 2. 3.	Meeting item	Standing instruc-tion	Project Manager
		Accident/ Incident/ NCR/Cor-rective action	Project leaders to re-port system failures leading to incidents/ non-conformances and subsequent corrective measures of last week. 1. 2. 3.	Meeting item	Standing instruc-tion	Project Manager

		Perform-ance measuring	Project leaders to ensure subordinates are performing to required standard as per safety plan	Enforce compliance	Standing instruction	Project Manager
3.2	Next week activities discussion	Presentation	Activities which need to perform on next week and its safe work procedure, additional precaution needs to be taken, approval, assignment and responsibilities of duties, etc. 1. 2. 3. 4. 5.	Meeting item	Standing instruction	Project Manager
3.3	Responsibilities		SHE management is the responsibility of all supervisory staff. Safety officers are required to offer support to the production team, to advice and to inspect for compliance. This means that hazards and non conformances must be reported and actioned immediate	Information briefings	Standard	Project Manager

4	Registers	Remarks	Action	Due by	Responsible
1					
2					
3					
4					
5					
6					
7					
8					
9					
10					

5		Name	Designation	Received Signature
	Copies Issued to Attending Members:			
	Cc Copies to non-attending members:	Designation	Denoted attendance

NEXT MEETING SCHEDULED FOR..............................ON NEXT WEEK. ALL ATTENDING MEMBERS AS PER ITEM 6, ABOVE TO ATTEND THE MEETING IN THE SITE MEETING ROOM. NON-ATTENDANCE NOT ALLOWED WITHOUT PROPER AUTHORIZATION AND NOTICE.

SDTP	7.2 Meeting Attendance Register	Document References		
		SDTP. SCM. 002		
		Rev. No	Doc. Date	
		0	1/1/2009	
		Pages		
		1 of 3		

Attendance Register for meeting   **NO:**   **Date:**

NAME	DESIGNATION	SIGNATURE

Emergency/Evacuation Drill Program 8

One of emergency evacuation drill program is defined

a. Emergency drill document.

Markings

As a part of developing safety, security and rescue purpose as well as in industry or construction site, evacuation drills needs to be conducted as a practice nature for the acknowledgement of staff and employees prior happening any emergency situations as per OSHA standards. Whenever alarm is heard, the staff and employees needs to stop there work activities, switch-off all electrical equipment and machineries and to close all windows and doors, and to march through emergency exit routes to assembly point. It should be practiced once in a week on hazardous area and at least once in six months on low hazard area.

What is the Infrastructure of Emergency/Evacuation Drill Program?

As per OSHA standards every company on project-wise select and assign a competent and expert person to act as the leader, control all site activities, correspond with all legal departments including media in case of Emergency situation until the site returns to the original stage. He has the authority and responsibility to select and train a group of persons as a team to perform under him in case of emergency. As a practice or drill program matter also he will be the chief and command the whole site.

The emergency/evacuation drill program is a training to be conducted either by announcing the date and time or without information, but both cases the whole company employees should respond to the situation hearing the emergency alarm according to the emergency/evacuation procedure. In this drill and training program any deviations happen from the actual procedure should be corrected in the next drill program. Every points including deviations while conducting emergency/evacuation drill program to be documented after the observer's signature.

What is the Importance of Emergency/Evacuation Drill Program?

The main emergency situations that can face on a company premises are fire, explosion, collapse of structures and beams, chemical and gas toxic condense on atmosphere, major accidents and natural disasters.

The importance of emergency and evacuation drill program is, each and every employee of the company premises should have the knowledge and practice of the Emergency and Evacuation procedure to follow prior in case of emergency situation for their own survival. This can be achieved only by periodical drill and training programs for their remembrance as well as practice

How is Emergency/Evacuation Drill Program Described in SDTP Document?

See the attached documents:

SDTP	**8.1 Emergency Evacuation and Fire Prevention Drill (Proforma)**	Document References	
		SDTP. ED. 001	
		Rev. No	Doc. Date
		0	1/1/2009
		Pages	
		1 of 1	

Project :				Project No. :	
Location :				Date of Exercise :	
Type of Drill :	Fire		Emergency	Time of Exercise :	

Observer's Name	Designation	Signature
	Chief Officer	
	Asst Officer	
	General Manager	

Scenario (Assumed) :	

Details of Response

Type of alarm :		Report No :	
Alarm raised at :		No. of persons reported missing:	
Time Evacuation completed :		Assembly point 1 :	
Duration of Evacuation :		Assembly point 2 :	

Search Party Information

Name :	Mobilized at :	Report back at :	Duration of search :

Learning Points :

Corrective Action Suggested:

	Information and Study Page	
Project:	ABC Power project	
Location:	Ground Floor, 1st and 2nd floor	
Type of Drill:	Emergency Alarm drill	
Type of Alarm:	False Alarm	
Report No.:	ED 001	
Alarm raised time:	HAm	
Date of Drill:	7/2/2009	
Project No.:	FXE1002	
No. of persons missing at ASS. 1:	One worker	
No. of persons missing at ASS. 2:	Nil	
Project:	11.40 AM	
Duration of drill	40 minutes	

Learning Point:

What happened to missing man:	Cannot find. At that time the employee went to the camp without informing to his supervisor.

Corrective actions:

Written warning given to missing employee. Strict instruction given to employee that after getting the permission from supervisor only he can go any where from site.

Search Party Information

Name:	Mobilized at:	Report back at:	Duration of search:
Sp 1	11.17 AM	11.33 AM	16 minutes
Sp 2	11.20 AM	11.35 AM	15 minutes
Sp 3	11.25 AM	11.40 AM	15 minutes

See the attach document how the Emergency Evacuation and Fire Prevention Drill program filled in:

SDTP	8.1 Emergency Evacuation and Fire Prevention Drill	Document References		
		SDTP. ED. 001		
		Rev. No	Doc. Date	
		0	1/1/2009	
		Pages		
		1 of 1		

Project :	ABC Power project		Project No. :	FXE1002
Location :	Ground Floor, 1st and 2nd floor		Date of Exercise :	7/2/2009
Type of Drill :	Fire	Emergency	yes	Time of Exercise :	11:00 AM

Observer's Name	Designation	Signature
Name of chief officer	Chief Officer	Name of chief officer
Name of Asst officer	Asst Officer	Name of Ass. officer
Name of GM	General Manager	Name of GM

Scenario (Assumed) :	An Emergency false alarm raised on 7th of February 2009 as a part of Emergency/Evacuation drill program.

Details of Response

Type of alarm :	False Alarm		Report No :	ED 001
Alarm raised at :	11. AM		No. of persons reported missing:	
Time Evacuation completed :			Assembly point 1 :	One
Duration of Evacuation :			Assembly point 2 :	Nil

Search Party Information

Name :	Mobilized at :	Report back at :	Duration of search :
Sp 1	11.17 AM	11.33 AM	16 minutes
Sp 2	11.20 AM	11.35 AM	15 minutes
Sp 3	11.25 AM	11.40 AM	11 minutes

Learning Points

a. Every employees participate in the Emergency/Evacuation drill.

b. One person missing on assembly point 1. Search party went to search him but in vain.

c. At that time the employee went to the camp without informing to his supervisor.

Corrective Action Suggested

a. Written warning to be given to missing employee. Strict instruction to be given to employee that after getting the permission from supervisor only he can go any where from site.

Safety Officer Weekly and Monthly Inspection Reports

9

Some of Inspection and Reports Program are Defined

a. Weekly site inspection.
b. Weekly statistic report.
c. Monthly statistic report

Markings

It is the duty of Safety Officer to prepare inspection and monthly and weekly reports. Signed reports should be filed and update in every week and month for the auditing purpose.

What is the Infrastructure of Weekly Statistic Report?

It is the prime duty of the Safety Advisor/Engineer/officer to prepare weekly site statistics to calculate the frequency and severity rate of accident/incident occurring, the site/industry condition which was satisfying the safety policy/plan and safe work procedure and to hand over the document to top management for further clarification and approval. This approved piece of document tells the exact position of the industry/site, and the safety personnel has to send a copy of it to clients and legal authorities (if necessary). For the preparation of the document, Safety personnel needs to collect information from time keeping department including subcontractors (if applicable) about the total man hours worked by employees and staff that current week including overtime and also collect information from first-aid center about the first-aid injuries including medical injuries and lost time accidents happen in that current week, including subcontractors (if applicable).

What is the Infrastructure of Monthly Statistic Report?

Though the calculations and application are same as of the weekly site statistics, monthly statistic report is to be prepared pertaining to that current month. In addition on monthly statistic report, the monthly disabling frequency rate is also calculated by the help of a common formula (i.e. Number of loss time accidents plus fatal, accidents happens in that current month multiplied by 100000 divided by total man hours worked in that month).

Disabling frequency rate shows that the company activities is on control or need to be improved with the agreed disabling frequency rate as per approved HSE Plan.

What is the Infrastructure of Weekly Site Inspection?

It is a document which defines systematic way of conducting work activity inspections including plant, machinery, equipment and welfare activities daily by an expert safety personnel in his work premises. Weekly site inspection will add merit to the method of "visible felt leadership" on working environment in implementing safe work practice as per OSHA standard. This document tells about any deviation of activities from the plan and polilcy as approved HSE plan and agreed safe work procedure.

If any deviations pointed out by the safety personnel in weekly site inspections, from the agreed HSE plan and safe work procedure, the project manager needs to countersign the document by spot checking the allegation area for the clearance of the deviation pointed out.

How is Weekly Site Inspection, Weekly Statistics Report and Monthly Statistics Report defined in SDTP Document?

See the attached documents:

A construction site is engaged in lot of activities involving different plants, machineries and equipments on daily basis. Safety personnel are also a mankind and they cannot inspect and cross check every activity on the site at day time or in a shift because of the lack of time. So everyday they are supposed to check certain work process or activities on construction site and also need to inspect and check high hazard nature work activities in the site on daily basis.

For this concern in the "weekly site Inspection" checklist, the day mentioning columns are filled with different colours. The "Items to be inspected" columns are equally divided for seven days in a week for easy inspection and cross checking. Each colour mentioned on the particular day column is also pasted to the divided "Items to be inspected" columns. On that particular day safety personnel must inspect and cross-check those activities as mentioned on "Items to be Inspected" column as colour coded and deviations found to be mentioned on "deviation" columns with photographs. These deviations shall be noticed to the concern Project Manager and shall take appropriate remedy measures to rectify the issue. Rest of the activities shall be checked in the balance days of the same week according to the colour coding and if any deviations found to be rectified as same as mentioned above.

SDTP	9.1 Safety Officer Weekly Site Inspection (Proforma)	Document References

Document References
SDTP. WSI. 001

Rev. No	Doc. Date
0	1/1/2009

Pages
1 of 7

SITE: **Period:**

Sr. No.	Items to be Inspected	Mark with "Yes" or "No" and if "No" write details of deviation on page 263						
		Mon	Tue	Wed	Thu	Fri	Sat	Sun
	Date Inspected							
1	**Public Safety**							
1.1	Signage "No unauthorized entry, Visitors to report to site office, work in progress, etc."							
1.2	Nets, canopies, screens, etc. to protect walkways pas Sign public							
2	**Personal Protective Equipment**							
2.1	Signage placed depicting required PPE to be worn on site and specific areas							
2.2	Are all persons (Visitors/Subcontractors included) wearing PPE?							
2.3	Are all persons and operators involved in hazardous tasks wearing the required PPE?							
3	**Housekeeping**							
3.1	Is the site clean and tidy, any clue of fire hazards?							
3.2	Rubble, spillage and Scrap being controlled, moved to the waste collecting area frequently							

3.3	Storage facilities provided, sort-by-sort material stacking facilitated in identified areas?							
3.4	Different wastes are collected in different marked skips. Are they removed regularly from site?							
3.5	Are subcontractors compiling with house-keeping standards?							
4	**Scaffolding**							
4.1	Does scaffolding comply with minimum standards?							
4.2	Are toe boards, hook-on—boards, hand railings being provided on scaffold platforms?							
4.3	Signage "scaffolding safe for use, scaffolding unsafe for use" displayed by scaffold inspector							
4.4	Clear and sound access from the bottom to top of scaffolds being provided							
4.5	Is tag system applied on scaffolds. Does scaffolds are inspected and signed on daily basis?							
5	**Ladders**							
5.1	Are ladders being used on site the correct type? (no make shift ladders?)							

5.2	Ladders numbered, hung up in specific allocated area, registers updated								
5.3	Fixed ladders have cages								
5.4	Ladders extend above platform level (1 meter) and landing secured at the base								
5.5	No painted wooden ladders, general condition acceptable, rungs in good condition, etc.								
6	**Electricity**								
6.1	Extension leads, DB boards numbered, registers updated								
6.2	DB board warning signs posted, locked, circuits marked								
6.3	DB placement accessible, safe, circuit-breakers fitted, no wet conditions, cables protected								
6.4	Electric cables feeding from DB boards being protected from vehicle/site traffic damage								
6.5	Electric power tools working through earth leakage unit? DB fitted with earth leakage								
7	**Fire Protection**								
7.1	Sufficient fire equipment available, equipment serviced, registers updated								
7.2	Equipment strategically placed, accessible, signage depicting placement								

7.3	Hazardous areas protected with sufficient, correct fire equipment							
7.4	Hazardous chemicals, gas cylinders, petroleum products are stacked in different compartments and protected with correct fire devices							
8	**Excavations**							
8.1	Inspected, barricaded, warnings posted, shored, braced and access provided, registers updated							
8.2	Permit to work system followed. Lighted after dark to prevent persons falling in							
9	**Edge Protection Barricading**							
9.1	Are all edges to decks being prepared and protected with sound materials?							
9.2	Construct of sound material, high enough, can prevent person falling through							
9.3	Demarcated, clearly visible, gated after dark							
10	**Tools**							
10.1	Hand tools correct type, serviced, maintained (NO MAKE SHIFT TOOLS). Registers updated							
10.2	Five percent of tools inspected weekly. Found any hand tools with mushroom head?							

11	Lifting Equipment, Mobile Cranes, Tower Cranes								
11.1	Operated by trained, authorized operator. All lifting tackles are strong enough and certified								
11.2	MML displayed, warnings posted, equipment checked, registers available and updated								
11.3	Slinging done correctly, rigger trained, signaling and communication procedure in place								
12	**Material/Man Hoist**								
12.1	Warning signs displayed, operator trained and appointed								
12.2	Landings and access barricading in place, moving parts guarded, Weekly inspection								
13	**Transport and Materials Handling Equipment/Plant and Machinery**								
13.1	Daily checklists being completed, trained licensed operators available								
13.2	No passengers on plant, no over speeding, no unauthorized operators								
13.3	All moving parts, nip points guarded/ unauthorized access in place checked								
14	**Health and Hygiene**								
14.1	Sufficient toilets (1/30 persons). Toilet paper, changing facility, water and soap								
14.2	Showers, eating facilities, drinking water available. Are eating and rest rooms cleaning on daily basis?								

14.3	Clean, hygienic drainage available. Are all food wastes being removed from site regularly?							
14.4	First-aid facilities, appointed first aider, signage depicting location of FA Room and box?							
14.5	Hazardous chemicals identified, MSDS and procedures on file?							
15	**Hot Works**							
15.1	Permit to work systems followed. Only authorised person entering in hot work area. Gas cylinders kept in trolleys. Flash back arresters are working properly							
15.2	Barricaded, Signage and Fire extinguishers placed. Fire watchers appointed. Fire blankets used in height works							
16	**Height Work**							
16.1	Are all scaffolds comply with standards? Tag systems are in place, inspected regularly?							
16.2	Lifting man-cages and lifting plant comply with standard, certified and authorised operator, registers updated							
16.3	Are all wearing safety harness, have ISO markings, numbered, and registers updated?							
16.4	Have life lines using in hazardous areas. Edge protection and Toeboards are in place?							

17	**Stacking**									
17.1	All hazardous material and chemicals identified, stacked in separate compartments, MSDS regulation followed and suitable fire extinguishers placed Signage's									
17.2	Sort-by-sort materials stacking provided and rack stacking system followed									
17.3	All compressed gas cylinders are stacked in separate compartments, chained and empty cylinders stacked separately in compartments									
18	**Cutting and Grinding using Electric Tools**									
18.1	Permit to system followed, all persons using approved PPEs as per safe work procedure. Suitable fire extinguishers and signage's placed									
18.2	Proper ventilation and lighting provided. Is the area kept free from tripping and slipping hazard? Metal scraps collected in closed containers and removed daily									
	Date Inspected:									
	Initialed by the Inspector (sign):									
	Spot Check by Project Manager (sign):									
	Date Spot Checked:									

Deviations and Corrective Actions						
Sr. No.	Description of Deviation	Corrective Action	Action by whom	Target Date	Comp. Date	Remark

SDTP	9.2 Weekly Site Safety Summary and Statistical Report	Document References	
		SDTP. WS. 001	
		Rev. No	Doc. Date
		0	1/1/2009
		Pages	
		1 of 5	

PROJECT/CONTRACT NAME:			
PROJECT/GENERAL MANAGER:			
PROJECT MANAGER:			
SAFETY OFFICER:		Week Ending	

DATE	Main Contractor Employees Present	Hours worked (including Overtime)	Subcontractor Employees Present	Hours worked (including Overtime)	TOTAL HOURS
MONDAY					
TUESDAY					
WEDNESDAY					
THURSDAY					
FRIDAY					
SATURDAY					
SUNDAY					
Total:					

1. Injury Statics

DESCRIPTION	Main contractor	Period	Subcontractor	Period	Comments
FATALITIES:					
LOST TIME ACCIDENTS:					
MEDICAL ONLY:					
First Aid:					
NEAR MISS:					
Remarks:					

2. Details of incident

Date	Name of Company	Severity	Detail of Incident

3. Property Damaged

Property Type	Cause of Damage	Department Responsible	Total Cost

4. Work Areas Stopped by Safety Officer

Work Areas	Section	Reason

5. Scheduled Work Area Inspections

Area Inspected	Hazardous Items Identified	Remedial Action Taken

6. Follow-up Actions

Sections	Percentage of previous identified hazards or potential hazards not corrected.	Action taken by safety officer to correct outstanding items.

7. Breaches of Legal Requirements

Sections	Regulations breached as identified by safety officer	Action required by the Line-Management

8. Determine Skills and Knowledge (Training)

a. Number of Person checked / Observed:		
b. Number of Persons showed lack of knowledge in there work knowledge:		

Areas identified as lacking by some employees and deviation founded:

Sr. No.	Date	Findings	Action by Whom	Target Date	Comple-tion Date

9. Training Conducted by Safety Department

Title of Training	Training given by	Number of Attendees
Induction Training for newcomers:		
Induction Training for Visitors:		
Toolbox Talk:		
Other Trainings		
a. Special Training:		
b. Third party Training:		
c. Emergency Training:		
d.		
e.		

10. General Comments

Sr. No.	Comments

This report is a true reflection of the Safety Condition and Actions on site:

NAME	POSITION	SIGNATURE	DATE
	SAFETY MANAGER/OFFICER		
	PROJECT MANAGER		
	GENERAL MANAGER		

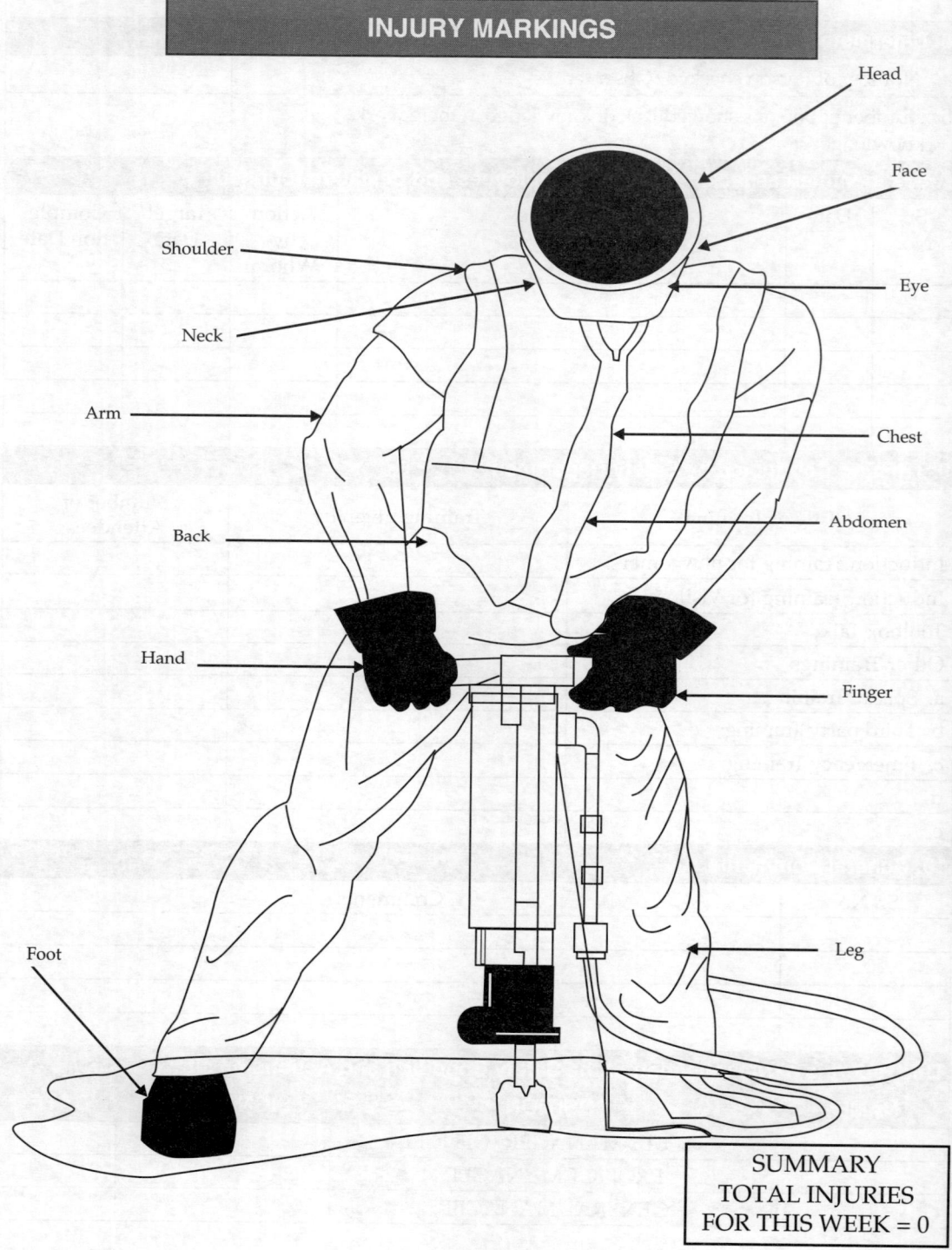

INJURY MARKINGS

Head

Face

Shoulder

Eye

Neck

Arm

Chest

Back

Abdomen

Hand

Finger

Foot

Leg

SUMMARY
TOTAL INJURIES
FOR THIS WEEK = 0

SDTP	9.3 Monthly Site Statistical Report	Document References	
		SDTP. MR. 001	
		Rev. No	Doc. Date
		0	1/1/2009
		Pages	
		1 of 3	

MONTHLY REPORT			
Project Name:		Month	
Project Number:			
Manpower Returns		Parent Company	Subcon-tractor
Total Number of Employees			
Number of Man-Hours Worked			
Total for the Month			
Injury Reporting			
Number of First Aid case Reported			
Number of Medical cases Reported (Non Lost Time)			
Number of Lost Time Incident cases			
Number of Fatalities			
Near-Miss Reporting			
Number of Near-Miss Reported			
Occupational Health			
Number of Occupational diseases Reported			
Environmental Incidents			
Number of Environmental diseases Reported			
Site Inspections/Audits			
Internal Audits			
Client Audits			
Standard Organization Audit/Inspection			
Department of Labor/Municipality Inspections (Health)			
Disabling Injury Frequency Rate		**Total for Month**	
(Number of Lost Time Incidents + Fatal) 100000/Total Man Hours worked			
MONTHLY DISEASE INJURY ANALYSES			

Parts of Body Effected				Type of Injury				Disease (Work Related)	
Title	First Aid	Medical	LTI	Title	First Aid	Medical	LTI	Title	Nos.
Head				Sprain/Strain				Skin	
Face				Contusion				Respiratory	
Neck				Cut				Muscular	
Eye				Laceration				Skeletal	
Back				Abrasion				Endemic	
Abdomen				Fracture				Hearing	
Arm				Burn				Neurological	
Hand				Amputation				Psychological	
Finger				Electric Shock				Other	
Leg				Asphyxia				Total	
Foot				Unconscious/ Fainting					
Toe				Poisoning					
Chest				Foreign Body					
Shoulder				Multiple Injury					
Internal				Total					
Multiple									
Total									

Date	Public Liability Incident Details	Potential Causes

NAME	POSITION	SIGNATURE	DATE
	SAFETY MANAGER/OFFICER		
	GENERAL MANAGER		

MONTHLY INJURY MARKINGS

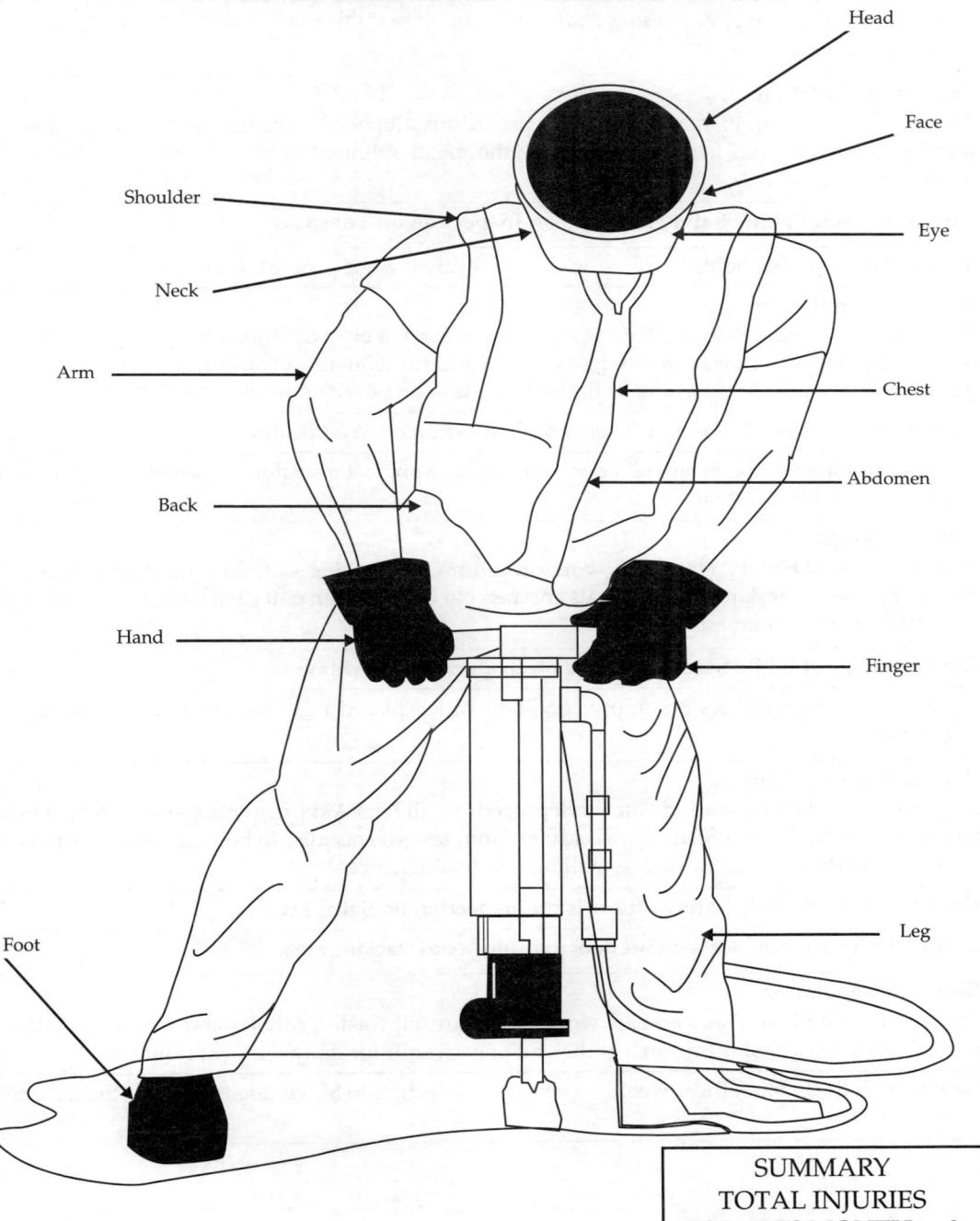

Head

Face

Shoulder

Eye

Neck

Arm

Chest

Back

Abdomen

Hand

Finger

Foot

Leg

SUMMARY
TOTAL INJURIES
FOR THIS MONTH = 0

Information Page
Deviation found while Safety officer is on Inspection on Monday:
a. Subcontractors doing pipe cutting works by means of portable electric cutters, not wearing mask and face shield.
Rule and Regulation: As per approved HSE Plan and Safe work procedure, all peoples engaged in pipe cutting work should wear mask and face shield because the metal splinters or dust can damage eyes and respiratory system.
Deviation found while Safety officer is on Inspection on Tuesday:
a. Found that flashing lights are not placed on excavation while people working on night.
Rule and Regulation: As per approved HSE Plan and Safe work procedure, all open excavations needs to provide solid barricading one meter away from edges of excavation, adequate and appropriate signage's to be placed and orange color warning or flashing lights to be provided for the visibility on night.
Deviation found while Safety officer is on Inspection on Wednesday:
a. Found painting workers on 2nd floor, working on a six meter scaffold which is red tagged and return incomplete scaffold.
Rule and Regulation: As per approved HSE Plan and Safe work procedure, all working scaffold should be constructed as per approved standard and materials and need to have a green entry tag which is signed daily by scaffold inspector after checking.
Deviation found while Safety officer is on Inspection on Friday:
a. Found fire extinguishers on empty condition which placed near 2nd floor chemical stacking compartment.
Rule and Regulation: As per approved HSE Plan and Safe work procedure, all First Aid fire fighting equipment should be checked daily for its efficiency, weekly random tested, checklist to be updated monthly and every six month needs compulsory refilling and maintenance.
Deviation found while Safety officer is on Inspection on Saturday:
a. Found a heap of rubbles near 3rd floor gypsum board stacking area.
Rule and Regulation: As per approved HSE Plan and Safe work procedure, all wastes, rubbles and scrapes should be moved to the waste collecting area at the end of each shift for daily removal from site.
Needs to fill the Safety officer weekly inspection according to SDTP document: See the attached Document sheet:

SDTP	9.1 Safety Officer Weekly Site Inspection		Document References SDTP. WSI. 001	
			Rev. No	Doc. Date
			0	1/1/2009
			Pages 1 of 7	

SITE: **Period:**

Sr. No	Items to be Inspected	Mark with "Yes" or "No" and if "No" Write details of deviation						
		Mon	**Tue**	**Wed**	**Thu**	**Fri**	**Sat**	**Sun**
	Date Inspected	1-Jan-09	2-Jan-09	3-Jan-09	4-Jan-09	5-Jan-09	6-Jan-09	7-Jan-09
1	**Public Safety**							
1.1	Signage "No un-authorized entry, Visitors to report to site office, work in progress, etc."	Yes						
1.2	Nets, canopies, screens, etc. to protect walkways pas Sign public	Yes						
2	**Personal Protective Equipment**							
2.1	Signage placed depicting required PPE to be worn on site and specific areas	Yes						
2.2	Are all persons (Visitors/Subcontractors included) wearing PPE?	No						
2.3	Are all persons and operators involved in hazardous tasks wearing the required PPE?	Yes						
3	**Housekeeping**							
3.1	Is the site clean and tidy, any clue of fire hazards?						Yes	

For rows 3.2-3.5, the Yes/No is near the right side (second to last column). For rows 4.1-4.4, the Yes/No is more toward the middle.

3.2	Rubble, spillage and Scrap being controlled, moved to the waste collecting area frequently						No	
3.3	Storage facilities provided, sort-by-sort material stacking facilitated in identified areas						Yes	
3.4	Different wastes are collected in different marked skips. Are they removed regularly from site						Yes	
3.5	Are subcontractors compiling with housekeeping standards?						Yes	
4	**Scaffolding**							
4.1	Does scaffolding comply with minimum standards?			Yes				
4.2	Are toe boards, hook-on-boards, hand railings being provided on scaffold platforms?			Yes				
4.3	Signage "scaffolding safe for use, scaffolding unsafe for use" displayed by scaffold inspector			No				
4.4	Clear and sound access from the bottom to top of scaffolds being provided			Yes				

4.5	Is tag system applied on scaffolds. Does scaffolds are inspected and signed on daily basis?			Yes				
5	**Ladders**							
5.1	Are ladders being used on site the correct type? (No make shift ladders?)			Yes				
5.2	Ladders numbered, hung up in specific allocated area, registers updated			Yes				
5.3	Fixed ladders have cages			Yes				
5.4	Ladders extend above platform level (1 meter) and landing secured at the base			Yes				
5.5	No painted wooden ladders, general condition acceptable, rungs in good condition, etc.			Yes				
6	**Electricity**							
6.1	Extension leads, DB boards numbered, registers updated				Yes			
6.2	DB board warning signs posted, locked, circuits marked				Yes			
6.3	DB placement accessible, safe, circuit-breakers fitted, no wet conditions, cables protected				Yes			

6.4	Electric cables feeding from DB boards being protected from vehicle/site traffic damage				Yes				
6.5	Electric power tools working through earth leakage unit? DB fitted with earth leakage				Yes				
7	**Fire Protection**								
7.1	Sufficient fire equipment available, Equipment serviced, registers updated					Yes			
7.2	Equipment strategically placed, accessible, signage depicting placement					Yes			
7.3	Hazardous areas protected with sufficient, correct fire equipment					Yes			
7.4	Hazardous chemicals, gas cylinders Petroleum products are stacked in different compartments and protected with correct Fire devises					No			
8	**Excavations**								
8.1	Inspected, barricaded, warnings posted, shored, braced and access provided, registers updated		Yes						

8.2	Permit-to-work System followed. Lighted after dark to prevent persons falling in		No						
9	**Edge Protection Barricading**								
9.1	Are all edges to decks being prepared and protected with sound materials?						Yes		
9.2	Construct of sound material, high enough, can prevent person falling through						Yes		
9.3	Demarcated, clearly visible, gated after dark						Yes		
10	**Tools**								
10.1	Hand tools correct type, serviced, maintained (NO MAKE SHIFT TOOLS). Registers updated								Yes
10.2	Five percent of tools inspected weekly. Found any hand tools with mushroom head								Yes
11	**Lifting Equipment, Mobile Cranes, Tower Cranes**								
11.1	Operated by trained, authorised operator. All lifting tackles are strong enough and Certified				Yes				
11.2	MML displayed, warnings posted, equipment checked, registers available and updated				Yes				

11.3	Slinging done correctly, rigger trained, signaling and communication procedure in place				Yes				
12	**Material/Man Hoist**								
12.1	Warning signs displayed, operator trained and appointed								Yes
12.2	Landings and access barricading in place, moving parts guarded, Weekly inspection								Yes
13	**Transport and Materials Handling Equipment/Plant and Machinery**								
13.1	Daily checklists being completed, trained licensed operators available					Yes			
13.2	No passengers on plant, no over speeding, no unauthorized operators					Yes			
13.3	All moving parts, nip points guarded/ unauthorized access in place checked					Yes			
14	**Health and Hygiene**								
14.1	Sufficient toilets (1/30 persons). Toilet paper, changing facility, water and soap							Yes	
14.2	Showers, eating facilities, drinking water available. Are eating and rest rooms cleaning on daily basis							Yes	

14.3	Clean, hygienic drainage available. Are all food wastes being removed from site regularly?						Yes	
14.4	First-aid facilities, appointed first aides, signage depicting location of FA Room and box						Yes	
14.5	Hazardous chemicals identified, MSDS and procedures on file						Yes	
15	**Hot Works**							
15.1	Permit-to-work systems followed. Only authorized person entering in hot work area. Gas cylinders kept in trolleys. Flash back arresters are working properly		Yes					
15.2	Barricaded, signage and Fire extinguishers placed. Fire watchers appointed. Fire blankets used in height works		Yes					
16	**Height Work**							
16.1	Are all scaffolds comply with standards. Tag systems are in place, inspected regularly?				Yes			
16.2	Lifting man-cages and lifting plant comply with standard, certified and authorised operator, registers updated				Yes			

					Yes			
16.3	Are all wearing safety harness, have ISO markings, numbered, and registers updated?				Yes			
16.4	Have life lines using in hazardous areas. Edge protection and Toeboards are in place?				Yes			
17	**Stacking**							
17.1	All Hazardous material and chemicals identified, stacked in separate compartments, MSDS regulation followed and suitable fire extinguishers placed signage's							Yes
17.2	Sort-by-sort materials stacking provided and rack stacking system followed							Yes
17.3	All compressed gas cylinders are stacked in separate compartments, chained and empty cylinders stacked separately in compartments							Yes
18	**Cutting and Grinding using Electric Tools**							
18.1	Permit to system followed, all persons using approved PPE's as per safe work procedure. Suitable fire extinguishers and signage's placed	Yes						
18.2	Proper ventilation and lighting provided. Is the area kept free	Yes						

from tripping and slipping hazard. Metal scraps collected in closed containers and removed daily							
Date Inspected:	1/1/ 2999	2/1/ 2009	3/1/ 2009	4/1/ 2009	5/1/ 2009	6/1/ 2009	7/1/ 2009
Initialed by the Inspector (sign):	Sign of safety Officer	Sign of safety Officer	Sign of safety Officer	Sign of safety Officer	Sign of safety Officer	Sign of safety Officer	Sign of safety Officer
Spot Check by Project Manager (sign):	Sign of Project Manger	Sign of Project Manger	Sign of Project Manger	Sign of Project Manger	Sign of Project Manger	Sign of Project Manger	Sign of Project Manger
Date Spot Checked:	1/1/ 2999	2/1/ 2009	3/1/ 2009	4/1/ 2009	5/1/ 2009	6/1/ 2009	7/1/ 2009

Deviations and Corrective Actions						
Sr. No.	Description of Deviation	Corrective Action	Action by whom	Target Date	Comp. Date	Remark
2.2	Subcontractors not wearing mask and face while doing pipe cutting works	Discipline and return warning to be given to subcontractor concern supervisor. Strict instruction to be given to wear all PPE's while working by Sr. Site Engineer.	Sr. Site Engineer	2/1/2009	2/1/2009	Done
3.2	Found a heap of rubbles near 3rd floor gypsum board stacking area.	Rubbles needs to be shifted to waste collecting area daily. Needs strict warning to be given to the 3rd floor area supervisor.	Sr. Site Engineer	8/1/2009		Completed on next week document
4.3	Found workers on 2nd floor, working on a six meter scaffold which is red tagged and return incomplete scaffold	Never allow to work on incomplete and red tagged scaffold. Needs return warning to be given to the painting supervisor.	Sr. Site Engineer	3/V2009	3/1/2009	Done
7.4	Found file extinguishers on empty condition which placed near 2nd floor chemical stacking compartment	All First Aid fire fighting equipment should be checked daily by a competent supervisors in there allotted areas. Empty extinguishers should send for re-filling.	Sr. Site Engineer	6/1/2009	6/1/2009	Done

8.2	Found that flashing lights are not placed on excavation while people working on night	Needs to erect flashing lights on night. Needs to given written warning to excavation supervisor.	Excavation Engineer	3/1/2009		Not Done/ Pending

Information and Study Page

Information of manhours worked during the current week got from Time keeper including staff and employees:

Company Staff and Employees

Day	Staff	Staff hours inc. over time	Employees	Employees hours including over time
Mon	36	390	1364	14844.50
Tue	34	420	1352	14534.50
Wed	31	360	1375	15263.00
Thu	32	390	1372	15146.50
Fri	34	400	1328	13965.00
Sat	28	320	1264	13262.50
Sun	10	190	382	3820.00

Subcontractor Staff and Employees

Day	Staff	Staff hours inc. over time	Employees	Employees hours including over time
Mon	8	88	75	820.50
Tue	10	106	79	854.00
Wed	12	134	84	912.50
Thu	9	103	84	917.25
Fri	11	122	84	914.25
Sat	12	132	82	908.00
Sun	3	36	32	368.50

Information of Injury Details of the Current Week of from the First-aid Center

Description	Company Employees	Period	Subcontractor Emp	Period
Fatality	Nil		Nil	
Lost Time Accident	Nil		One	5/1/2009
medical Accidents	One	2/1/2009	Nil	
First-aid cases	One	4/1/2009	Nil	
Near Miss	Nil		One	3/1/2009

Detail of Injury happenings

Date	Company Detail	Severity	Details of Incident
2/1/2009	Parent Company	Head Cut, three stitches on head	While lifting steel clamps by means of hand in a scaffold, a steel clamp slips down and struck on helmet. Then helmet slip down and the edges of the clamp touches on the head and cut him.
3/1/2009	Subcontractor	Near miss	While conducting pressure text on fire line in 4th floor, a 3 meter fire pipe line breaks and fall down where four person standing with out touching them.
4/1/2009	Parent Company	First Aid, hand cut	While cutting a steel pipe by means of a portable electric cutter, he slips and the edge of the cutting blade touched on his hand and cut him.
5/1/2009	Subcontractor	Lost time Accident, Fracture on lower leg	While lifting manually a box of steel couplers, he slips and box fall to his leg and fractured him.

Work Area Stopped by Safety Personnel (See safety officer weekly inspection)

1. Stopped the pipe cutting work of Subcontractor where workers not wearing mask and face shield. (Date 1/12009)

2. Stopped the painting work on 2nd floor where workers standing on a red tagged scaffold. (Date 2/1/2009

Scheduled Work Area Inspection by Safety Personnel (*see* safety officer weekly inspection)

Area Inspected	Hazardous items identified	Remedial action Taken
Ground Area	Found subcontracting workers engaged in pipe cutting works by means of electric cutter with out wearing FFE.	Stop the work, Informed Project Manager
Ground Area	Found an excavation with out erecting flashing lights while on night.	Stop the work, Informed Project Manager
2nd Floor	Found painting workers on six meter scaffold which is red tagged.	Stop the work, Informed Project Manager
2nd Floor	Found fire extinguishers on empty condition near chemical stacking compartment.	Informed Project Manager
3rd Floor	Found a heap of rubbles on gypsum board stacking area	Informed Project Manager

Follow-up Actions (*see* safety officer weekly inspection)

Section	Percentage of previous identified hazards or potential hazards not corrected.	Action taken by safety officer to correct outstanding items.
Excavation	Flashing Lights till not erected.	Informed General Manager

Determine Skills and Knowledge (*see* safety officer weekly inspection)	
a. Number of persons checked/observed:	340
b. Number of persons showed lack of knowledge in there work knowledge:	2

Areas Identified as Lacking by some Employees and Deviation Founded: (See weekly inspection)

Sr. No.	Date	Findings	Action by Whom	Target Date	Completion Date
1	1/12009	Subcontractor workers engaged in pipe cutting work not wearing mask and face shield.	Sr. Site Engineer	2/1/2009	2/1/2009
2	2/1/2009	Painting workers standing on a red tagged scaffold while working.	Sr. Site Engineer	3/1/2009	3/1/2009
3	2/1/2009	Found an excavation with out erecting flashing lights while on night.	Excavation Engineer	3/1/2009	Not yet Finished
4	5/1/2009	Found fire extinguishers on empty condition near chemical stacking compartment.	Sr. Site Engineer	6/1/2009	fyl/ 2009
5	7/1/2009	Found a heap of rubbles on gypsum board stacking area.	Sr. Site Engineer	8/1/2009	Continue on next

Training Conducted by Safety Department

Title of Training	Training given by	Number of Attendees
Induction Training for newcomers:	Safety Personnel	28
Induction Training for Visitors:	Safety Personnel	4
Toolbox Talk:	Site Supervisors	91267

Needs to fill the Weekly Site Statistics Report according to SDTP document: *see* the attached Document sheet:

SDTP	9.2 Weekly Site Safety Summary and Statistical Report	Document References	
		SDTP. WS. 001	
		Rev. No	Doc. Date
		0	1/1/2009
		Pages 1 of 5	

PROJECT/CONTRACT NAME:	FXE 1002	
PROJECT/GENERAL MANAGER:	Name of GM	
PROJECT MANAGER:	Name of PM	
SAFETY OFFICER:	Name of Safety Officer	Week Ending

DATE	Main Contractor Employees Present	Hours worked (including Overtime)	Subcontractor Employees Present	Hours worked (including Overtime)	TOTAL HOURS
MONDAY	36 + 1364	390 + 14844.50	8 + 75	88 + 820.50	88 + 820.50
TUESDAY	34 + 1352	420 + 14534.50	10 + 79	106 + 854.00	106 + 854.00
WEDNESDAY	31 +1375	360 + 15263.00	12 + 84	134 + 912.50	134 + 912.50
THURSDAY	32 + 1372	390 + 15146.50	9 + 84	103 + 917.25	103 + 917.25
FRIDAY	34 + 1328	400 + 13965.00	11 +84	122 + 914.25	122 + 914.25
SATURDAY	28 + 1264	320 + 13262.5	12 + 82	132 + 908.00	132 + 908.00
SUNDAY	10 + 382	190 + 3820.00	3 + 32	36 + 368.50	36 + 368.50
Total:	36 + 1364	2470 + 90836.00	12 + 84	721 + 5695.00	721 + 5695.00

1. Injury Statics

DESCRIPTION	Main contractor	Period	Subcontractor	Period	Comments
FATALITIES:	Nil		Nil		
LOST TIME ACCIDENTS:	Nil		One	5/1/2009	
MEDICAL ONLY:	One	2/1/2009	Nil		

First Aid:	One	4/1/2009	Nil		
NEAR MISS:	Nil		One	3/1/2009	

Remarks:

2. Details of Incident

Date	Name of Company	Severity	Detail of Incident
2/1/2009	Parent Company	Head cut, three stitches on head	While lifting steel clamps by means of hand in a scaffold, a steel clamp slips down and struck on helmet Then helmet slip down and the edges of the clamp touches on the head and cut him.
3/1/2009	Sub-contractor	Near miss	While conducting pressure text on fire line in 4th floor, a 3 meter fire pipe line breaks and fall down where four person standing without touching them.
4/1/2009	Parent Company	First Aid, hand cut	While cutting a steel pipe by means of a portable electric cutter, he slips and the edge of the cutting blade touched on his hand and cut him.
5/1/2009	Sub-contractor	Lost time accident, fracture on lower leg	While lifting manually a box of steel couplers, he slips and box fall to his leg and fractured him.

3. Property Damaged

Property Type	Cause of Damage	Department Responsible	Total Cost
Nil			

4. Work Areas Stopped By Safety Officer

Work Areas	Section	Reason
Ground Area	Hot work, Pipe cutting work	Stopped the pipe cutting work of Subcontractor where workers not wearing mask and face shield.
2nd Floor	Painting Work	Stopped the painting work on 2nd floor where workers standing on a red tagged scaffold.

5. SCHEDULED WORK AREA INSPECTIONS

Area Inspected	Hazardous Items Identified	Remedial Action Taken
Ground Area	Found subcontracting workers engaged in pipe cutting works by means of electric cutter with out wearing PPE.	Stop the work, informed Project Manager
Ground Area	Found an excavation with out erecting flashing lights while on night.	Informed Project Manager
2nd Floor	Found painting workers on six meter scaffold which is red tagged.	Stop the work, informed Project Manager
2nd Floor	Found fire extinguishers on empty condition near chemical stacking compartment	Informed Project Manager
3rd Floor	Found a heap of rubbles on gypsum board stacking area	Informed Project Manager

6. Follow-up Actions

Sections	Percentage of previous identified hazards or potential hazards not corrected.	Action taken by safety officer to correct outstanding items.
Excavation	Flashing Lights till not erected.	Remembered Project Manager and informed General Manager

7. Breaches of Legal Requirements

Sections	Regulations breached as identified by safety officer	Action required by the Line-Management
Nil		

8. Determine Skills and Knowledge (Training)

a. Number of Person checked/Observed:	340
b. Number of Persons showed lack of knowledge in there work knowledge:	2

Areas identified as lacking by some employees and deviation founded:

Sr. No.	Date	Findings	Action by Whom	Target Date	Completion Date
1	1/1/2009	Subcontractor workers engaged in pipe cutting work not wearing mask and face shield.	Sr. Site Engineer	2/1/2009	2/1/2009

2	2/1/1009	Painting workers standing on a red tagged scaffold while working	Sr. Site Engineer	3/1/2009	3/1/2009
3	2/1/2009	Pound an excavation with out erecting flashing lights while on night	Excavation Engineer	3/1/2009	Not yet Finished
4	5/1/2009	Found fire extinguishers on empty condition near chemical stacking Compartment.	Sr. Site Engineer	6/1/2009	6/1/2009
5	7/1/2009	Found a heap of rubbles on gypsum board stacking area.	Sr. Site Engineer	8/1A009	Continue on next

9. Training Conducted by Safety Department

Title of Training	Training given by	Number of Attendees
Induction Training for newcomers:	Safety personnel	28
Induction Training for Visitors:	Safety personnel	4
Toolbox Talk:	Site Supervisor	91267
Other Trainings		
(a) Special Training:	Nil	
(b) Third party Training:	Nil	
(c) Emergency Training:	Nil	
(d)		
(e)		

10. General Comments

Sr. No.	Comments
	Nil

This report is a true reflection of the Safety Condition and Actions on site:

NAME	POSITION	SIGNATURE	DATE
Name of safety Manager/ Officer	SAFETY MANAGER/OFFICER	Sign of Safety Manager/Officer	Date Sign
Name of PM	PROJECT MANAGER	Sign of PM	Date Sign
Name of GM	GENERAL MANAGER	Sign of GM	Date Sign

INJURY MARKINGS

Head

Face

Eye

Shoulder

Neck

Arm

Chest

Abdomen

Back

Hand

Finger

Foot

Leg

SUMMARY
TOTAL INJURIES
FOR THIS WEEK = 3

Information and Study Page							
Project Name:		ABC Company		Project No.:		FXE1002	
Information from Time Keeper :							
Total No. of Employees		1425 (Parent Company)			410 (Subcontractors)		
Total Man hours worked		369721 hours (Parent Company)			24112 hours (Subcontractors)		
Information from First Aid center:							
No. of First Aid cases			No. of Medical cases		No. of Lost Time		No. of Near miss
Head (cut)	1 (sc), 2 (pc)	3	Head (cut)	1 (Pc)	1	Leg (Fracture)	1 (sc)	1	SC	1
Hand (cut)	2 (sc), 2 (pc)	4	Hand (cut)	1 (Sc)	1				
Leg (strain)	1 (sc), 1 (pc)	2							
Information from Safety Department:							
No. of Internal Audits:		4 (Pc)	4 (Sc)				
No. of Client Audits:		2 (Pc)	2 (Sc)				
No. of Municipality Inspections:		1 (Pc)	1 (Sc)				
Calculation of Disabling Injuries:							
Formula:		(Number of Lost Time Incidents + Fatal) 100000/Total Man Hours worked					
No. of Lost Time Accidents:		One					
No. of Fatalities:		Nil					
Total Man Hours worked:		369721 + 24112 = 393833					
Calculation:		1 100000/393833 = 0.25					
Needs to fill the Weekly Site Statistics Report according to SDTP document: *See* the attached Document sheet:							

SDTP	9.3 Monthly Site Statistical Report	Document References	
		SDTP. MR. 001	
		Rev. No	Doc. Date
		0	1/1/2009
		Pages	
		1 of 3	

MONTHLY REPORT

Project Name:	ABC Company	Month	Jan-09
Project Number:	FXE 1002		

Manpower Returns	Parent Company	Subcon-tractor
Total Number of Employees	1425	410

Number of Man-Hours Worked		
Total for the Month	369721	24112

Injury Reporting		
Number of First Aid case Reported	5	4
Number of Medical cases Reported (Non Lost Time)	1	1
Number of Lost Time Incident cases	0	1
Number of Fatalities	0	0

Near-Miss Reporting		
Number of Near-Miss Reported	0	1

Occupational Health		
Number of Occupational diseases Reported	0	0

Environmental Incidents		
Number of Environmental diseases Reported	0	0

Site Inspections/Audits		
Internal Audits	4	4
Client Audits	2	2
Standard Organization Audit/Inspection	0	0
Department of Labour/Municipality Inspections (Health)	1	1
Disabling Injury Frequency Rate	Total for Month	
(Number of Lost Time Incidents + Fatal) 100000/Total Man-Hours worked	0.25	

MONTHLY DISEASE INJURY ANALYSES									
Parts of Body Effected				Type of Injury				Disease (Work Related)	
Title	First Aid	Medical	LTI	Title	First Aid	Medical	LTI	Title	Nos.
Head	3	1	0	Sprain/ Strain	2	0	0	Skin	0
Face	0	0	0	Contusion	0	0	0	Respira- tory	0
Neck	0	0	0	Cut	7	2	0	Muscular	0
Eye	0	0	0	Laceration	0	0	0	Skeletal	0
Back	0	0	0	Abrasion	0	0	0	Endemic	0
Abdomen	0	0	0	Fracture	0	0	1	Hearing	0
Arm	0	0	0	Burn	0	0	0	Neurologi- cal	0
Hand	4	1	0	Amputation	0	0	0	Psycho- logical	0
Finger	0	0	0	Electric Shock	0	0	0	Other	0
Leg	2	0	1	Asphyxia	0	0	0	Total	0
Foot	0	0	0	Unconscious/ Fainting	0	0	0		
Toe	0	0	0	Poisoning	0	0	0		
Chest	0	0	0	Foreign Body	0	0	0		
Shoulder	0	0	0	Multiple Injury	0	0	0		
Internal	0	0	0	Total	9	2	1		
Multiple	0	0	0						
Total	9	2	1						

Date	Public Liability Incident Details	Potential Causes
	Nil	

NAME	POSITION	SIGNATURE	DATE
Name of SM/ Officer	SAFETY MANAGER/OFFICER	Sign of SM/Officer	Sign Date
Name of GM	GENERAL MANAGER	Sign of GM	Sign Date

MONTHLY INJURY MARKINGS

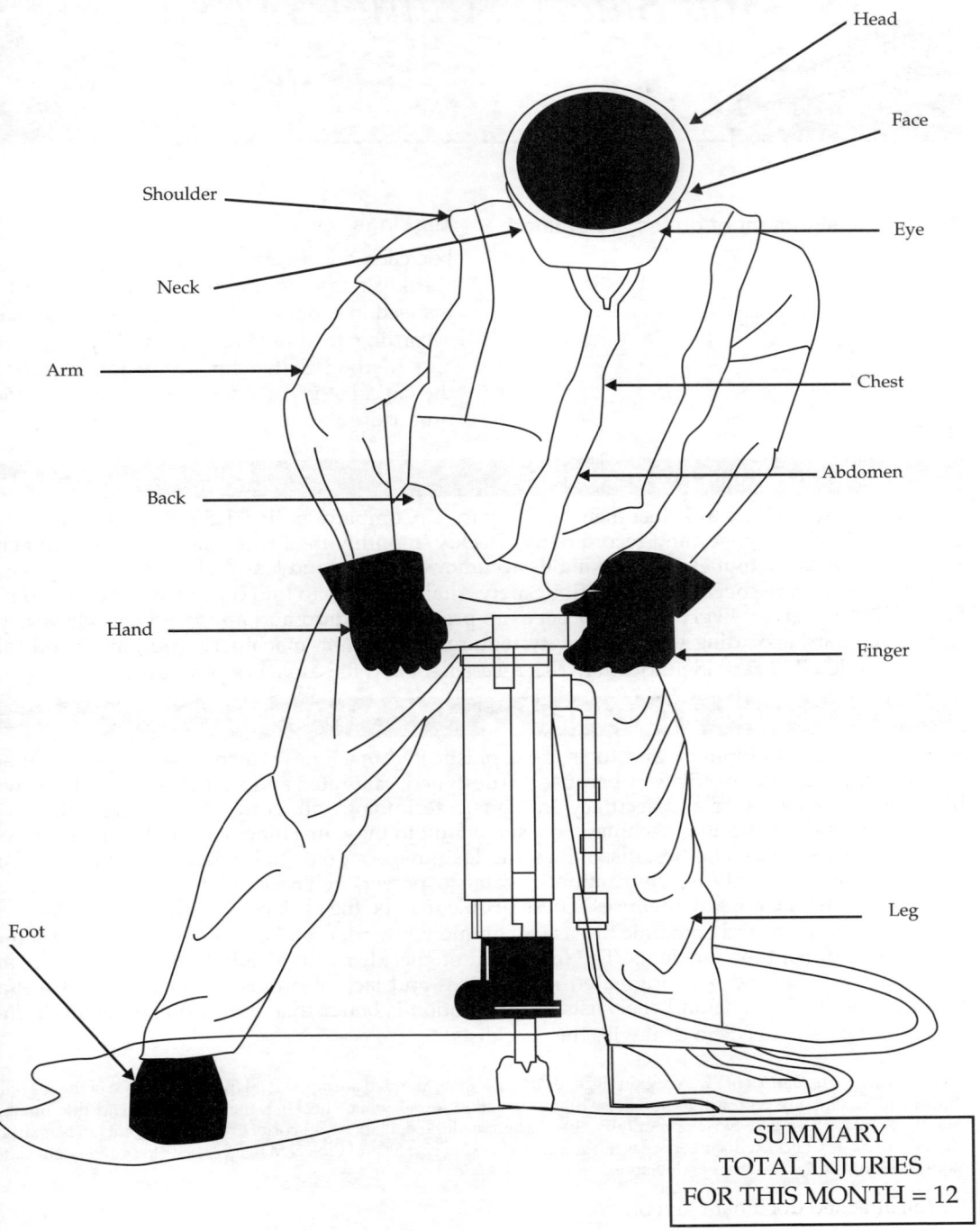

SUMMARY
TOTAL INJURIES
FOR THIS MONTH = 12

Job Safety Analyses (JSA) Checklist 10

Some of assignment of duties are defined

a. Excavation
b. Hand tool
c. Portable electric tool
d. Gas cutting
e. Hygiene

Markings

For each and every job and welfare activities checklists are to be prepared as a part of inspection program. All checklists are prepared according to the company's policy, HSE plan and Method Statement. Deviation marked in the checklist is considered as points for review procedure.

What is the Infrastructure of Checklists?

It is an operational control document and a part of Company policy, HSE Plan and Safe work procedure to visibly check and record daily/weekly/monthly the Plant/Machinery/Equipment/ work and welfare activities for checking its accuracy and wear and tear for better efficiency.

In this procedure checklists or JSA (Job Safety Analyses) help to find out any deviations from its Safe work procedure. Every company needs to prepare standard and approved checklists as per OSHA standard according to their activity requirements, plant, machinery, equipment and tools used, after detail hazard identification, risk assessment and its safe work procedure.

What is the Importance of Checklists?

Plant, machinery, equipment and tools are sophisticated machines which are very powerful and stronger beyond human efficiency because it is designed, calibrated and strengthened to do job with high speed, performance and accuracy. In modern technology all work activities are linked with these types of sophisticated machines. So a small fault to these machines can be deadly dangerous and can proceed with fatalities, disabilities and lost time accidents to humans. Lot of people from different countries, culture, civilization, resistance power, behavior, attitude and lifestyle are grouping under a company premises. These people may be the back bone of their families.

The occupational and epidemic disease is common in work premises which can cause fatalities, disability and lost time incidents. The important of checklist is to visibly find and eliminate any deficiency in wear and tear originated in machines and lack of provisions in work and welfare activities. So the logic behind checklists is "Prevention is better than cure". A competent, trained and expert person needs to deal with the checklists.

Note on checklists to filled in : The checklist legends or deviations founded are marked on the right column of the checklist document accompany with a title numbers. If any deviation found while checking, the concern legend title number should be marked on daily wise or monthly wise column, and the deviation and corrective actions points to be marked on the 2nd page. Checklist documents has to be sign by checking person and Project Manager needs to countersign it after spot checking for the clearance of deviations.

See the attached document given:

SDTP

10.1 Excavation Checklists and Deviations

Document References		
SDTP.CL.001		
Rev. No	Doc. Date	
0	1/1/2009	
	Pages	
	1 of 2	

⚠ WARNING

OPEN HOLE

Site: _____ Date/Year: _____

Do Not "tick" Write "OK", if faulty use legend number.

Date of Inspection

Sr. No.	Depth of Excavation	Location of Excavation	1/1	1/2	1/3	1/4	1/5	1/6	1/7	1/8	1/9	1/10	1/11	1/12	1/13	1/14	1/15
Date Inspected:																	
Initialed by the Inspector:																	
Spot Check by Project Manager:																	
Date Spot Checked:																	

Checklist/Legend

E 1	Are all working faces secure/all shoring materials secure?
E 2	Timber condition good; signs of rot: Wedges tight?
E 3	Soil seeping through gap of shoring.
E 4	All examinations properly recorded and communicated?
E 5	Are soil heaps far enough back, and are they posing danger for men in the excavation below?
E 6	Are warning signs posted during day and night time?
E 7	Are pipes, bricks, stones or tools too near to the edges?
E 8	Is excavation properly guarded(1M from face) with solid materials?
E 9	Is the excavation properly guarded and lighted at night?
E 10	Gangways guard railed and toe boards fixed at edges?
E 11	Is regular testing for harmful gas carried out?
E 12	Is all PPE's wearing. Are existing structures sufficiently protected?
E 13	Do employees know what to do if evacuation is necessary?
E 14	Is the site tidy? Excavations are the most difficult to keep tidy. Tidiness is essential for safety 1
E 15	Ladders used are correct type and secure at the bottom and top.

Contd...

Deviations and Corrective Actions

Date	Description of Deviations	Corrective Actions	Action by whom	Target Date	Completion Date

			Document References		
			SDTP.CL.0012		
			Rev. No	Doc. Date	
			0	1/1/2009	
			Pages		
			1 of 4		

SDTP — 10.2 Hand tool Checklists and Deviations

Do Not "tick" Write "OK", if faulty use legend number.

Checklist/Legend

H 1	Blunting?
H 2	Mushrooming/Split Heads?
H 3	Handles? (No Make Shift)
H 4	Right Tool/Right Job?
H 5	All Wearing PPE?
H 6	Using any Wear and Tear Tools?
H 7	Storage Facility Adequate?
H 8	Broken, Cracked Tool
H 9	Other

Sr. No	Location of Excavation	Type of Tool	Date/Year:												
			Jan	Feb	Mar	Apr	May	Jun	July	Aug	Sep	Oct	Nov	Dec	

Date Inspected:

Initialed by the Inspector:

Spot Check by Project Manager:

Date Spot Checked:

Contd....

Deviations and Corrective Actions

Month	Description of Deviations	Corrective Actions	Action by whom	Target Date	Completion Date
JAN					
FEB					
MAR					
APR					

Deviations and Corrective Actions

Month	Description of Deviations	Corrective Actions	Action by whom	Target Date	Completion Date
MAY					
JUN					
JUL					
AUG					

Contd...

Deviations and Corrective Actions

Month	Description of Deviations	Corrective Actions	Action by whom	Target Date	Completion Date
SEP					
OCT					
NOV					
DEC					

Contd...

SDTP | **10.3 Portable Electric Tools Checklists and Deviations**

Document References
SDTP.CL.005
Rev. No 0 | Doc. Date 1/1/2009
Pages 1 of 4

Do Not "tick" Write "OK", if faulty use legend number.

Sr. No.	Location of Excavation	Type of Tool	Site:								Date/Year:					Checklist/Legend
			Jan	Feb	Mar	Apr	May	Jun	July	Aug	Sep	Oct	Nov	Dec	
															PE 1 Loose connections?
															PE 2 Continuous earthing?
															PE 3 Cable free from bad joints?
															PE 4 Cracked or broken insulation?
															PE 5 Polarity of extension cables?
															PE 6 Any switches faulty, electric shock hazard
															PE 7 Home made sockets, Plug in order?
															PE 8 Guards missing or damaged?
															PE 9 Earth wire not connected?
															PE 10 Extension overcrowded?
															PE 11 Unusual noise, vibration
															PE 12 Using on wet, vapor or flammable condition
															PE 13 Trained/expert Operatives
															PE 14 Operatives wearing lose cloths
															PE 15 Using wear and tear parts
															PE 16 Other

Date Inspected:

Initialed by the Inspector:

Spot Check by Project Manager:

Date Spot Checked:

Contd...

Deviations and Corrective Actions

Month	Description of Deviations	Corrective Actions	Action by whom	Target Date	Completion Date
JAN					
FEB					
MAR					
APR					

Contd...

Deviations and Corrective Actions

Month	Description of Deviations	Corrective Actions	Action by whom	Target Date	Completion Date
MAY					
JUN					
JUL					
AUG					

Contd...

Deviations and Corrective Actions

Month	Description of Deviations	Corrective Actions	Action by whom	Target Date	Completion Date
SEP					
OCT					
NOV					
DEC					

SDTP

10.4 Gas Cutting and Welding Checklists and Deviations

Document References

SDTP.CL006	
Rev. No	Doc. Date
0	1/1/2009
	Pages
	1 of 4

Do Not "tick" Write "OK", if faulty use legend number.

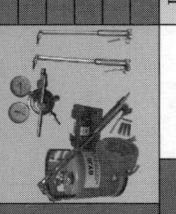

Site: **FXE 1002**   Date/Year: **JAN-08**

Checklist/Legend

CYLINDERS
- GS 1 Secured, valve guards?
- GS 2 Valves undamaged, closed if not
- GS 3 Spindle key on each?

REGULATORS
- GS 4 Clean and no contamination?
- GS 5 Identifiable gas. Correct range capacity?
- GS 6 Is outlet pressure steady. Gauges
- GS 7 Pressure adjusting screw turns freely?

FLASH BACK ARRESTORS
- GS 8 Reset lever is in good condition?
- GS 9 Body and connection nuts undamaged?

HOSES
- GS 10 ISO Standards. Color identification?
- GS 11 Free of oil, grease, cuts, burns, cracks?
- GS 12 Hose secured with " o-clip" ?.
- GS 13 Flashback arrestors fitted on torch end?

TORCHES, NOZZLES, FLAME
- GS 14 Free of visual damage,
- GS 15 Valves free to Open and close-off?
- GS 16 Cutting oxygen valve closes completely?
- GS 17 Nozzle nut in good condition?
- GS 18 Nozzle correct size. No leaks?
- GS 19 Friction type lighters used?

GOGGLES
- GS 20 ISO approved type.
- GS 21 Filter lens shade correct for type of work.
- GS 22 Frame and head band in good condition.
- GS 23 Every operator own goggles?

OTHER
- GS 24 PTW System followed?
- GS 25 Cylinder trolley, Extinguisher Provided?

Gas Set ID No.	Cylinder Capacity	Location of the Gas set	Cylinder Description	Jan	Feb	Mar	Apr	May	Jun	July	Aug	Sep	Oct	Nov	Dec

Date Inspected:

Initialed by the Inspector:

Spot Check by Project Manager:

Date Spot Checked:

Contd...

Deviations and Corrective Actions

Month	Description of Deviations	Corrective Actions	Action by whom	Target Date	Completion Date
JAN					
FEB					
MAR					
APR					

Contd...

Deviations and Corrective Actions

Month	Description of Deviations	Corrective Actions	Action by whom	Target Date	Completion Date
MAY					
JUN					
JUL					
AUG					

Contd...

Deviations and Corrective Actions

Month	Description of Deviations	Corrective Actions	Action by whom	Target Date	Completion Date
SEP					
OCT					
NOV					
DEC					

SDTP — 10.5 Hygiene Checklists and Deviations

Document References		
SDTP.CL.007		
Rev. No	Doc. Date	
0	1/1/2009	
Pages		
1 of 4		

Do Not "tick" Write "OK", if faulty use legend number.

Location of the Facility — Site: _____ Date/Year: _____

ID. No.	Checklist/Legend	Jan	Feb	Mar	Apr	May	Jun	July	Aug	Sep	Oct	Nov	Dec
	SHOWER/WASHING AREA												
H 1	Area clean, hygienic, good drainage?												
H 2	Area demarcated screened off?												
H 3	Hot water, detol, soap available.?												
	KITCHEN												
H 4	Floors clean, non slip?												
H 5	Tables and chairs good condition?												
H 6	Utensils clean, kitchen clean												
H 7	Hot water, good ventilation and lighting?												
	EATING AREAS												
H 8	Sufficient seating, dust free.?												
H 9	Clean, hygienic, dustbins available?												
	DRINKING WATER												
H 10	Drinking water (Hot and Cold) available?												
H 11	Are Water cooler filter changing regularly?												
H 12	Clean drainage, waste removal program.												
H 13	Dust controlling are in practice?												
	TOILETS												
H 14	Sufficient and cleaning regularly?												
H 15	Separate men / women												
H 16	No damages, partitions for privacy?												
H 17	Wash basins, soap, no damages and leaks?												
H 18	Urinals clean, no blockages and leaks?												
H 19	General condition good, no strange Water												
	CHANGE AREA												
H 20	Sufficient, lighting, ventilation?												
H 21	No changing in stores, stacking												
H 22	Area clean, disinfected, sufficient bins?												
	TUCK-SHOP OUTLET												
H 23	Well placed, tidy, hygienic?												
H 24	No stagnant water used?												
H 25	Pest control system is in practice?												

Date Inspected: _____
Initialed by the Inspector: _____
Spot Check by Project Manager: _____
Date Spot Checked: _____

Contd...

Deviations and Corrective Actions

Month	Description of Deviations	Corrective Actions	Action by whom	Target Date	Completion Date
JAN					
FEB					
MAR					
APR					

Deviations and Corrective Actions

Month	Description of Deviations	Corrective Actions	Action by whom	Target Date	Completion Date
MAY					
JUN					
JUL					
AUG					

Contd...

Deviations and Corrective Actions

Month	Description of Deviations	Corrective Actions	Action by whom	Target Date	Completion Date
SEP					
OCT					
NOV					
DEC					

Information and Study Page

Sr.No.	GF. Exc. 01	Depth of Excavation:	7 Meters	Site:	FXE 1002	Date/Year:	January-09
Checking Conducted:	Dailywise	Location of Excavation:	D2 Oil Storage tank, 'A' module ground floor				

While on checking excavation on 1st January:

a. While checking excavation found that the gap between the shoring material and edges of excavation not back filled. (E3)

b. While checking excavation found that warning flashing lights are not erected (E 9).

c. While checking excavation found ladders used as assess are not secured properly (E 15).

While on checking excavation on 2nd January:

a. While checking excavation found that the gap between the shoring material and edges are back filled.

b. While checking excavation found that warning flashing lights are not yet posted (E 9).

c. While checking excavation found ladders used as assess are secured properly.

d. While checking excavation found that toe boards on the two sides of excavation are removed (E 10).

While on checking excavation on 3rd January:

a. While checking excavation found that warning flashing lights are not yet posted (E 9).

b. While checking excavation found that toe boards on the four sides of excavation is fixed.

While on checking excavation on 8th January:

a. While checking excavation found that warning flashing lights are posted.

After 10th and up to 14th, deviations nil and on 15th excavation work completed and the excavation was permanently closed. *See the attached document how the excavation checklist filled in on STDP documents:*

SDTP — 10.1 Excavation Checklists and Deviations

Document References

SDTP.CL.001	
Rev. No	Doc. Date
0	1/1/2009
	Pages
	1 of 2

⚠ WARNING — OPEN HOLE

Do Not "tick" Write "OK", if faulty use legend number.

Site: FXE 1002 **Date/Year:** Jan-09

Sr. No.	Depth of Excavation	Location of Excavation
G.F. Exc. 01	7 Meters	D2 Oil Storage tank, 'A' module ground floor

Date of Inspection

Ref	Checklist/Legend	1/1	1/2	1/3	1/4	1/5	1/6	1/7	1/8	1/9	1/10	1/11	1/12	1/13	1/14	1/15
E 1	Are all working faces secure/all shoring materials secure?	E3	OK	OK	OK	OK	OK	OK	OK	OK	OK	OK	OK	OK	OK	-
E 2	Timber condition good; signs of rot: Wedges tight?	E9	E9	E9	E9	E9	E9	E9	E9	OK	OK	OK	OK	OK	OK	-
E 3	Soil seeping through gap of shoring.	E15	OK	OK	OK	OK	OK	OK	OK	OK	OK	OK	OK	OK	OK	-
E 4	All examinations properly recorded and communicated?	OK	E10	OK	OK	OK	OK	OK	OK	OK	OK	OK	OK	OK	OK	-
E 5	Are soil heaps far enough back, and are they posing danger for men in the excavation below?															
E 6	Are warning signs posted during day and night time?															
E 7	Are pipes, bricks, stones or tools too near to the edges?															
E 8	Is excavation properly guarded (1 M from face) with solid materials?															
E 9	Is the excavation properly guarded and lighted at night?															
E 10	Gangways guard railed and toe boards fixed at edges?															
E 11	Is regular testing for harmful gas carried out?															
E 12	Is all PPE's wearing. Are existing structures sufficiently protected?															
E 13	Do employees know what to do if evacuation is necessary?															
E 14	Is the site tidy? (Excavations are the most difficult to keep tidy). Tidiness is essential for safety 1															
E 15	Ladders used are correct type and secure at the bottom and top.															
	Date Inspected:	1/1	1/2	1/3	1/4	1/5	1/6	1/7	1/8	1/9	1/10	1/11	1/12	1/13	1/14	1/15
	Initialed by the Inspector:	Sign EI	Sign EI	Sign EI	Sign n EI	Sign EI	Sign EI	Sign EI	Sign EI	Sign EI	Sign EI	Sign EI	Sign EI	Sign EI	Sign EI	Sign EI
	Spot Check by Project Manager:	Sign PM	Sign PM	Sign PM	n	Sign PM	Sign PM	Sign PM	Sign PM	Sign PM	Sign PM	Sign PM	Sign PM	Sign PM	Sign PM	Sign PM
	Date Spot Checked:	1/1	1/2	1/3	1/4	1/5	1/6	1/7	1/8	1/9	1/10	1/11	1/12	1/13	1/14	1/15

Deviations and Corrective Actions

Date	Description of Deviations	Corrective Actions	Action by whom	Target Date	Completion Date
1-Jan-09	Found the gap between the shoring material and edges of excavation not back filled (E 3).	The gap to be back filled.	Excavation Supervisor	2/1/2009	2/1/2009
1-Jan-09	The warning flashing lights are not erected (E 9).	Needs to erect warning flashing lights.	Excavation Supervisor	2/1/2009	8/1/2009
1-jan-09	Ladders used as assess are not secured properly (E 15).	Needs to secure ladders.	Excavation Supervisor	2/1/2009	2/1/2009
2-jan-09	Toe boards on the two sides of excavation are removed (E 10).	Needs to fix toe boards.	Excavation Supervisor	3/1/2009	3/1/2009

Contd...

Information and Study Page

Sr. No.	Toolbox (TB 001 to TB 004)	Date/Year:	1/1/2008	**Site:**	FXE 1002	Type of Tools	Hammer	Chisel
							Screwdriver	Axie
							Cutting Player	Pipe Ranger
							Adjustable Spanners	
Checking Conducted:	Monthly wise	Location of Tools:	Random checking the Toolbox (TB)					

While on checking 1st floor on January: (Date: 1/1/2008)

(TB 4) A chisel founded with mushroom head (H 2).   (TB 2) Found dismantling a charged generator wire by a player

(TB 3) Found a screwdriver with a broken handle (H 3).   Which has partially damaged insulation grip (H 6).

While on checking ground floor on April: (Date: 2/4/2008)

(TB 4) Found tightening a pipe using a broken spanner.   (TB 1) Nailing a box using adjustable spanner (H 4).

While on checking 2nd floor on September: (Date: 2/9/2008)

(TB 1) Found a hammer with a split head (H 2).   (TB 3) Found a worker holding a chisel on a concrete braking

(TB 2) Found cutting a aluminum sheet using broken handle axle (H 3).   Works with out wearing gloves (H 5).

See the attached document how the hand tool checklist filled in on STDP documents:

SDTP — 10.2 Hand tool Checklists and Deviations

Document References

SDTP.CL.0012	
Rev. No	Doc. Date
0	1/1/2009
Pages	
1 of 4	

Do Not "tick" Write "OK", if faulty use legend number.

Checklist/Legend

H 1	Blunting?
H 2	Mushrooming/Split Heads?
H 3	Handles? (No Make Shift)
H 4	Right Tool/Right Job?
H 5	All Wearing PPE?
H 6	Using any Wear and Tear Tools?
H 7	Storage Facility Adequate?
H 8	Broken, Cracked Tool
H 9	Other

Site: FXE 1002 Date/Year: 2008

Sr. No.	Location of Excavation	Type of Tool	Jan	Feb	Mar	Apr	May	Jun	Jul	Aug	Sep	Oct	Nov	Dec
TB 1	Ground floor, 1st Floor, 2nd Floor	Hammer	OK	OK	OK	OK	OK	OK	OK	OK	H2	OK	OK	OK
		Pipe Ranger	OK	OK	OK	OK	OK	OK	OK	OK	OK	OK	OK	OK
		Screwdriver	OK	OK	OK	OK	OK	OK	OK	OK	OK	OK	OK	OK
		Adjustable Spanner	OK	OK	OK	H 4	OK	OK	OK	OK	OK	OK	OK	OK
		Cutting player	OK	OK	OK	OK	OK	OK	OK	OK	OK	OK	OK	OK
		Chisel	OK	OK	OK	OK	OK	OK	OK	OK	OK	OK	OK	OK
		Axle	OK	OK	OK	OK	OK	OK	OK	OK	OK	OK	OK	OK
TB 2	Ground floor, 1st Floor, 2nd Floor	Hammer	OK	OK	OK	OK	OK	OK	OK	OK	OK	OK	OK	OK
		Pipe Ranger	OK	OK	OK	OK	OK	OK	OK	OK	OK	OK	OK	OK
		Screwdriver	OK	OK	OK	OK	OK	OK	OK	OK	OK	OK	OK	OK
		Adjustable Spanner	OK	OK	OK	OK	OK	OK	OK	OK	OK	OK	OK	OK
		Cutting player	H6	OK	OK	OK	OK	OK	OK	OK	OK	OK	OK	OK
Date Inspected:			1-Jan	2-Feb	2-Mar	2-Apr	2-May	2-Jun	1-Jul	2-Aug	2-Sep	1-Oct	1-Nov	2-Dec
Initialed by the Inspector:			Sign Ins.	Sign Ins.	Sign Ins.	Sign Ins.	Sign Ins.	Sign Ins.	Sign Ins.	Sign Ins.	Sign Ins.	Sign Ins.	Sign Ins.	Sign Ins.
Spot Check by Project Manager:			Sign PM	Sign PM	Sign PM	Sign PM	Sign PM	Sign PM	Sign PM	Sign PM	Sign PM	Sign PM	Sign PM	Sign PM
Date Spot Checked:			1-Jan	2-Feb	2-Mar	2-Apr	2-May	2-Jun	1-Jul	2-Aug	2-Sep	1-Oct	1-Nov	2-Dec

SDTP — Hand tool Checklists and Deviations

Document References		
SDTP.CL.0012		
Rev. No		Doc. Date
0		1/1/2009
Pages		
1 of 4		

Site: FXE 1002 **Date/Year:** 2008

Checklist/Legend

Do Not "tick" Write "OK", if faulty use legend number.

H 1	Blunting?
H 2	Mushrooming/Split Heads?
H 3	Handles? (No Make Shift)
H 4	Right Tool/Right Job?
H 5	All Wearing PPE?
H 6	Using any Wear and Tear Tools?
H 7	Storage Facility Adequate?
H 8	Broken, Cracked Tool
H 9	Other

Sr. No.	Location of Excavation	Type of Tool	Jan	Feb	Mar	Apr	May	Jun	Jul	Aug	Sep	Oct	Nov	Dec
		Chisel	OK	OK	OK	OK	OK	OK	OK	OK	OK	OK	OK	OK
		Axle	OK	OK	OK	OK	OK	OK	OK	OK	OK	H3	OK	OK
		Hammer	OK	OK	OK	OK	OK	OK	OK	OK	OK	OK	OK	OK
TB 3	Ground floor, 1st Floor, 2nd Floor	Pipe Ranger	OK	OK	OK	OK	OK	OK	OK	OK	OK	OK	OK	OK
		Screwdriver	H3	OK	OK	OK	OK	OK	OK	OK	OK	OK	OK	OK
		Adjustable Spanner	OK	OK	OK	OK	OK	OK	OK	OK	OK	OK	OK	OK
		Cutting player	OK	OK	OK	OK	OK	OK	OK	OK	OK	OK	OK	OK
		chisel	OK	OK	OK	OK	OK	OK	OK	OK	H5	OK	OK	OK
		Axle	OK	OK	OK	OK	OK	OK	OK	OK	OK	OK	OK	OK
TB 4	Ground floor, 1st Floor, 2nd Floor	Hammer	OK	OK	OK	OK	OK	OK	OK	OK	OK	OK	OK	OK
		Pipe Ranger	OK	OK	OK	OK	OK	OK	OK	OK	OK	OK	OK	OK
		Screwdriver	OK	OK	OK	OK	OK	OK	OK	OK	OK	OK	OK	OK
Date Inspected:			1-Jan	2-Feb	2-Mar	2-Apr	2-May	2-Jun	1-Jul	2-Aug	2-Sep	1-Oct	1-Nov	2-Dec
Initialed by the Inspector:			Sign Ins.	Sign Ins.	Sign Ins.	Sign Ins.	Sign Ins.	Sign Ins.	Sign Ins.	Sign Ins.	Sign Ins.	Sign Ins.	Sign Ins.	Sign Ins.
Spot Check by Project Manager:			Sign PM	Sign PM	Sign PM	Sign PM	Sign PM	Sign PM	Sign PM	Sign PM	Sign PM	Sign PM	Sign PM	Sign PM
Date Spot Checked:			1-Jan	2-Feb	2-Mar	2-Apr	2-May	2-Jun	1-Jul	2-Aug	2-Sep	1-Oct	1-Nov	2-Dec

SDTP — 10.3 Hand tool Checklists and Deviations

Document References	
SDTP.CL.0012	
Rev. No	Doc. Date
0	1/1/2009
Pages	
1 of 4	

Site: FXE 1002 **Date/Year:** 2008

Do Not "tick" Write "OK", if faulty use legend number.

Checklist/Legend

H 1	Blunting?
H 2	Mushrooming/Split Heads?
H 3	Handles? (No Make Shift)
H 4	Right Tool/Right Job?
H 5	All Wearing PPE?
H 6	Using any Wear and Tear Tools?
H 7	Storage Facility Adequate?
H 8	Broken, Cracked Tool
H 9	Other

Sr. No.	Location of Excavation	Type of Tool	Jan	Feb	Mar	Apr	May	Jun	Jul	Aug	Sep	Oct	Nov	Dec
		Adjustable Spanner	OK	OK	OK	H 8	OK	OK	OK	OK	OK	OK	OK	OK
		Cutting Player	OK	OK	OK	OK	OK	OK	OK	OK	OK	OK	OK	OK
		Chisel	H 2	OK	OK	OK	OK	OK	OK	OK	OK	OK	OK	OK
		Axle	OK	OK	OK	OK	OK	OK	OK	OK	OK	OK	OK	OK
Date Inspected:			1-Jan	2-Feb	2-Mar	2-Apr	2-May	2-Jun	1-Jul	2-Aug	2-Sep	1-Oct	1-Nov	2-Dec
Initialed by the Inspector:			Sign Ins.	Sign Ins.	Sign Ins.	Sign Ins.	Sign Ins.	Sign Ins.	Sign Ins.	Sign Ins.	Sign Ins.	Sign Ins.	Sign Ins.	Sign Ins.
Spot Check by Project Manager:			Sign PM	Sign PM	Sign PM	Sign PM	Sign PM	Sign PM	Sign PM	Sign PM	Sign PM	Sign PM	Sign PM	Sign PM
Date Spot Checked:			1-Jan	2-Feb	2-Mar	2-Apr	2-May	2-Jun	1-Jul	2-Aug	2-Sep	1-Oct	1-Nov	2-Dec

Contd...

Deviations and Corrective Actions

Month	Description of Deviations	Corrective Actions	Action by whom	Target Date	Completion Date
JAN					
FEB					
MAR					
APR					
MAY					

Contd

Deviations and Corrective Actions

Month	Description of Deviations	Corrective Actions	Action by whom	Target Date	Completion Date
JUN					
JUL					
AUG					
SEP					
OCT					

Contd...

Deviations and Corrective Actions

Month	Description of Deviations	Corrective Actions	Action by whom	Target Date	Completion Date
NOV					
DEC					

Contd.

Information and Study Page

Sr. No.	PET 001 TO PET 004	Date/Year:	Site:	FXE 1002	Type of Tools	Portable drills	Portable cutter
		1/1/2008				Portable grinder	
Checking Conducted:	Monthly wise	Location of Tools:	check the information below			Portable saw	

While on checking 1st floor on January: Date: 1/1/2008

(PG 2) Found a portable grinder, which has home made joint cables (PE 3).

(PC 1) Found cutting a alloy pipe using cracked insulated handle portable cutter (PE 4).

(PS 2) Found a portable saw missing safety guards (PE 8).

While on checking ground floor on April: Date: 2/4/2008

(PD 1) Found a metal beam drilling work using portable drill where its cables touching with water and the insulation of cable is partially damaged (PE 12).

While on checking 2nd floor on August: Date: 1/8/2008

(PG 1) Found a grinding work using partially cracked grinding wheel (PE 15).

(PC 2) Found cutting a aluminum sheet by an electric cutter where its cables connected to the electric source openly with out plugs (PE 1).

Description of Tool Coding

Item	Description of Tool	No. of Tools	
PD	Portable drills	PD 1, PD 2, PD 3	3
PG	Portable grinder	PG 1, PG 2	2
PS	Portable saw	PS 1, PS 2, PS 3	3
PC	Portable cutter	PC 1, PC 2, PC 3	3

See the attached document how the Portable Electric Tool checklist filled in on STDP documents:

Contd...

SDTP — 10.3 Portable Electric Tools Checklists and Deviations

Document References: SDTP.CL.005 — Rev. No 0 — Doc. Date 1/1/2009 — Pages 1 of 4

Site: FXE 1002 **Date/Year:** Jan-08

Do Not "tick" Write "OK", if faulty use legend number.

Checklist/Legend

PE 1	Loose connections?
PE 2	Continuous earthing?
PE 3	Cable free from bad joints?
PE 4	Cracked or broken insulation?
PE 5	Polarity of extension cables?
PE 6	Any switches faulty, electric shock hazard
PE 7	Home made sockets, Plug in order?
PE 8	Guards missing or damaged?
PE 9	Earth wire not connected?
PE 10	Extension overcrowded?
PE 11	Unusual noise, vibration
PE 12	Using on wet, vapor or flammable condition
PE 13	Trained/Expert Operatives
PE 14	Operatives wearing lose cloths
PE 15	Using wear and tear parts
PE 16	Other

Sr. No.	Location of Excavation	Type of Tool	Jan	Feb	Mar	Apr	May	Jun	July	Aug	Sep	Oct	Nov	Dec
PD 1	Ground floor, 1st floor and 2nd floor	Portable drill	OK	OK	OK	PE 12	OK	OK	OK	OK	OK	OK	OK	OK
PD 2	Ground floor, 1st floor and 2nd floor	Portable drill	OK	OK	OK	OK	OK	OK	OK	OK	OK	OK	OK	OK
PD 3	Ground floor, 1st floor and 2nd floor	Portable drill	OK	OK	OK	OK	OK	OK	OK	OK	OK	OK	OK	OK
PG 1	Ground floor, 1st floor and 2nd floor	Portable drill	OK	OK	OK	OK	OK	OK	OK	PE 15	OK	OK	OK	OK
PG 2	Ground floor, 1st floor and 2nd floor	Portable drill	PE 3	OK	OK	OK	OK	OK	OK	OK	OK	OK	OK	OK
PS 1	Ground floor, 1st floor and 2nd floor	Portable drill	OK	OK	OK	OK	OK	OK	OK	OK	OK	OK	OK	OK
PS 2	Ground floor, 1st floor and 2nd floor	Portable drill	PE 8	OK	OK	OK	OK	OK	OK	OK	OK	OK	OK	OK
PS 3	Ground floor, 1st floor and 2nd floor	Portable drill	OK	OK	OK	OK	OK	OK	OK	OK	OK	OK	OK	OK
PC 1	Ground floor, 1st floor and 2nd floor	Portable drill	PE 4	OK	OK	OK	OK	OK	OK	OK	OK	OK	OK	OK
PC 2	Ground floor, 1st floor and 2nd floor	Portable drill	OK	OK	OK	OK	OK	OK	OK	PE 1	OK	OK	OK	OK
PC 3	Ground floor, 1st floor and 2nd floor	Portable drill	OK	OK	OK	OK	OK	OK	OK	OK	OK	OK	OK	OK
Date Inspected:			1-Jan	2-Feb	2-Mar	2-Apr	2-May	2-Jun	1-Jul	2-Aug	2-Sep	1-Oct	1-Nov	2-Dec
Initialed by the Inspector:			Sign Ins.	Sign Ins.	Sign Ins.	Sign Ins.	Sign Ins.	Sign Ins.	Sign Ins.	Sign Ins.	Sign Ins.	Sign Ins.	Sign Ins.	Sign Ins.
Spot Check by Project Manager:			Sign PM	Sign PM	Sign PM	Sign PM	Sign PM	Sign PM	Sign PM	Sign PM	Sign PM	Sign PM	Sign PM	Sign PM
Date Spot Checked:			1-Jan	2-Feb	2-Mar	2-Apr	2-May	2-Jun	1-Jul	2-Aug	2-Sep	1-Oct	1-Nov	2-Dec

Deviations and Corrective Actions

Month	Description of Deviations	Corrective Actions	Action by whom	Target Date	Completion Date
JAN	(PG 2). Found a portable grinder, which has home made joint cables (PE 3).	Fix the cable with closed socket	Electrical Supervisor	2/1/2008	2/1/2008
	(PS 2). Found a portable saw missing safety guards (PE 8).	Fix safety guard immediately. Needs to give written warning to the operator who removes the safety guard.	Electrical Engineer	2/1/2008	2/1/2008
	(PC 1). Found cutting a alloy pipe using cracked insulated handle portable cutter (PE 4).	Replace the tool with new one.	Electrical Supervisor	2/1/2008	2/1/2008
FEB					
MAR					
APR	(PD 1). Found a metal beam drilling work using portable drill where its cables touching with water and the insulation of cable is partially damaged (PE 12).	Stop the work. Instruct to remove or dry the water from portable electric tool using area and also Instructed to replace cable with new one.	Work Supervisor	3/4/2008	3/4/2008
MAY					

Contd...

Deviations and Corrective Actions

Month	Description of Deviations	Corrective Actions	Action by whom	Target Date	Completion Date
JUN					
JUL					
AUG	(PG 1). Found a grinding work using partially cracked grinding wheel (PE 15).	Stop the work and instructed to change the grinding!. wheel.	Work Supervisor		2/8/2008
AUG	(PC 2). Found cutting a aluminum sheet by an electric cutter where its cables connected to the electric source openly without plugs (PE 1).	Stop the work. Written warning to be issued to the work supervisor. A special induction training to be given to the total crew about the electric safety and portable electric tools safety by concern Engineer.	Concern Engineer	2/8/2008	2/8/2008
SEP					
OCT					

Deviations and Corrective Actions

Month	Description of Deviations	Corrective Actions	Action by whom	Target Date	Completion Date
NOV					
DEC					

Contd...

Information and Study Page

Sr. No.	WS 1 to WS 3	Date/Year:	1/1/2008	**Site:**	FXE 1002	Type of Cylinders	20 kg Oxygen Cylinders
Checking Conducted:	Monthly wise	Location of Tools:	on floors in trolley while working				20 kg Acetylene Cylinders

While on checking 1st floor on March: Date: 2/3/2008

a. Oxygen valve of WS 3 set seen tight to open and close (GS 15).

b. Oxygen pressure cage vale glass of WS 1 set seen in dirty and scratched condition (GS 4).

While on checking ground floor on June: Date: 2/6/2008

a. The hose of WS 1 set seen in damage condition (GS 11).

While on checking 2nd floor on September: Date: 1/9/2008

a. Found a welder welding with WS 2 set without wearing welding shield (GS 23).

Description of Welding set Coding

Item	Description of Cylinder	Markings	
WS 1	Oxygen and Acetylene cylinders	OXY 1 and ACY 1	1
WS 2	Oxygen and Acetylene cylinders	OXY 2 and ACY 2	1
WS 3	Oxygen and Acetylene cylinders	OXY 3 and ACY 3	1

See the attached document how the cutting and welding set checklist filled in on STDP documents:

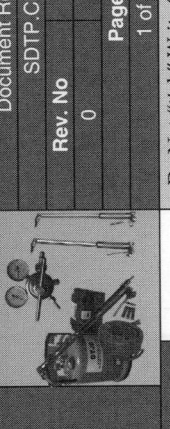

SDTP — Document References — SDTP.CL.006

Rev. No	Doc. Date
0	1/1/2009

Pages 1 of 4

10.4 Gas Cutting and Welding Checklists & Deviations

Site: FXE 1002 **Date/Year:** JAN-08

Gas Set ID No.	Cylinder Capacity	Location of the Gas set	Cylinder Description	Jan	Feb	Mar	Apr	May	Jun	July	Aug	Sep	Oct	Nov	Dec
WS 1	20 kg each	On floors in trolley while in working	OXY 1 and ACY 1	OK	OK	GS 4	OK	OK	OK	GH 11	OK	OK	OK	OK	OK
WS 2	20 kg each	On floors in trolley while in working	OXY 2 and ACY 2	OK	OK	OK	OK	OK	OK	OK	OK	GH 23	OK	OK	OK
WS 3	20 kg each	On floors in trolley while in working	OXY 2 and ACY 2	OK	OK	GH 15	OK	OK	OK	OK	OK	OK	OK	OK	OK
Date Inspected:				1-Jan	2-Feb	2-Mar	2-Apr	2-May	2-Jun	1-Jul	2-Aug	2-Sep	1-Oct	1-Nov	2-Dec
Initialed by the Inspector:				Sign Ins.	Sign Ins.	Sign Ins.	Sign Ins.	Sign Ins.	Sign Ins.	Sign Ins.	Sign Ins.	Sign Ins.	Sign Ins.	Sign Ins.	Sign Ins.
Spot Check by Project Manager:				Sign PM	Sign PM	Sign PM	Sign PM	Sign PM	Sign PM	Sign PM	Sign PM	Sign PM	Sign PM	Sign PM	Sign PM
Date Spot Checked:				1-Jan	2-Feb	2-Mar	2-Apr	2-May	2-Jun	1-Jul	2-Aug	2-Sep	1-Oct	1-Nov	2-Dec

Do Not "tick" Write "OK", if faulty use legend number.

Checklist/Legend

CYLINDERS
- GS 1 Secured, valve guards?
- GS 2 Valves undamaged, closed if not
- GS 3 Spindle key on each?

REGULATORS
- GS 4 Clean and no contamination?
- GS 5 Identifiable gas. Correct range capacity?
- GS 6 Is outlet pressure steady. Gauges
- GS 7 Pressure adjusting screw turns freely?

FLASH BACK ARRESTORS
- GS 8 Reset lever is in good condition?
- GS 9 Body and connection nuts undamaged?

HOSES
- GS 10 ISO Standards. Color identification?
- GS 11 Free of oil, grease, cuts, burns, cracks?
- GS 12 Hose secured with "o-clip"?
- GS 13 Flashback arrestors fitted on torch end?

TORCHES, NOZZLES, FLAME
- GS 14 Free of visual damage,
- GS 15 Valves free to Open and close-off?
- GS 16 Cutting oxygen valve closes completely?
- GS 17 Nozzle nut in good condition?
- GS 18 Nozzle correct size. No leaks?
- GS 19 Friction type lighters used?

GOGGLES
- GS 20 ISO approved type.
- GS 21 Filter lens shade correct for type of work.
- GS 22 Frame and head band in good condition.
- GS 23 Every operator own goggles?

OTHER
- GS 24 PTW System followed?
- GS 25 Cylinder trolley, Extinguisher Provided?

Contd...

Deviations and Corrective Actions

Month	Description of Deviations	Corrective Actions	Action by whom	Target Date	Completion Date
JAN					
FEB					
MAR	Oxygen valve of WS 3 set seen tight to open and close (GS 15).	Send to the workshop for changing the valve.	Storekeeper	3/3/2008	3/3/2008
	Oxygen pressure cage vale glass of WS 1 set seen in dirty and scratched condition (GS 4).	Send to the workshop for changing the pressure cage valve glass.	Storekeeper	3/3/2008	3/3/2008
APR					
MAY					

Deviations and Corrective Actions

Month	Description of Deviations	Corrective Actions	Action by whom	Target Date	Completion Date
JUN					
JUL					
AUG					
SEP	Found a welder welding with WS 2 set without wearing welding shield (GS 23).	Warning letter plus cutting some hours to be issued, Re-induction to be given to the total gang.	Site Engineer	2/9/2008	2/9/2008
OCT					

Contd...

Deviations and Corrective Actions

Month	Description of Deviations	Corrective Actions	Action by whom	Target Date	Completion Date
NOV					
DEC					

Contd...

Information and Study Page

Sr. No.	RA, EA, TL 1, TL 2, WA	Date/Year:	1/1/2008	Site:	FXE 1002
Checking Conducted:	Monthly wise	Location of Tools:		Ground Floor	

While on checking on March: Date: 2/3/2008

a. Found the partition screen on the toilet (TL 1) in damage condition (H 16).

b. Found water cooler filter in dirty condition (H 11).

While on checking on June: Date: 2/6/2008

a. Complaints got from the employees that hot water is not in supply in washing area (H 3).

While on checking on September: Date: 1/9/2008

a. Found the drinking water cooler area drainage in overflowing condition (H 12).

Description of Welding set Coding

Item	Description of Cylinder	No of units
RA	Rest Area	One
EA	Eating Area	One
TL	Toilet	TL 1, TL 2
WA	Washing Area	One
DW	Drinking water	One

See the attached document how the Hygiene checklist filled in on STDP documents:

SDTP — 10.5 Hygiene Checklists and Deviations

Document References	
SDTP.CL.007	Doc. Date
Rev. No	1/1/2009
0	
Pages	
1 of 4	

Site: FXE 1002 **Date/Year:** JAN-08

ID. No.	Location of the facility	Jan	Feb	Mar	Apr	May	Jun	July	Aug	Sep	Oct	Nov	Dec
TL.1	Ground Floor	OK	OK	H 16	OK	OK	OK	OK	OK	OK	OK	OK	OK
TL.2	Ground Floor	OK	OK	OK	OK	OK	OK	OK	OK	OK	OK	OK	OK
RA	Ground Floor	OK	OK	OK	OK	OK	OK	OK	OK	OK	OK	OK	OK
EA	Ground Floor	OK	OK	OK	OK	OK	OK	OK	OK	OK	OK	OK	OK
WA	Ground Floor	OK	OK	OK	OK	OK	H 3	OK	OK	OK	OK	OK	OK
DW	Ground Floor	OK	OK	H 11	OK	OK	OK	OK	OK	H 12	OK	OK	OK
Date Inspected:		1-Jan	2-Feb	2-Mar	2-Apr	2-May	2-Jun	1-Jul	2-Aug	2-Sep	1-Oct	1-Nov	2-Dec
Initialed by the Inspector:		Sign Ins.	Sign Ins.	Sign Ins.	Sign Ins.	Sign Ins.	Sign Ins.	Sign Ins.	Sign Ins.	Sign Ins.	Sign Ins.	Sign Ins.	Sign Ins.
Spot Check by Project Manager:		Sign PM	Sign PM	Sign PM	Sign PM	Sign PM	Sign PM	Sign PM	Sign PM	Sign PM	Sign PM	Sign PM	Sign PM
Date Spot Checked:		1-Jan	2-Feb	2-Mar	2-Apr	2-May	2-Jun	1-Jul	2-Aug	2-Sep	1-Oct	1-Nov	2-Dec

Checklist/Legend

Do Not "tick" Write "OK", if faulty use legend number.

SHOWER/WASHING AREA

- H 1 — Area clean, hygienic, good drainage?
- H 2 — Area demarcated screened off?
- H 3 — Hot water, detol, soap available.?

KITCHEN

- H 4 — Floors clean, non slip?
- H 5 — Tables and chairs good condition?
- H 6 — Utensils clean, kitchen clean
- H 7 — Hot water, good ventilation and lighting?

EATING AREAS

- H 8 — Sufficient seating, dust free.?
- H 9 — Clean, hygienic, dustbins available?

DRINKING WATER

- H 10 — Drinking water (Hot and Cold) available?
- H 11 — Are Water cooler filter changing regularly?
- H 12 — Clean drainage, waste removal program.
- H 13 — Dust controlling are in practice?

TOILETS

- H 14 — Sufficient and cleaning regularly?
- H 15 — Separate men/women
- H 16 — No damages, partitions for privacy?
- H 17 — Wash basins, soap, no damages and leaks?
- H 18 — Urinals clean, no blockages and leaks?
- H 19 — General condition good, no strange Water

CHANGE AREA

- H 20 — Sufficient, lighting, ventilation?
- H 21 — No changing in stores, stacking
- H 22 — Area clean, disinfected, sufficient bins?

TUCK-SHOP OUTLET

- H 23 — Well placed, tidy, hygienic?
- H 24 — No stagnant water used?
- H 25 — Pest control system is in practice?

Contd

Deviations and Corrective Actions

Month	Description of Deviations	Corrective Actions	Action by whom	Target Date	Completion Date
JAN					
FEB					
MAR	Found the partition screen on the toilet in damage condition (H 16).	Fix the partition correctly.	Maintenance Supervisor	3/3/2008	3/3/2008
	Found water cooler filter in dirty condition (H 11).	Change the water filter with new one	Maintenance Supervisor	2/3/2008	2/3/2008
APR					
MAY					

Contd....

Deviations and Corrective Actions

Month	Description of Deviations	Corrective Actions	Action by whom	Target Date	Completion Date
JUN	Complaints got from the employees that hot water is not in supply in washing area (H 3).	Repair or replace the water heater.	Maintenance Supervisor	3/6/2008	3/6/2008
JUL					
AUG					
SEP	Found the drinking water cooler area drainage in overflowing condition (H 12).	Needs to immediately clear the waste accumulated on drainage.	Housekeeping Supervisor	2/19/2008	2/19/2008
OCT					

Deviations and Corrective Actions

Month	Description of Deviations	Corrective Actions	Action by whom	Target Date	Completion Date
NOV					
DEC					

Permit-to-Work (PTW) Systems

11

Some of PTW program are defined

a. Hot work permit.
b. Confined space entry permit.
c. Excavation permit.

Markings

For every hazardous jobs like hot work, confined space entry, excavation works, working near to live lines, road works, work on electricity and pressurised systems, etc. Permit to work systems should be adopted for taking additional precautionary measures against the risk and hazards involved in it.

What is the Infrastructure of Permit-to-Work Systems?

It is an operational document prepared by a responsible and competent person after a detailed assessment of hazards and danger involved in the process of preparing precautions and control measures prior starting two activities. This document is a permit or authorization given in writing after a detailed checking, describing that the concern area is safe to enter/can start work/way in which the sequence of work is to be done/the duties and responsibilities of the persons involved in work activities/the safety checks and precaution to be taken while work in progress/and the time when work must stop.

The permit-to-work systems is valid only, if both the permit issuing authority and the responsible person who is in charge of work activity including the concern technician signs the document after understanding and agreeing the purpose and condense of the document. The permit issuing authority is assigned and selected by top management after client approval, for preparation of safe working environment task only for the matter of high risk and hazard involved work activities.

What is the Importance of Permit-to-Work System?

Some work activities and area of work has high accidental and incidental nature because in course of carrying; out work in such areas and activities which can produce high disaster such as fatalities, disabilities, multiple loss times and heavy damages regularly. For eliminating hazards and controlling risk on hazardous and danger zone, the procedure of Permit-to-Work specific-job systems come into part.

The Permit-To-Works system procedure activities mainly in the circumstances of carrying out jobs in confined space entry/hot works/working in the area where gas and vapor presents in atmosphere/ working with or near chemical and compressed gas cylinders/deep excavation/working on charged electric system working near or overhead live services/working with pressurized—pneumatic and hydraulic systems/cold works working in assess restricted areas/road works, etc.

How is Permit-to-Work Systems Defined in SDTP Document ?

See the attached documents given:

SDTP	**11.1 Hot Work Permit**	Document References

		Document References
		SDTP. PTW. 001
		Rev. No — Doc. Date
		0 — 1/1/2009
		Pages
		1 of 2

Project :		Project No. :	
Type of hot work :		PTW requested by :	
Exact location :		Date :	

Section 1

Description of Work :

Duration

Start Date :		Finish Date :		Expected time of closure :	

Section 2

Safety Precautions

Sr. No.	Description	Yes	No
1.	The above location has been inspected		
2.	There are no combustible liquids, vapours, gases or dusts		
3.	All combustible materials have been either been removed or suitably protected against heat and sparks		
4.	Ensure a man standing with an extinguisher/hose reel while the operation is in progress		
5.	The operatives have the nearest fire alarm/telephone line point		
6.	The operatives have been inducted/training completed		
7.	Areas barricaded and signage's placed to avoid unauthorised personnel's		
8.	Strong enough and standard scaffolding provided for working at heights		

PPE's Used: (Daily checks to be made by Foreman and random checks by safety officer)

Helmet	Coverall	Goggles	Ear plugs	Face mask	Welding Shield	Gloves	Safety shoes	Chin strap	Breathing apparatus	Safety glass

Any additional PPE and safety elements required :

Section 3			
Authorization			
With the above conditions checked authorise the undersigned to carry-out the works as stated.			
Name	**Designation**	**Signature**	**Date**
	Authorized permit issuing authority		
	Project Manager		
Client/Consultant representative			
Name	**Designation**	**Signature**	**Date**
	Consultant Resident Engineer		
	Client Resident Engineer		

Section 4			
Acknowledgement			
I understand the hazards and the precautions to be taken whilst executing this work			
Name	**Designation**	**Signature**	**Date**
	Field Engineer/Supervisor		
	Welder/Technician		
	Fire Watcher		

Section 5			
Cancellation			
The work detailed in this permit is completed and the area restored to a safe and orderly condition.			
Name	**Designation**	**Signature**	**Date**
	Authorised permit issuing authority		
	Field Engineer/Supervisor		
	Fire watcher		

SDTP	11.2 Confined Space Entry Permit	Document References	
		SDTP. PTW. 002	
		Rev. No	Doc. Date
		0	1/1/2009
		Pages	
		1 of 2	

Project No. :		Date :	
Time required :		Name and signature of person who requested the permit :	
Exact location :			

Section 1

Description of Work :	

Duration

Start Date :		Finish Date :		Expected time of closure :	

Section 2

Safety Precautions

Sr. No.	Description	Yes	No
1.	The above location has been inspected, tests for gas concentration done		
2.	Is the client required to test for gas concentration?		
3.	Are there combustible/toxic liquids and gases present?		
4.	Are all combustible/toxic liquids and gases evacuated or removed completely?		
5.	The operatives have the nearest fire alarm/telephone line point		
6.	The operatives have been inducted/training completed		
7.	Areas barricaded and signage's placed to avoid unauthorized personnel's		
8.	Is security personnel's provided at the doors for emergency rescue purpose?		

PPE's Used: (Daily checks to be made by Foreman and random checks by safety officer)

Helmet	Coverall	Goggles	Ear plugs	Face mask	Welding Shield	Gloves	Safety shoes	Chin strap	Breathing apparatus	Safety glass

Any additional PPE and safety elements required :	

Section 3			
Authorization			
With the above conditions checked authorize the undersigned to carry-out the works as stated.			
Name	**Designation**	**Signature**	**Date**
	Authorised permit issuing authority		
	Project Manager		
Client/Consultant representative			
Name	**Designation**	**Signature**	**Date**
	Consultant Resident Engineer		
	Client Resident Engineer		

Section 4			
Acknowledgement			
I understand the hazards and the precautions to be taken whilst executing this work			
Name	**Designation**	**Signature**	**Date**
	Field Engineer/Supervisor		
	Technician		
	Security Person		
	Fire Watcher		

Section 5			
Cancellation			
The work detailed in this permit is completed and the area restored to a safe and orderly condition.			
Name	**Designation**	**Signature**	**Date**
	Authorised permit issuing authority		
	Field Engineer/Supervisor		
	Security Person		
	Fire watcher		

SDTP	11.3 Excavation Permit	Document References	
		SDTP. PTW. 003	
		Rev. No	Doc. Date
		0	1/1/2009
		Pages	
		1 of 2	

Project :		Project No. :	
Type of hot work :		Drawing No. :	
Exact location :		Date :	

Section 1

Equipment to be used :	

Maximum Depth :				Maximum width :				Maximum Length :		

Shoring required		Yes		Shoring required		Yes		Shoring required		Yes
		No				No				No

Shoring Details Attached:		Batter Details Attached:		Risk Assessment Attached:		Risk Assessment Sr.. No. :	Method Statement Attached:		M/S Sr. No.
Yes	No	Yes	No	Yes	No		Yes	No	

Duration

Start Date :		Finish Date :		Expected time of closure :	

Section 2

Safety Precautions

Sr. No.	Description	Yes	No
1.	Underground services marked on attached plan's		
2.	Approval received to remove redundant services		
3.	Routes for underground services identified and marked		
4.	Trial pits dug to expose underground services		
5.	Underground services detection survey carried out		
6.	All over head power line services disconnected/removed		
7.	All existing structures/buildings near by, well supported		

PPE's Used: (Daily checks to be made by Foreman and random checks by safety officer)

Helmet	Coverall	Goggles	Ear plugs	Face mask	Welding Shield	Gloves	Safety shoes	Chin strap	Breathing apparatus	Safety glass

Any additional PPE and safety elements required :	

Section 3			
Authorization			
With the above conditions checked authorize the undersigned to carry-out the works as stated.			
Name	**Designation**	**Signature**	**Date**
	Authorised permit issuing authority		
Client/Consultant representative			
Name	**Designation**	**Signature**	**Date**
	Consultant Resident Engineer		
	Client Resident Engineer		

Section 4			
Acknowledgement			
I understand the hazards and the precautions to be taken whilst executing this work			
Name	**Designation**	**Signature**	**Date**
	Field Engineer/Supervisor		

Section 5			
Cancellation			
The work detailed in this permit is completed and the area restored to a safe and orderly condition.			
Name	**Designation**	**Signature**	**Date**
	Authorised permit issuing authority		
	Field Engineer/Supervisor		

Information and Study Page					
Project	ABC Power project	Project No.	FXE 1002	Location	2nd Floor, gas turbine room
Permit req. by:	Mechanical Engineer	Date	5/2/2009	Type of work	welding work

Description of work

Joining permanently the alloy metal pipe line by the means of gas welding set.

Start Date.:	5/2/2009	Finish Date.:	5/2/2009	Expected time of finishing	7 PM	

Description of the legends

The above location has been inspected?	Yes
There are no combustible liquids, vapors, gases or dusts?	Yes
All combustible materials have been either been removed or suitably protected against heat and sparks?	Yes
Ensure a man standing with an extinguisher/Hose reel while the operation is in progress?	Yes
The operatives have the nearest Fire Alarm/Telephone line point?	Yes
The operatives have been inducted/training completed?	Yes
Areas barricaded and signage's placed to avoid unauthorized personnel's?	Yes
Strong enough and standard scaffolding provided for working at heights?	Yes

PPEs used:

Coverall	Welding Helmet	Gloves	Safety shoes	Any additional PPEs and safety elements required	Warning signs, fire extinguishers

Note : The cancellation columns only be sign one hour after the completion of the hot work activity because the fire watcher to be stand there at least one hour in search of any evident of fire in the area of hot work been performed. If the work is not completed on time this permit need to be cancelled and new permit to be issued considering as a new hot work activity.

See the attached document how the hot work permit is filled in:

SDTP	11.1 Hot Work Permit	Document References
		SDTP. PTW. 001

Document References	
SDTP. PTW. 001	
Rev. No	Doc. Date
0	1/1/2009
Pages	
1 of 2	

Project :	ABC Power	Project No. :	FXE 1002
Type of hot work :	Welding work	PTW requested by :	Mechanical
Exact location :	2nd Floor, Gas	Date :	5/2/2009

Section 1

Description of Work :

Joining permanently the alloy metal pipe line by the means of gas welding set.

Duration

Start Date :	5/2/2009	Finish Date :	5/2/2009	Expected time of closure :	7 PM

Section 2

Safety Precautions

Sr. No.	Description		Yes	No
1.	The above location has been inspected		Yes	
2.	There are no combustible liquids, vapors, gases or dusts		Yes	
3.	All combustible materials have been either been removed or suitably protected against heat and sparks		Yes	
4.	Ensure a man standing with an extinguisher/hose reel while the operation is in progress		Yes	
5.	The operatives have the nearest fire alarm/telephone line point		Yes	
6.	The operatives have been inducted/training completed		Yes	
7.	Areas barricaded and signage's placed to avoid unauthorized personnel's		Yes	
8.	Strong enough and standard scaffolding provided for working at heights		NA	

PPEs Used: (Daily checks to be made by Foreman and random checks by safety officer)

Helmet	Coverall	Goggles	Ear plugs	Face mask	Welding shield	Gloves	Safety shoes	Chin strap	Breathing apparatus	Safety glass
	X				X	X	X			

Any additional PPE and safety elements required :	Warning Signs, Fire extinguishers

Section 3			
Authorization			
With the above conditions checked authorize the undersigned to carry-out the works as stated.			
Name	Designation	Signature	Date
Name of the issuing authority	Authorised permit issuing authority	Sign of the issuing authority	Signed date
Name of PM	Project Manager	Sign of PM	Signed date
Client/Consultant representative			
Name	Designation	Signature	Date
Name of the person	Consultant Resident Engineer	Sign	Signed date
Name of the person	Client Resident Engineer	Sign	Signed date

Section 4			
Acknowledgments			
I understand the hazards and the precautions to be taken whilst executing this work			
Name	Designation	Signature	Date
Name of Engineer or Supervisor	Field Engineer/Supervisor	Sign	Signed date
Name of welder	Welder/Technician	Sign	Signed date
Name of Fire watcher	Fire Watcher	Sign	Signed date

Section 5			
Cancellation			
The work detailed in this permit is completed and the area restored to a safe and orderly condition.			
Name	Designation	Signature	Date
Name of the issuing authority	Authorised permit issuing authority	Sign	Signed date
Name of Engineer or Supervisor	Field Engineer/Supervisor	Sign	Signed date
Name of Fire watcher	Fire watcher	Sign	Signed date

Information and Study Page					
Project	ABC Power project	Project No.	FXE 1002	Location	2nd Floor, gas turbine room
Permit req. by:	Electromechanical Engineer	Time req.	7 hours	Type of work	Reinstall work

Description of work

Reinstalling the pressure cage unit which was damaged.

Start Date.:	7/2/2009	Finish Date.:	7/2/2009	Expected time of finishing	2 PM	

Description of the legends

The above location has been inspected, tests for gas concentration done	Yes	
Is the client required to test for gas concentration	No	
Are there combustible/toxic liquids and gases present	No	
Are all combustible/Toxic liquids and gases evacuated or removed completely	Yes	
The operatives have the nearest Fire Alarm/Telephone line point	Yes	
The operatives have been inducted/Training completed	Yes	
Areas barricaded and signage's placed to avoid unauthorized personnel's	Yes	
Is security personnel's provided at the doors for emergency rescue purpose	Yes	

PPEs used:

Coverall	Breath-ing helmet	Gloves	Safety shoes	Any additional PPE's and safety elements required	Warning signs, fire extinguishers

Note : The cancellation columns only be sign after the completion of the confined space work activity, if not completed on the mention time this permit needs to be cancelled and new permit to be issued after detailed checking of legends considering all points as a new activity.

See the attached document how the hot work permit is filled in:

SDTP	11.2 Confined Space Entry Permit	Document References	
		SDTP. PTW. 002	
		Rev. No	Doc. Date
		0	1/1/2009
		Pages	
		1 of 2	

Project No. :	FXE 1002	Date :	7/2/2009
Time required :	7 hours	Name and signature of person who requested the permit :	Electromechanical Engineer
Exact location :	1st Floor, Gas turbine		Sign of EM Engineer

Section 1

Description of Work :

Re-installing the pressure cage unit which was damaged.

Duration

Start Date :	7/3/2009	Finish Date :	7/3/2009	Expected time of closure :	

Section 2

Safety Precautions

Sr. No.	Description	Yes	No
1.	The above location has been inspected, tests for gas concentration done	Yes	No
2.	Is the client required to test for gas concentration?		No
3.	Are there combustible/toxic liquids and gases present?		
4.	Are all combustible/toxic liquids and gases evacuated or removed completely?	Yes	
5.	The operatives have the nearest fire alarm/telephone line point	Yes	
6.	The operatives have been inducted/training completed	Yes	
7.	Areas barricaded and signage's placed to avoid unauthorised personnel's	Yes	
8.	Is security personnel's provided at the doors for emergency rescue purpose?	Yes	

PPEs Used (Daily checks to be made by Foreman and random checks by safety officer)

Helmet	Coverall	Goggles	Ear plugs	Face mask	Welding shield	Gloves	Safety shoes	Chin strap	Breathing apparatus	Safety glass
	X					X	X		X	

Any additional PPE and safety elements required :	Warning Signs, Fire extinguishers

Section 3			
Authorization			
With the above conditions checked authorize the undersigned to carry-out the works as stated.			
Name	**Designation**	**Signature**	**Date**
Name of the issuing authority	Authorised permit issuing authority	Sign of the issuing authority	Signed date
Name of PM	Project Manager	Sign of PM	Signed date
Client/Consultant representative			
Name	**Designation**	**Signature**	**Date**
Name of the person	Consultant Resident Engineer	Sign	Signed date
Name of the person	Client Resident Engineer	Sign	Signed date

Section 4			
Acknowledgements			
I understand the hazards and the precautions to be taken whilst executing this work			
Name	**Designation**	**Signature**	**Date**
Name of Engineer or Supervisor	Field Engineer/Supervisor	Sign	Signed date
Name of Technician	Technician	Sign	Signed date
Name of Main Security Person	Security Person	Sign	Signed date
Name of Tire watcher	Fire Watcher	Sign	Signed date

Section 5			
Cancellation			
The work detailed in this permit is completed and the area restored to a safe and orderly condition.			
Name	**Designation**	**Signature**	**Date**
Name of the issuing authority	Authorized permit issuing authority	Sign	Signed date
Name of Engineer or Supervisor	Field Engineer/Supervisor	Sign	Signed date
Name of Main Security Person	Security Person	Sign	Signed date
Name of Fire watcher	Fire watcher	Sign	Signed date

Information and Study Page					
Project	ABC Power project	Project No.	FXE 1002	Date :	26/1/2009
Excavation for:	Under ground diesel storage tank	Drawing No.	DST 009	Location:	Ground Floor 'c' module
Description of work	Excavating Shovel, tripping trucks, hand shovel, hydraulic jack hammer.				
Excavation Depth:	24 Meters	Excavation Width:	12 Meters	Excava-tion Length:	14 Meters	
Shoring/ stepping/sloping required	Shoring details attached —Yes	Risk assessment attached—Yes		RA No.	HIRA 13.2
	Batter details attached— Yes	Method statement attached—Yes		MS No.	HIRA 13.2
Start Date:	26/1/2009	Finish Date:	27/1/2009	Expected time of finishing	5 PM	

Description of the legends

Underground services marked on attached plan (s)	Yes
Approval received to remove redundant services	Yes
Routes for underground services identified and marked	Yes
Trial pits dug to expose underground services	Yes
Underground services detection survey carried out	Yes
All over head power line services disconnected/removed	NA
All existing structures/buildings near by, well supported	Yes

PPEs used:

Coverall	Goggles	Ear Plugs	Face Mask		
Coverall	Helmet	Gloves	Safety shoes	Any additional PPEs and safety elements required :	Warning Signs, Fire extinguishers, warning lights

Note : The cancellation columns only be sign after the completion of the excavation work activity, if not completed on the mention time this permit can be extended by taking the permission of authorised permit issuing authority. The authority will give permission after a detailed checking of excavation work area, if he satisfied with the condition.

See the attached document how the hot work permit is filled in:

SDTP	11.3 Excavation Permit	Document References

Document References
SDTP. PTW. 003

Rev. No	Doc. Date
0	1/1/2009

Pages
1 of 2

Section 1

Project :	ABC Power project	Project No. :	FXE 1002
Type of hot work :	Under ground diesel storage tank	Drawing No. :	DST 009
Exact location :	Ground Floor 'c' module	Date :	26/1/2009

Equipment to be used : Excavating Shovel, tripping trucks, hand shovel, hydraulic jack hammer.

Maximum Depth :	24 Meters	Maximum width :	12 Meters	Maximum Length :	12 Meters
Shoring required	Yes / No	Shoring required	Yes / No	Shoring required	Yes / No

Shoring Details Attached:		Batter Details Attached:		Risk Assessment Attached:		Risk Assessment Sr. No. :	Method Statement Attached:		M/S Sr. No.
Yes	No	Yes	No	Yes	No	HIRA 13.2	Yes	No	HIRA 13.2

Duration

Start Date :	26/1/2009	Finish Date :	27/1/2009	Expected time of closure :	5 PM

Section 2

Safety Precautions

Sr. No.	Description	Yes	No
1.	Underground services marked on attached plan's	Yes	
2.	Approval received to remove redundant services	Yes	
3.	Routes for underground services identified and marked	Yes	
4.	Trial pits dug to expose underground services	Yes	
5.	Underground services detection survey carried out	Yes	
6.	All over head power line services disconnected/removed	NA	
7.	All existing structures/buildings near by, well supported	Yes	

PPE's Used: (Daily checks to be made by Foreman and random checks by safety officer)

Helmet	Coverall	Goggles	Ear plugs	Face mask	Welding shield	Gloves	Safety shoes	Chin strap	Breathing apparatus	Safety glass
X	X	X	X	X		X	X			

Any additional PPE and safety elements required :	Warning Signs, Fire extinguishers, warning lights

Section 3			
Authorization			
With the above conditions checked authorize the undersigned to carry-out the works as stated.			
Name	**Designation**	**Signature**	**Date**
Name of the issuing authority	Authorised permit issuing authority	Sign	Signed date
Client/Consultant representative			
Name	**Designation**	**Signature**	**Date**
Name of the person	Consultant Resident Engineer	Sign	Signed date
Name of the person	Client Resident Engineer	Sign	Signed date

Section 4			
Acknowledgment			
I understand the hazards and the precautions to be taken whilst executing this work			
Name	**Designation**	**Signature**	**Date**
Name of Engineer or Supervisor	Field Engineer/Supervisor	Sign	Signed date

Section 5			
Cancellation			
The work detailed in this permit is completed and the area restored to a safe and orderly condition.			
Name	**Designation**	**Signature**	**Date**
Name of the issuing authority	Authorized permit issuing authority	Sign	Signed date
Name of Engineer or Supervisor	Field Engineer/Supervisor	Sign	Signed date

Accident/Incident Investigation Report

12

Some of the investigation reports program are defined

a. Preliminary report.

b. Accident investigation report.

Markings

Each and every incident and accident should be investigated and root cause of incident/accident needs to be found out and remedial measures to be taken, filed and its training should be conducted among the work force.

What is the Infrastructure of Accident/Incident Investigation Program?

Each and every accident, incident and near miss in the company premises should be investigated to find out the direct, indirect and root causes of the accident/incident happened, to find out the failures in operating safe working procedure and remedial measures to be taken against the failures happened according to OSHA standard. The bottom line of accident/incident investigation is to stop the recurrence of the accident/incident/Accident/Incident investigation to be done by a group of members assigned by the top management.

This group of members need to visit the accident/incident place and find out the actual reason of accident by clarifying all witnesses and activities and need to note all points regarding accident and to fill investigation report, by considering all relevant details founded and by communicating with each members of the group.

Whom should be Selected as the Members of Accident/Incident Investigation Team?

As per OSHA standards the selection of accident/incident members to be:

a. Project Manager—head of Investigation team.

b. Site Engineer—who implements safe site work activities.

c. Site Supervisor—incharge of workers.

d. Safety personnel—to find any defects in activating safe work procedure.

e. Quality Engineer—to find any defects in quality of plant, machinery and materials used.

What is Preliminary Report and its Difference from Investigation Report?

Preliminary report is the first investigation report and the accident/incident investigation report is the detail investigation report.

Preliminary report condense only mark the details of the injured person, near miss or property damage, concerned supervisor details, first aid center harm and consequence report, the description of accident, incident near miss and property damage happened, the direct/immediate cause findings and remedial measures preferred by safety personnel and the decision of Project Manager to proceed or not to proceed with detailed investigation by signing officially.

But for each accident and incident, Preliminary report needs to be filled in.

How is Accident/Incident Investigation and Preliminary Reports Defined in SDTP Document?

See the attached documents given:

SDTP	12.1 Preliminary Accident Report	Document References	
		SDTP. AI. 001	
		Rev. No	Doc. Date
		0	1/1/2009
		Pages	
		1 of 2	

(This Section to be completed by the Responsible Supervisor/General Foreman)
Note: To be filled in for all incidents

Injured Person/ Damage

Name		Age	
Occupation		Supervisor	

What Happened

Brief Description (Severity) of Incident/Accident/Damage

This Section is to be Filled by Safety Officer

Date			Site				
Time			Department				
Classification			Location				
Actual Injury Outcome	Near Miss		Medical Time Accident		First Aid		Lost Time Accident
Potential Injury Outcome	Medical Time Accident		Lost Time		Fatality		
Equipment Damage	Yes				No		

Primary Root Cause	Immediate Corrective Actions
Hazard Identification/Risk Assessment Completed. YES/NO	If NO do not continue work until HIRA completed
Standard Procedure available. YES/NO	If NO do not continue work until SP issued!
Hazards Identified? YES/NO	If NO do not continue work - Revise Risk Assessment.
Written Daily Safe Task Instruction. YES/NO	If NO—Discipline supervisor/Instruct to complete before work continues.
Risks and precautions communicated? YES/NO	If NO do not continue work until communicated!
Adequate and competent Supervision? YES/NO	If NO do not continue work until appointed!
Direct/ Immediate Causes	**Remedial Actions**
Safety Officer Signature	
Site Supervisor Signature	
Site Nurse Signature	

Instruction:

The Site Safety Officer must ensure this PIR is submitted to the responsible Project Manager and e-mailed faxed to the HSE Manager within 3 hours of the injury/incident occurred.

Project Manager Signature		
Full Incident Investigation Required	Yes	No
HIRA = Hazard Identification/Risk Assessment		
SP = Standard Procedure		
DSTI = Written Daily Safe Task Instruction		

SDTP	12.2 HSE Accident/Incident Investigation Reports	Document References	
		SDTP. AI. 001	
		Rev. No	Doc. Date
		0	1/1/2009
		Pages	
		1 of 2	

Near Miss		Injury		Disease		Damage		Report No.	

Company Name:		Section/Department:	

A Part

Date of Accident/ Incident:		Time of Accident/ Incident:		Date Reported:	
Name of Affected Person:			Present Job Title:		

Time Employed		Experience		Expected Time Off Work					
yrs:	Mt-hs:	yrs:	Mt-hs:	0 days:	0 to 14 days:		> 14 days:		

Accident/Incident Reported to Legal Department/Government:	Yes		No		Date:	

B Part

Machine/process involved/type of work performed:	

Brief Description of Accident/Incident

Supervisor Name :		Supervisor Signature		Date:	

C Part

Injury		Disease		Effect		Damage		Estimated Cost	
Head		Skin		Sprain/ Strain		Buildings		Damages	
Neck		Respiratory		Contusion		Machinery		Medical	
Eye		Skeletal		Laceration		Equipment		Salary	

Back		Muscular		Fracture		Tools		Investigations	
Trunk		Hearing		Burn		Vehicles		Other	
Arm		Neurological		Amputation		Product		Total	
Hand		Psychological		Electric Shock		Safety			
Finger		Endemic		Asphyxiate		Other			
Leg		Other		Unconscious					
Foot				Poisoning					
Toe				Fatal					
Internal				Multiple					

D Part

Direct Causes	Fall/ Slip/Trip		Fire			Dust	
	Moving Machinery		Environment			Noise	
	Motorized Transport		Chemical			Ergonomics	
	Electrical		Fume/vapor			Flying/Falling objects.	

Indirect cause (At Risk behaviours	Operating without authority		Horse play	
	Working on moving machinery		Not using P.P.E	
	Taking up unsafe position		Other	
	Not using safety device			

Unsafe Condition (At Risk Condition)	Inadequate guarding		Poor layout	
	Unnatural environment		Housekeeping	
	Poor lighting		Other	
	Ventilation			

E Part

Root Cause Of Accident/Incident

Hardware			Error Enforcing Condition	
Design			Housekeeping	

Organization			Maintenance Management	
Incompatible Goals			Training	
Procedures			Communication	

Conclusions:

Recommendations to Prevent Reoccurrence:

Prevention—Control Steps	Person Responsible	Action Date	Follow-up completed
Review equipment supply and design			
Review Maintenance process			
Review work procedures			
Employee—counsel/warn			
Housekeeping controls			
Refer to Group meetings			
Define responsibilities			
Training			
Other			

F part		
Head of Investigator	**HSE Official**	**General Manager**
Name:	Name:	Name:
Sign:	Sign:	Sign:
Date:	Date:	Date:

Legal Authority/ Client Comments:

G Part		
Detailed Accident Investigation Required	Yes	No

General Manager					
Name:		Sign:		Date:	

Why Questioning Technique			
Company Name		Time	
Location		Notify/Work order Number	
Incident investigation number			
Date			

Participants			
Safety officer		Project Manager	
Medical Officer		General Foreman	
Injured (if available)		Responsible Engineer	
Direct Supervisor		Jr. Engineer	
Line Supervisor		Other	

1. Incident: What was observed—before, during and after the problem occurred. Attach all necessary prints, recordings, data sheets, photos, etc.

2. Brainstorm: Possible Immediate Cause

3. Most Probable Cause

4. Ask Why of most Probable Cause (ticked item in 3. above)

Why:	
Why:	
Why:	
Why:	
Why:	
Why:	

Why:	

5. Possible Remedy

6. Feedback to Review Panel

Yes		No	

PARTICIPANTS NAME	DESIGNATION	SIGNATURE

Case Study			
Company Name:	ABC Company	Company Branch No	FXE1002

Injured Person Details:

Name of person:	Ravi Yadav		Trade/Occupation		Mason	
Age:	32 years	Total Experience:		6 years	Experience in current Company	3 years

Details of concern Supervisor:

Supervisor Name	Hassan Mohammad	Supervisor direct reporting Engineer's Name:	Tony Martin

Detail of Accident Happen:

While shifting hollow blocks for masonry work on 1st floor area by means of hand on a scaffold, a stacked block piece on the platform fall down to his helmet. Then helmet slip down and the edge of the block touches on his head and cut him.

Date of Accident:	28/08/2007	Time of Accident:	4:00 PM	Location:	1st Floor, Electric Room

Details of harm and consequence report from First Aid center:

Harm:	Deep Head Injury	Consequence:	Deep Cut, Send to hospital for Stitching

Details found in the course of Safety Personnel's First Investigation :

Direct Causes	Remedial Actions
1. Struck by falling block piece.	Nets or proper garding to be erected on scaffolds.
2. Incomplete Scaffold which is red tagged.	Allow only to work on green tagged scaffolds which was signed byscaffold inspector.

So needs to fill the Preliminary report according to SDTP Preliminary report document
See the attached Document sheet:

SDTP	12.1 Preliminary Accident Report	Document References	
		SDTP. AI. 001	

Rev. No	Doc. Date
0	1/1/2009

Pages
1 of 2

(This section to be completed by the Responsible Supervisor/General Foreman)
Note: To be filled in for all incidents

Injured Person/ Damage

Name	Ravi Yadav	Age	32 years
Occupation	Mason	Supervisor	Hassan Mohammad

What Happened

While shifting hollow blocks for masonry work on 1st floor area by
means of hand on a scaffold, a stacked block piece on the platform fall
down to his helmet. Then helmet slip down and the edge of the block
touches on his head and cut him.

Brief Description (Severity) of Incident/Accident/Damage

Deep head cut, send to hospital for stitching.

This Section is to be Filled by Safety Officer

Date	28/08/2007	Site	FXE1002
Time	4:00 PM	Department	Masonry
Classification	Mason	Location	1st Floor, Electric Room

Actual Injury Outcome	Near Miss	Medical Time Accident		First Aid	Lost Time Accident
Potential Injury Outcome	Medical Time Accident		Lost Time		Fatality

Equipment Damage	Yes	No

Primary Root Cause	Immediate Corrective Actions
Hazard Identification/Risk Assessment Complected. YES/NO	If NO do not continue work until HIRA completed
Standard Procedure available. YES/NO	If NO do not continue work until SP issued!
Hazards Identified? YES/NO	If NO do not continue work—Revise Risk Assessment.
Written Daily Safe Task Instruction. YES/NO	If NO—Discipline supervisor/Instruct to complete before work continues.
Risks and precautions communicated? YES/NO	If NO do not continue work until communicated!
Adequate and competent Supervision? YES/NO	If NO do not continue work until appointed!
Direct/ Immediate Causes	**Remedial Actions**
1. Struck by falling block piece.	Nets or proper garding to be erected on scaffolds.
2. Incomplete Scaffold which is red tagged.	Allow only to work on green tagged scaffolds which was signed by scaffold inspector.
Safety Officer Signature	*Sign of safety officer*
Site Supervisor Signature	*Sign of site supervisor*
Site Nurse Signature	*Sign of site Nurse*

Instruction:

The Site Safety Officer must ensure this PIR is submitted to the responsible Project Manager and e-mailed faxed to the HSE Manager within 3 hours of the injury/ incident occurred.

Project Manager Signature		
Full Incident Investigation Required	Yes	No
HIRA = Hazard Identification/Risk Assessment		
SP = Standard Procedure		
DSTI = Written Daily Safe Task Instruction		

Case Study				
Company Name:	ABC Company	Company Branch No		FXE 1002
Injured Person Details:				
Name of person:	Ravi Yadav	Trade/Occupation		Mason
Age:	32 years	Total Experience:	6 years	Experience in Current Company	3 years

Details of concern Supervisor

Supervisor Name	Hassan Mohammad	Supervisor direct reporting Engineer's Name:		Tony Martin

Detail of Accident Happen:
While shifting hollow blocks for masonry work on 1st floor area by means of hand on a scaffold, a stacked block piece on the platform fall down to his helmet. Then helmet slip down and the edge of the block touches on his head and cut him.

Date of Accident:	28/08/2007	Time of Accident:	4:00 PM	Location:	1st Floor, Electric Room

Details of harm and consequence report from First-aid center:

Harm:	Deep Head Injury	Consequence:	Deep Cut, Send to hospital for stitching

Details found in the course of Safety Personnel's First Investigation :

Direct Causes

1. Struck by falling block piece.

2. Incomplete Scaffold which is red tagged.

Details found in the course of Accident/Incident Investigation:

1. Injured person was standing under the suspended load.

2. Supervisor is not on the work area while accident happens.

3. Daily written safe task instruction or Toolbox talk is not conducted by supervisor.

4. Found that the scaffold was ridged due to the lack of space and work planned on incomplete scaffolds.

5. Found that blocks are stacked on the top platform which was not properly guarded (lack of toe boards).

6. Found that workers were lifting blocks by means of hand which was neglecting safe work procedure.

7. Found evident of failure by supervisor in obeying instruction given by site Engineer to implement safe work activity.

So needs to fill the Accident Investigation report according to SDTP document:
See the attached Document sheet:

SDTP	**12.2 HSE Accident/Incident Investigation Reports**	Document References	
		SDTP. AI. 001	
		Rev. No	Doc. Date
		0	1/1/2009
		Pages	
		1 of 2	

Near Miss		Injury	X	Dise-ase		Damage		Report No.	

Company Name:	ABC Company	Section/Department:	Masonry

A Part

Date of Accident/Incident:	28/08/2007	Time of Accident/Incident:	4:00 PM	Date Reported:	28/08/2007
Name of Affected Person:		Ravi Yadav		Present Job Title:	Mason

Time Employed				Experience				Expected Time Off Work						
yrs:	6	Mths:	0	yrs:	3	Mths:	0	0 days:	0 to 14 days:	2 days	> 14 days:		Perma-nent:	

Accident/Incident Reported to Legal Department/Government:	Yes		No	X	Date:	

B Part

Machine/process involved/type of work performed:	Block work
Brief Description of Accident/Incident	While shifting hollow blocks for masonry work on

1st floor area by means of hand on a scaffold, a stacked block piece on the platform fall down to his helmet. Then helmet slip down and the edge of the block touches on his head and cut him.

Supervisor Name :	Hassan Mohammad	Supervisor Signature :	Sign of supervisor	Date:	Date Signed

C Part

Injury		Disease	Effect	Damage	Estimated Cost	
Head	X	Skin	Sprain/ Strain	Buildings	Dama-ges	0
Neck		Respiratory	Contusion	Machinery	Medical	

Eye		Skeletal		Laceration/cut	X	Equipment		Salary	
Back		Muscular		Fracture		Tools		Investigations	
Trunk		Hearing		Burn		Vehicles		Other	
Arm		Neurological		Amputation		Product		Total	
Hand		Psychological		Electric Shock		Safety			
Finger		Endemic		Asphyxiate		Other			
Leg		Other		Unconscious					
Foot				Poisoning					
Toe				Fatal					
Internal				Multiple					

D Part								
Direct Causes	Fall/Slip/Trip		Fire		Dust			
	Moving Machinery		Environment		Noise			
	Motorised Transport		Chemical		Ergonomics			
	Electrical		Fume/Vapor		Flying/Falling objects			

Indirect cause (At Risk behaviours	Operating without authority		Horse play	
	Working on moving machinery		Not using PPE	
	Taking up unsafe position	X	Other	
	Not using safety device			

Unsafe Condition (At Risk Condition)	Inadequate guarding	X	Poor layout	
	Unnatural environment		Housekeeping	
	Poor lighting		Other	
	Ventilation			

E Part				
Root Cause Of Accident/Incident				
Hardware			Error Enforcing Condition	
Design			Housekeeping	
Organization			Maintenance Management	
Incompatible Goals			Training	
Procedures	X		Communication	X

Conclusions:

1. Inadequate supervision.
2. Failure to follow instruction.

Recommendations to Prevent Reoccurrence:

1. Allow workers only in a green tag scaffolds which was signed by scaffold inspector.
2. Needs proper guarding.
3. Needs adequate supervision

Prevention—control Steps	Person Responsible	Action Date	Follow-up completed
Review equipment supply and design			
Review Maintenance process			
Review work procedures	Sr. Site Engineer	29/08/2007	
Employee—counsel/warn			
Housekeeping controls			
Refer to Group meetings	Safety Officer	6/9/2007	
Define responsibilities	Site Engineer	29/08/2007	
Training			
Other			

F part					
Head of Investigator		**HSE Official**		**General Manager**	
Name:	Name of Project Manager	Name:	Name of safety officer	Name:	Name of General Manager
Sign:	Sign of Project Manager	Sign:	Sign of safety officer	Sign:	Sign of General Manager
Date:	Date Signed	Date:	Date Signed	Date:	Date Signed

Legal Authority/Client Comments:

G Part					
Detailed Accident Investigation Required		Yes		No	
General Manager					
Name:	Name of General Manager	Sign:	Sign of General Manager	Date:	Date Signed
Why Questioning Technique					
Company Name	ABC Company		Time	4 PM	
Location	1st Floor, Electric room		Notify/Work order Number		
Incident investigation number	AI 001				
Date	28/7/2008				
Participants					
Safety officer	X		Project Manager	X	
Medical Officer			General Foreman	X	
Injured (if available)			Responsible Engineer	X	
Direct Supervisor	X		Jr. Engineer	X	
Line Supervisor	X		Other		

1. Incident: What was Observed—before, during and after the problem occurred. Attach all necessary prints, recordings, data sheets, photos, etc.

- Working on unsafe scaffold which was red tagged.
- Workers were shifting blocks by means of hand.
- Blocks were placed on the top of the scaffold without proper guarding.
- Person on the top level accidentally hit his leg on the block piece and block piece fall down to
- The injured persons head, who was working below.
- Worker was wearing helmet.

2. Brainstorm: Possible Immediate Cause

- Working on not tagged and improper scaffold.
- Unsafe shifting and placing of materials.
- Lack of supervision.
- Congested Area

3. Most Probable Cause

- Unsafe placing of materials.

4. Ask Why of Most Probable Cause (ticked item in 3. above)

Why:	• Insufficient space to place materials.

Why:	• Congested area.
Why:	• Stacking more materials.
Why:	• Lack of supervision.
Why:	
Why:	
Why:	

5. Possible Remedy

- Needs strict supervision.
- Only use tagged and proper scaffolds.
- Specific job training.

6. Feedback To Review Panel

Yes		No	
PARTICIPANTS NAME	**DESIGNATION**	**SIGNATURE**	
Name of Safety officer	Safety officer	Date Signed	
Name of Direct Supervisor	Supervisor	Date Signed	
Name of General Foreman	General Foreman	Date Signed	
Name of Line Supervisor	Jr. Supervisor	Date Signed	
Name of Site Engineer	Site Engineer	Date Signed	
Name of Quality Engineer	Quality Control Engineer	Date Signed	
Name of Project Manager	Project Manager	Date Signed	

Hazard Identification, Risk Assessment and Safe Work Procedure

13

Some of HIRA program are defined
a. Electric power tools.
b. Excavation.
c. Lifting operation with mobile crane.
d. Gas welding.
e. Working on scaffolds.

Markings
For each and every work activities hazard identification, risk assessment and safe work procedure to be made, approved and procedures to be trained to the low management and activated according to OSHA standard. If there any deviation found it should be revised, reviewed and new HIRAS to be re-made according to the hazards and risks found further. The low management to restrict unsafe act and unsafe conditions. For this disciplinary actions Safety violation Slip and Stop work order Notice come to part.

What is the infrastructure of Hazard Identification, Risk Assessment and Safe Work Procedure?

It is an operational document activated for each and every work activities on company premises prior designing work methods to identify hazards, to analyze its harm and consequence and to find out it control and additional control remedies for bringing down the high exposure limit of hazards to low exposure limit. The hazard identification, risk assessment and safe work procedure is designed to reduce to rate the frequency of exposure with consequence to a safer workable level (i.e. high or medium exposure level to low level).

These documents should be send for client and legal authority's approval (if necessary) and to be filled in till the end of the project. If any hazard identification further find on committing these activities, then the document needs to revised and new control measures to be added of safe working environment as per OSHA standards.

In this document, the hazard identification of which work activity, the equipment and tools to be used, the activity task in one by one step, hazard identified in each step, harm and consequences can happen by these hazards, the exposure limit ratings, the control and additional control measures, the rate of exposure limit bring down to which level and the name and signature of the persons who conduct the done should be clearly marked.

What is the importance of Hazard Identification, Risk Assessment and Safe Work Procedure?

Every activity on any company premises co-link with high risk and hazard nature and so planning and designing is crucial to avoid accident and incidents because of using of high sophisticated equipment, machineries, tools and different atmospheric conditions. The hazard identification, risk assessment and safe work procedure document is a tool to bring down the actual exposure limit to low exposure limit by finding the precaution of the risks and hazards that identified in each and every activity.

According to the control and additional control remedies of each activity, the top management needs to conduct safety training and safety talks prior starting work to the line and low management for good work practice.

How is Hazard Identification, Risk Assessment and Safe Work Procedure done?

See the attached documents for some generic HIRA's for chosen activities:

Document References		
SDTP.RA 001		
Rev.No		Doc. Date
0		1/1/2009
	Pages	
	1 of 2	

SDTP 13 Hazard Identification and Risk Assessment

Project : ABC power
Client: XYZ power
Date: Date Made

HIRA No.

Task / Activity Description:

Risk Assessment:

Tools / Equipment in use:

Sr. No.	Task in one-by-one step	Hazard Identification in each step	Harm/Consequences (without controls)	Risk Assessment				Who could be effected (mark on right column)	With Controls		A	B	C	
				A Fq.	B Cq.	C Risk Store	Risk Rating		Current Control	Additional Control	Fq.	Cq.	Risk Store	Risk Store

Who could be effected (mark on right column):
- Client
- Project Manager
- Control Manager
- Supervisor
- Employees
- Operator
- Banksman
- Contractor
- Public
- Welder
- Fabricator
- Cutters
- Engineers
- Site Agent
- Foreman
- Charge Hand

Contd...

Consultant	Safety Officer	Store Keeper	Painter	Cleaner	Steel Fitter	Mason	Carpenter	Helper

PPE Needed: Coverall, Goggles, Hard Hat, Vest, Safety Shoes.

Additional PPE Needed: Safety Guards, Earplug, Breath Mask

Average Rounded:

Frequency of Exposure (Fq.)

20	Continuously (daily)
15	Frequently (weekly)
10	From time to time (monthly)
5	Sometimes/occasional (yearly)
1	Realy (less than once per year)

Consequence (Cq.)

20	Disaster	Several Detail
15	Very serious	One Death
10	Serious	Disability/Permanent Injury
5	Signification	Lost Time Injury
1	Minimal	Minor Injury

Risk Rating

Low (L)	1 To 30
Medium (M)	31 To 150
Medium High (H)	151 To 200
High (H)	201 and above

Contd...

Risk Rating= Frequency X consequence

Frequency of Exposure (F)	CONSEQUENCE (Cq)				
	20	15	10	5	1
20	400 - High	300 - High	200 - Medium	100 - Medium	20- Low
15	300 - High	325 - High	150 - Medium	75 - Medium	15 - Low
10	200 - Medium	150 - Medium	100 - Medium	50 - Medium	10 - Low
5	100 - Medium	75 - Medium	50 - Medium	25 - Low	5 - Low
1	20 - Low	15 - Low	10 - Low	5 - Low	1 - Low

Risk Assessment Committee Members

Name	Designation	Signature	Name	Designation	Signature

SDTP — 13.1 Hazard Identification and Risk Assessment

Document References		
SDTP.RA 001		
Rev.No 0	Doc. Date	1/1/2009
Pages		
1 of 2		

HIRA No. Project: ABC power

Task / Activity Description: Hand held electrical tools are to be used for the purpose they have been designed for, Drilling, cutting grinding. — Client: XYZ power — Date: Date Made

Risk Assessment: Use of Electrical Power Tools Drills, Cutters and grinders

Tools / Equipment in use: Drill bits: High speed steel and masonry. Discs: Metal and masonry. Tools: Chuck keys, spanners and retaining wash

Sr. No.	Task in one-by-one step	Hazard Identification in each step	Harm/ consequences (without controls)	A Fq.	B Cq.	C Risk Store	Risk Rating	Who could be affected (mark on right column)	Current Control	Additional Control	A Fq.	B Cq.	C Risk Store	Risk Rating
1	Energising (connecting to power source)	Electrical Shock	Burns, injury and electro-cution	20	10	200	M	Client; Project Manager; Control Manager; Supervisor (y)	Instruction and triaining given about the use of tool circuit breakers to be fitted with all circuits. Only double insulated tools to be used.	Operatives to be tested for there for the competence in use of tool. Ensure supply voltages conforms operating voltage. All default tools to be red tagged as "Tools Damaged" and not to be used.	20	1	20	L
2	Connecting extension leads	Electrical shock	Burns, injury and electro-cution	20	10	200	M	Employees (y); Operator (y); Banks man; Contractor	Use only color coded plugs and sockets. Regular checks for the polarity and serviceability	Ensure that 420 V, 220 V, 110 V supply line include func-tioning with circuit breakers. Protect extension leads from vehicle wheel damage.	20	1	20	L
3	Drilling	Drill bit grab	Muscular skeletal injuries	20	10	200	M	Public; Welder (y); Fabricator (y)	Hold drill firmly with one hand on drill grip and	User to be in a good and comfor-table distance in order to absorb	20	1	20	L

Contd...

No.	Activity	Hazard	Consequence	Average Rounded	Fq	Cq	Cq Rating	Persons Exposed	Additional PPE Needed / Control Measures	Average Rounded	P	Value	Risk Rating
4	Cutting (circular saw)	Flying particles and rotating blade	Eye and body injury	20	5	100	M	Cutters, Engineers, Site Agent, Foreman, Charge Hand	Permit-to-work systems to be followed. Guards to fitted and eye protection to be worn. Other hand on auxiliary grip arm. Standard PPE to be worn. Trigger switch lock not to be used. Sudden shock. Trigger switch is to be unlocked.	20	1	20	L
5	Cutting and grinding (grinders)	Flying particles and rotating blade. Sparks generated. Disc disintegration	Eye and body injury. Start a fire	20	5	200	M	Consultant, Safety Officer, Storekeeper, Painter	Guards to fitted and eye protection to be worn. Keep work area clean of combustible materials. Place suitable fire extinguishers. Only use good quality discs. Avoid disk in contact with the surroundings when not cutting or grinding.	20	1	20	L
6	Drilling, cutting and grinding	Dust generation	Lung damage and eye injury	20	5	100	M	Cleaner, Steel Fitter, Mason, Carpenter, Helper	Operatives to be issued with breathing mask and eye protection. Use of PPE to be enforced.	20	1	20	L
		Noise generation	Loss of hearing	20	5	100	M		Operative to be issued ear plug. Noise level to check at regular intervals.	20	1	20	L
PPE Needed:	Coverall, Goggles, Hard hat, Vest, safety shoes		Average Rounded	20	10	200		Additional PPE Needed	Safety Guards, Ear plug Breath marks	Average Rounded	20	1	20	L

Frequency of Exposure (Fq.)

| 20 | Continuously (Daily) |
| 15 | Frequently (weekly) |

Consequence (Cq)

Disaster	Very serious	Several Detail	20
	One Death	15	

Risk Rating

| Low (L) | 1 To 30 |
| Medium (M) | 31 To 150 |

Contd

CONSEQUENCE (Cq.)

Frequency of Exposure (F)			Serious	Disability/Permanent Injury		151 To 200	Medium High (MH)
			Signification	Lost Time Injury		201 and above	High (H)
			Minimal	Minor Injury			
10	From time to time (monthly)			10	10	5	1
5	Sometimes/occasional (yearly)		20	15	10	5	1
1	Relay (less than once per year)						

Risk Rating = Frequency × consequence

Frequency of Exposure (F)	Cq → 20	15	10	5	1
20	400 - High	300 - High	200 - Medium	100 - Medium	20 - Low
15	300 - High	325 - High	150 - Medium	75 - Medium	15 - Low
10	200 - Medium	150 - Medium	100 - Medium	50 - Medium	10 - Low
5	100 - Medium	75 - Medium	50 - Medium	25 - Low	5 - Low
1	20 - Low	15 - Low	10 - Low	5 - Low	1 - Low

Risk Assessment Committee Members

Name	Designation	Signature	Name	Designation	Signature

SDTP | 13.2 Hazard Identification and Risk Assessment

Document References	
SDTP:RA 002	
Rev.No 0	Doc. Date 1/1/2009
Pages	1 of 3

Task/Activity Description: Excavating by means of Excavator, Excavation safety, etc.

HIRA-002 — Excavation Operations

Client: ABC power / XYZ power
Date: Date Made

Tools/Equipment in use: Shovel (manual/mechanical), crane, barricading, equipment, traffic cones safety equipment and gadgets as required.

Sr. No.	Task in one-by-one step	Hazard Identified in each step	Harm/Consequences (without controls)	A Fq.	B Cq.	C Risk Store	Risk Rating	Who could be effected (mark on right column)	Current Control	Additional Control	A Fq.	B Cq.	C Risk Store	Risk Rating
1	Existing under-ground Condition	Touching any existing under ground services	Injury, Fatality, Disability, Damage to equip-ment/tools, loss of services supply to surrounding areas, Electric shock, Short circuit, Pressure hammer.	20	15	200	H	Client; Project Manager; Contact Manager; Supervisor (y)	Permission and area clearance certificate from municipality/client and local authorities needs prior designing excavation work.	Check with under-ground service detector prior starting excavating work. Approved drawings of the existing round services of site premises also be considered before designing excavation.	20	1	20	L
		Toxic or inflammable contaminants	Asphyxiation, Lung disease, unconscious-ness, shock, irritating, fire, explo-sion, disabi-lity and fatality.	20	20	400		Employees (y); Operator (y); Banksman (y); Contractor (y)	Collect data's of ancient work activities that followed on the site premises to know the hazardous chemicals or materials used. Workers need to wear breathing apparatus compulsory in doubtful areas.	Appropriate fire extinguishers to be placed. Always check with toxic gas detector in every consecutive 2 meters depth of digging by competent persons. Unauthorised entry of persons should be prohibited.	20	1	20	L

Contd

#	Activity	Hazard	Effects	L	C	RPN		Persons Exposed	Control Measures	Further Controls	L	C	RPN	
		Under ground water flow	Drowning, fatality, Unconsciousness, Shock	20	15	300	H	Public (y), Welder (y), Fabricator (y), Cutter (y)	Main contractor must collect all relevant details about the atmospheric and underground geographic condition to recognize the possible natural hazards.	De-watering to be done if there is presence of water using de-watering pumps up to the safe level.	20	1	20	L
2	Excavating	Loose soil Condition	Side wall collapse, Fatality, Shock, Disability, Injury, Unconsciousness.	20	15	300	H	Engineers (y), Site Agent (y), Foreman (y), Charge Hand (y)	Permit-to-work system should be followed. Shoring, stepping, slopping (which is applicable) shall be enforced. Restrict workers to dig manually in excavation of lose soil type.	Stop guards to be provided 1 meter around the excavation to prevent excavating equipment and batter running through edges. Special induction training to be conducted prior To start work about the hazards involved in excavation activities.	20	1	20	L
		Equipment abuse.	Accidents, Fatality, damage to property, Injury, Lost time Injury.	20	15	300	H	Cleaner, Steel Fitter, Mason, Carpenter, Helper (y)	Only trained, certified and experienced operators should allowed to operate plant. Strict instruction to be given to operatives about site traffic rules. MML display units to be work in order.	Operator should be inducted by the Site Engineer about the issued to Operator. Both must sign to acknowledge they understand the full requirements. The Operator will monitored by the Engineer or foreman during the full excavation process	20	1	20	L
		Truck failure while collecting digging batter	Overturning of Truck, Accidents while Truck stopping to close to	20	15	300	H	Consultant, Safety Officer, Storekeeper, Painter	Only trained, certified and experienced operators should allowed to operate	MML display units to be work in order. All Trucks to be oadedevenly to prevent overturning when tip	20	1	20	L

Contd....

No.	Activity	Hazard	Consequence	S	P	Risk	Level	Control Measures	S	P	Risk	Level
			Excavator, break failure Fatality, damage to property, Injury, Lost time Injury.	20	15	300	H	trucks. Always provide bankman with flag, whistle and reflective vest to control the truck movements on site. ping. Trucks to be parked level during loading. Use stop blocks for trucks while parkingnear excavator.	20	1	20	L
		Overhead electric wires while crane or shove is on work	Damage to property, and or injury/disability and death to personnel	20	15	300	H	Permit system requires to follow while working close to electric wires. Needs Supervision in all times basis while work in progress. Provide "Goal post" system is protect crane boom or shovel boom, while touching with charged electric line ahead.	20	1	20	L
3	Excavation Protection	Depth of Excavation	Man/materials/tools to fall in, damage to property, disability, lost time injury, fracture, Injury, Fatality	20	15	300	H	Solid barricading to provide around the excavation (1.5 m away from edges). Toe boards also to be provided (at least 900 mm height) on the bottom side barriers to prevent loose materials and tools to fall inside the excavation. Daily checklist to be updated and filled. Warning signage's "Deep excavation work in progress, keep away", "Danger of falling, keep away" to be place on the barriers. Flashing yellow lights to be provided on nights for easy identification of excavation. All small excavations (less than 1.2 m depth) to be well covered with solid materials at the end of each shift.	20	1	20	L
4	Excavation Access	Unsafe Access inside excavation	Slipping/Falling, disability, lost time injury, fracture, Injury, Fatality	20	15	300	H	Access ladders should fixed 1 m above zero level and to be secured firmly at both ends. Keep always ladder and stairways free A special Toolbox talk and training about the excavation hazards involved, to be conducted prior to start work on excavation by supervisor. Safety	20	1	20	L

Contd

No.	Activity	Hazard	Effect	L	S	P	Risk	Control Measures	PPE	L	S	P	Risk
								from grease, oil and dirt. Only allow authorized personnel's to go inside the excavation by appointing security. All standard PPE to be worn.	harness attached with life line or fall arrester to be used while stepping in and climbing out of deep excavations by workers. Ensure good illumination facilities especially on assess area				
		Protruding parts, sharp edges	Lacerations, deep cuts, shock, unconsciousnes	20	15	300	H	Regular inspection needs to carry-out for sharp edges and nails, if anything found and cannot be removed, capping system to be adopted. Ensure good housekeeping at all times.	Hard gloves, coverall, hard hat and safety shoes to be worn by the workers while working on excavation	20	1	20	L
5	Excavation on Roads	Traffic jam, vehicle collision	Damage to property, and or injury/disability and death to personnel.	20	15	300	H	Should be barricade or market with safety cones and warning tapes. Flashing yellow lights to be provided on nights for easy identification of excavation.	The personnel's working in these premises should always wear yellow/orange color jackets or reflective vest. These excavations should be backfilled as soon as possible.	20	1	20	L

Contd...

PPE Needed:	Coverall, Goggles, Hard hat, Vest, safety shoes	Average Rounded	20	20	1	20	L
	Additional PPE Needed	Safety Guards, Earplug, Breath Mask				M	100	20

Frequency of Exposure (Fq)

20	Continuously (daily)
15	Frequently (weekly)
10	From time to time (monthly)
5	Sometimes/occasional (yearly)
1	Realy (less than once per year)

Consequence (Cq)

20	Disaster	Several Detail
15	Very serious	One Death
10	Serious	Disability / Permanent Injury
5	Signification	Lost Time Injury
1	Minimal	Minor Injury

Risk Rating

1 To 30	Low (L)
31 To 150	Medium (M)
151 To 200	Medium High (H)
201 and above	High (H)

Risk Rating = Frequency × cosequence

Frequency of Exposure (F)	CONSEQUENCE (Cq)				
	20	15	10	5	1
20	400 - High	300 - High	200 - Medium	100 - Medium	20 - Low
15	300 - High	325 - High	150 - Medium	75 - Medium	15 - Low
10	200 - Medium	150 - Medium	100 - Medium	50 - Medium	10 - Low
5	100 - Medium	75 - Medium	50 - Medium	25 - Low	5 - Low
1	20 - Low	15 - Low	10 - Low	5 - Low	1 - Low

Risk Assessment Committee Members

Name	Designation	Signature	Name	Designation	Signature

SDTP — 13.3 Hazard Identification and Risk Assessment

Document References		
SDTP.RA 002		
Rev.No	Doc. Date	
0	1/1/2009	
	Pages	
	1 of 3	

HIRA NO.	HIRA-003	Risk Assessment:	Mobile crane Lifting operation	Project:	ABC power
Task / Activity	Mobile (Crawler) crane operations	Tools/Equipment in use:	Sling: Nylon, wire rope and chains shackles spreader bars, beams, man cages, etc.	Client:	XYZ power
Description:				Date:	Date Made

Sr. No.	Task in one-by-one step	Hazard Identification in each step	Harm/consequences (without controls)	A Fq.	B Cq.	C Risk Store	Risk Rating	Who could be affected (mark on right column)		With Controls – Current Control	With Controls – Additional Control	A Fq.	B Cq.	C Risk Store	Risk Rating
1	Mobilizing	Moving Equipment.	Collision, crashing, fatality, damage to property.	20	20	400	H	Client	y	Toolbox talk is required to emphasize tti dangers of moving plant before starting the work. MML systems in the crane are to be work in order.	Restrict all non-work related equipment/vehicle/persons entering in crane work area. Banksmans with flags to be provided with the mobile crane for giving standard hand signals.	20	1	20	L
								Project Manager	y						
								Contact Manager	y						
								Supervisor	y						
		Equipment abuse.	Accidents, Fatality, damage to property, Injury, Lost time Injury.	20	20	400	H	Employees	y	Only trained, certified and experienced operators should allowed to operate plan Strict instruction to be given about site traffic rules.	Supervisors and management persons should familiar with crane's application and operative capabilities. Communication system to be sustain by means of walky-talky service.	20	1	20	L
								Operator	y						
								Banksman	y						
								Contractor	y						
		Fire	Burns, fatalities, damage of property Injury,	5	20	100	M	Public	y	Mobile crane checklist to be updated on daily basis.	Charged fire extinguisher is to be carried in Mobile crane at all times.	5	1	5	L
								Welder							
								Fabricator							
								Cutters							

Contd...

No.	Activity	Hazard	Consequence					Persons exposed	Control Measures					L
2	Lifting	Exceeding Crane capacity	Damage to property, and or injury/disability and death to personnel, public, etc.	20	20	400	H	Engineers (y), Site Agent (y), Foreman (y), Charge Hand (y)	Load movement devises always work/stability. No lifts to done without a banksman. Lifting to be done only by experienced operator.	Banksman hand signal to be well-understood by operator and banksman should always wear orange vest for visibility. For long distance communication, provide whistle or walky-talky.	20	1	20	L
		Load slipping/falling/dropping crushing limbs, etc.	Damage to property, and or injury/disability and death to personnel, public, etc.	20	10	200	M	Consultant (y), Safety Officer (y), Storekeeper, Painter	Toolbox talk is needed to given according to the principals of slinging and rigging. Lot should always kept steady without slldln; to either sides.	Special instruction to be given about the crar ls design only for lifting not for lifting. Standard PPE and hard glove to be worn all times at operation. Tag lines to be erected or loads for trie control of suspended load.	20	1	20	L
		Other Cranes operating within crane/boom radius.	Damage to property, and or injury/disability and death to personnel.	10	20	200	M	Cleaner (y), Steel Fitter, Mason, Carpenter, Helper (y)	Banksman and operator should be madi aware of other cranes within sticking distance all times. Walky-talkies to be provided to both cranes operatives and banksman for communication purpose.	Anti collision sensors to be fitted on each cranes where necessary.	10	1	10	L
		Tandem Lifts	Damage to property, and or injury/disability and death to personnel.	10	20	200	M	Crane operator (y)	Use similar weight capacity cranes. Requirements for both cranes before lifting.	Special Toolbox talk should be done among the employees about the slinging and riggings planed before prior to working two cranes in a Tandem	10	1	10	L
		Working under hook path	Damage to property, and or injury/	20	20	400	H		Special instruction should be given to employees	Special training should given to supervisor, charge hand and employees	20	1	20	L

Cont'd

Hazard	Consequence	P	S	Risk	Rating	Control Measures	P	S	Risk	Rating
(continued)	...disability and death to personnel, public, etc.					about suspended load path way as well as employees and operatives and banksman. / about the risks and hazards involved in lifting operation especially when the crane is on suspended load.				
Working close to electric wires	Damage to property, and or injury/disability and death to personnel, public, etc.	20	15	300	H	Permit system requires to follow while working close to electric wires. Needs Supervision in all times basis / Use nylon slings only. Ensure crane is electrically earthed	20	1	20	L
Failure: Slings, Chains, buckets	Damage to property, and or injury/disability and death to personnel	10	20	200	H	Only use standard (ISO marked) lifting equipment. Checklists to be completed and updated at all times. All lifting equipment to be load tested and certified with visible identification marks. / Ensure that lifting equipment capacity not exceeding the limit than it designed.	20	1	20	L
Uncontrolled and spinning loads on hook	Damage to property, and or injury/disability and death to personnel.	20	20	400	H	Instruction "do not operate" needs to given while heavy wind is blowing. Tag lines to be erected always on loads for the control of suspended load. / Special instruction and training to be given to lifting team employees that the crane is design only for lifting not used for pulling.	20	1	20	L
Weather: Wind, rain, lightning	Damage to property, and or injury/disability and death to personnel.	15	20	300	H	Shutdown the crane while on windy and lightning condition. / Ground level checking to be conducted on rainy atmospheric condition in the case of mudslide for the placement of outriggers.	15	1	15	L

Contd...

	Activity	Hazard	Consequence					Additional PPE Needed	Control 1	Control 2				
3	Moving with load	Derailing of crane, Overturning	Damage to property, and or injury/disability and death to personnel.	10	20	200	H		Observe manufactures charts and procedures whether the crane is design to move with the suspended load. Instruct to follow strictly safe work procedure and method statement.	Proceed at extremely slow speed with banks men or flag men at back and front and boom fully retracted over the cabin.	10	1	10	L
4	Welding on items suspended from hook	Electronic equipment malfunction	Damage to electronic measuring and control systems	10	20	200	H		Never allow welding while suspended load on hook.	If welding cannot be avoided, use only nylon slings for lifting purpose. Hot work permit system to be followed. Beware of fire hazards on nylon slings.	10	1	10	L
5	Shutdown	Parking up equipment	Injury, fatalities and property damage	20	15	300	H		Follow manufactures shutdown procedures. Park crane with park brakes applied and low/reverse gear also applied.	Remove keys immediately and place in safe. Prevent all unauthorised entries	20	1	20	L
PPE Needed:	Coverall, Goggles, Hard hat, Vest, safety shoes		Average Rounded		20	200	M	Safety Guards, Earplug, Breath Maks		Average Rounded	20	1	20	L

Frequency of Exposure (Fq)

Continuously (daily)	20
Frequently (weekly)	15

Very Serious	20	Several Detail	20
Disaster		One Death	15

Risk Rating

1 To 30	Low (L)
31 To 150	Medium (M)

Contd...

10	From time-to-time (monthly)	Serious	Disability/Permanent Injury	10	151 To 200	Medium High (H)
5	Sometimes/Occasional (yearly)	Significant	Lost Time Injury	5	201 and above	High (H)
1	Realy (less than once per year)	Minima	Minor Injury	1		

CONSEQUENCE (Cq)

Risk Rating = Frequency consequence

Frequency of Exposure (F)	20	15	10	5	1
20	400 - High	300 - High	200 - Medium	100 - Medium	20 - Low
15	300 - High	225 - High	150 - Medium	75 - Medium	15 - low
10	200 - Medium	150 - Medium	100 - Medium	50 - Medium	10 - Low
5	100 - Medium	75 - Medium	50 - Medium	25 - Low	5 - Low
1	20 - Low	15 - low	10 - Low	5 - Low	1 - Low

Risk Assessment Committee Members

Name	Designation	Signature	Name	Designation	Signature

SDTP — 13.4 Hazard Identification and Risk Assessment

Document References

	SDTP.RA 002
Rev.No	0
Doc. Date	1/1/2009
Pages	1 of 3

HIRA No.	HIRA-004		
Task/Activity Description:	Gas Welding		
Risk Assessment:	Using working with welding set.		
Tools/Equipment in use:	Chipping Hammer, Wire Brush, Welding Goggles. Welding wire, gas cylinders, Full set of nozzles and nozzle cleaner		
Project:	ABC power		
Client:	XYZ power		
Date:	Date Made		

Sr. No.	Task in one by one step	Hazard Identification in each step	Harm/consequences (without controls)	A Fq.	B Cq.	C Risk Store	Risk Rating	Who could be affected (mark on right column)	Current Control (With Controls)	Additional Control	A Fq.	B Cq.	C Risk Store	Risk Rating
1	Startup	Explosion and fire	Burns, bodily injury, disability, fatality and property damage	10	20	200	M	Client / Project Manager / Contracts Manager / Supervisor	Hot work permit system to be followed. Welding area to be free from flammable materials Area to be barricaded and signage's to be placed.	At least two trained and competent fire watchers to be placed in that area. Fire blankets to be erect underneath while working at heights. Emergency alarm and fire point should be accessible very near.	10	1	10	L
2	Welding	Fire	Burns, bodily injury, disability, fatality and property damage	20	20	400	H	Employees / Operator / Banksman / Contractor	Hot work permit system to be followed. Welding area to be free from flammable materials Area to be barricaded and signage's to be placed.	At least two trained and competent fire watchers to be placed in that area. Fire blankets to be erect underneath while working at heights. Emergency alarm and fire point should be accessible very near.	20	1	20	L
		Hot metal spatter	Eye injury, swellings and skin burns	20	10	200	M	Public / Welder / Fabricator / Cutters	Use appropriate PPEs, including leather aprons, gloves and UV rays filters in welding in goggles or	Surrounding area should be wetted to diffuse hot spatters. First Aider with First Aid kit also needed to be appointed.	20	1	20	L

#	Hazard	Effect				Responsible	Control Measures				
3	Infra red and visible light emission	Eye injury, cancer and skin burns	20	10	200	M	Engineers, Site Agent, Foreman, Charge Hand	welding mask. Unauthorized entry of personnels to be prohibited. Special induction training to be given to the operators and co-workers about the dangers involved and precaution to be taken by competent engineer.	20	1	20	L
	Toxic gases	Asphyxiation, Lung disease, unconsciousness, shock, irritating and fatality	10	20	200	M	Consultant, Safety Officer, Storekeeper, Painter	Ventilation to be provided as per standards. Use of welding mask for filter air for breathing. Emergency response procedure to be remembered before starting the work. Breathing apparatus to be provided in the hazardous areas for workers.	10	1	10	L
	Explosion and fire	Burns, bodily injury, disability, fatality and property damage	10	20	200	M	Steel Fitter, Mason, Carpenter, Helper	Ensure shut-off valve in cylinders needs to closed first and then of the nozzle. Working condition of flash back arrestors to be checked by a competent person before starting the start up work. Special induction training to be given to the operators about the dangers involved and precaution to be taken by competent engineer. Third party inspection and certification should be done regularly as per HSE plan.	10	1	10	L
	Cylinder Falling	Bodily injury, fracture, cut, sprain, swellings and disability	20	10	200	M	Scaffold Inspector	Always keep the cylinders in trolleys and needs to secure the trolleys firmly. Tie-up and coil the After welding work, welding set including cylinders and trolley to be returned to the stacking.	20	1	20	L

Contd...

PPE Needed: Coverall, Goggles, Hard hat, Vest, safety shoes

Additional PPE Needed: Safety Guards, Earplug, Breath Mask / horses after the welding activity.

Frequency of Exposure (Fq)

Fq	Description
20	Continuously (daily)
15	Frequently (weekly)
10	From time to time (monthly)
5	Sometimes/Occasional (yearly)
1	Realy (less than once per year)

Consequence

Value	Severity	Description
20	Disaster	Several Detail
15	Very Serious	One Death
10	Serious	Disability/Permanent Injury
5	Significant	Lost Time Injury
1	Minima	Minor Injury

Risk Rating

Average Rounded	Risk Rating
1 To 30	Low (L)
31 To 150	Medium (M)
151 To 200	Medium High (H)
201 and above	High (H)

Risk Rating = Frequency × consequence

Frequency of Exposure (F)	CONSEQUENCE (C)				
	20	15	10	5	1
20	400- High	300- High	200- Medium	100- Medium	20- Low
15	300 - High	225 - High	150 - Medium	75 - Medium	15 - Low
10	200 - Medium	150 - Medium	100 - Medium	50 - Medium	10 - Low
5	100 - Medium	75 - Medium	50 - Medium	25 - Low	5 - Low
1	20 -Low	15- Low	10 - Low	5 - Low	1- Low

Risk Assessment Committee Members

Name	Designation	Signature	Name	Designation	Signature

SDTP — 13.5 Hazard Identification and Risk Assessment

Document References		
SDTP.RA 002		
Rev.No	Doc. Date	
0	1/1/2009	
Pages		
1 of 3		

HIRA No.	HIRA-005	Project: ABC power
Task/Activity Description:	Scaffolding to carry-out tasks	Client: XYZ power
Risk Assessment:	Using, working from, walking on scaffolding.	Date: Date Made
Tools/Equipment in use:	Hand tools, Portable electric tools, Forklift, etc.	

Sr. No.	Task in one-by-one step	Hazard Identification in each step	Harm/ consequences (without controls)	A Fq.	B Cq.	C Risk Store	Risk Rating	Who could be effected (Mark on right column)		Current Control	Additional Control	A Fq.	B Cq.	C Risk Store	Risk Rating
	All tasks requiring the use of scaffolding	Scaffolding Collapse	Fatalities, Disability, Injury, Damage to property.	15	20	300	H	Client		Use only standard (ISO marked) materials for scaffold erection. Only authorised and competent persons allowed to erect scaffolds. Supports and ties to designed and checked by competent personnels.	Safety personnels needs to inspect scaffold prior to any work and once in a day. Only authorized scaffolders allow to alter the scaffold. Scafftag systems to follow and to be signed by scaffold inspector daily.	15	1	15	L
								Project Manager							
								Contracts Manager							
								Supervisor	y						
		Access	Falling, Slipping, Injury, injury from Congestion	20	5	100	M	Employees	y	Ladders and stair ways to secured firmly on both ends. Ensure good illumination facilities for scaffolds particularly on access area.	Entry and exit points to be well marked. Access required landings facilities at regular intervals.	20	1	20	L
								Operator	y						
								Banksman	y						
								Contractor	y						
		Falling from heights	Lacerations, cuts, fracture, back pain, disability and death.	20	15	300	H	Public		Platforms to be boarded and hand railed. Toeboards required in all platform edges. All wor-	Regular Toolbox talk regarding the "use of scaffolds" needs to taken by supervisor. Manual handling techniques to be	20	1	20	L
								Welder	y						
								Fabricator	y						
								Cutters	y						

Contd...

#	Activity	Hazard	Consequences	P	S	R	Rating	Responsible Persons	Control Measures	P	S	R	Rating
		Falling objects	Lacerations, cuts, fracture, back pain, disability and death.	20	15	200	H	Engineers (y), Site Agent (y), Foreman (y), Charge Hand (y)	...kers needs to wear and hook safety harness above 2 meters heights. All taught and trained to the workers. Minimise people in the scaffolding area. Signage such as "Danger people working overhead" to be placed. Erect catch platforms and nets were practicable. Keep platforms and access clear and free from materials and obstacles. Erect toe boards at the edges of platforms. Use tool retainers or lanyards for preventing tools falling down from scaffolds.	20	1	20	L
		Nails, sharp objects / edges, etc	Lacerations, deep cuts, shock,	20	10	200	M	Consultant, Safety Officer, Storekeeper, Painter (y)	Regular inspection needs to carry-out for sharp edges and nails, if anything found and cannot be removed, capping system to be adopted. Ensure good housekeeping at all times. Hard gloves, coverall, hard hat and safety shoes to be worn by the workers while working on scaffolds.	20	1	20	L
2	Lifting by means of crane close to scaffold	Crane Collision with scaffold	Fatalities, Disability, Injury, Damage to property	15	20	300	H	Cleaner (y), Steel Fitter (y), Mason (y), Carpenter (y), Helper (y)	All lifts to be controlled only by the presents of authorised banksman. Banksmen to be in full view of scaffolding within the crane's sphere of work. Site specific induction about the "use of scaffold" and "safe lifting operation" to be conducted among the workers by safety personnels and supervisors. Standard PPEs to be worn by workers.	15	1	15	L
3	Manual Material / Tools Lifting	Falling / Flying objects	Fatalities, Disability, Cut, Lost Time Injury, Damage to property.	20	15	300	H	Scaffold Inspector (y)	Prohibit hand-by-hand material lifting. Ropes or pulleys attached with bucket and rope can use for the purpose. Needs to barricade the area which is planed for manual lifting with solid materials and warning tape. Signage's," Danger, Manual lifting in	20	1	20	L

PPE Needed:	Average Rounded:			M			Additional PPE Needed		Average Rounded			L
Coverall, Goggles, Hard hat, Vest, safety shoes		20	20	200			Safety Guards, Earplug, Breath Maks			20	1	20

No one is permitted to stand under the suspended load while lifting is in progress. Supervisors need to take special Toolbox talk prior starting working on scaffolds. "...progress, keep away", "Danger of Flying/Falling objects. Keep away" to be placed for the identification of area.

Frequency of Exposure (Fq)

20	Continuously (daily)
15	Frequently (weekly)
10	From time-to-time (monthly)
5	Sometimes/Occasional (yearly)
1	Realy (less than once per year)

CONSEQUENCE (Cq)

20	Disaster	Several Detail
15	Very Serious	One Death
10	Serious	Disability/Permanent Injury
5	Significant	Lost Time Injury
1	Minima	Minor Injury

Risk Rating

Average Rounded	
1 To 30	Low (L)
31 To 150	Medium (M)
151 To 200	Medium High (H)
201 and above	High (H)

Risk Rating = Frequency × cosequence

Frequency of Exposure (F) \ CONSEQUENCE (Cq)	20	15	10	5	1
20	400 - High	300 - High	200 - Medium	100 - Medium	20 - Low
16	400 - High	300 - High	150 - Medium	75 - Medium	15 - Low
10	300 - High	220 - High	100 - Medium	50 - Medium	10 - Low
5	200 - Medium	150 - Medium	50 - Medium	25 - Low	5 - Low
1	20 - Low	15 - Low	10 - Low	5 - Low	1 - Low

Risk Assessment Committee Members

Name	Designation	Signature	Name	Designation	Signature

Safety Punishment Notice

14

Some of PTW Program are defined

a. Safety default notice.

b. Stop work notice.

Markings

As a part of developing safety culture and activate visible felt leadership, safety inspections to be conducted regularly on site work activities to found any deviations from health and safety procedure and legal requirements. If any deviations find, site top management should discipline the low management to restrict unsafe act and unsafe conditions. For this disciplinary actions safety violation Slip and Stop work order notice come to part.

What is the Infrastructure of Safety Violation Default Notice?

It is the prime duty of the safety personnel to correct unsafe working condition and unsafe working acts to avoid accidents. While on inspection any activities found deviating from the actual safe work procedure, safety personnel and top management need to discipline the workers as well as supervisor to prevent accident occurrence.

In this document, the deviating condition and acts from the part of employee, the penalty enforced as fine, name of violation slip issuer, default employee and concern supervisors details including identification numb to be clearly marked. For final approval this document should be counter sign by Project Manager or General Manager. A copy of the document should send to the HR Manager for the deduction of amount.

What is the Infrastructure of Stop Work Order Notice?

As same as safety violation slip procedure, the stop work order notice to be issued where the work activities and working condition found dangerous which were totally against safe work procedure. If the work had continued in that fashion it would cause major accidents that can turn to fatalities, disabilities, lost times and multiple injuries. In this document also the deviating conditions and acts from the part of employer or employee, the identified risks and hazards points, the sign of the issuer and acceptors name and signature should be clearly marked. After correcting the identified risk and hazards according to the safe work procedure, the notice issuer may re-check the concern area and permission granted to start work by his own signing, then only work can re-start.

What is the Importance of Safety Violation Slip and Stop Work Order Notice?

Training, safety talks and safety awareness will help to prevent accident to a great extend. But the human mind can forget everything they had been inducted in a second due to various reasons like tension, negligence, horseplay, production pressure, lack of knowledge, amnesia, absence of mind, habit, etc. The fault of one can also cause accidents to others. These human errors may lead to major accidents and damages and need to be disciplined. It is the basic nature of the workforce that, if they are warned and they lose a part of wages as penalty for committing their own unsafe work or mistakes, they won't forget in their life time and didn't commit that activity again. For this concept the safety violation slip and stop work order notice are implemented.

How is Safety Violation Default Notice and Stop Work Order Notice Described?

See the attached documents:

SDTP	**14.1 Safety Violation Default Notice**	Document References

Document References	
SDTP. SVN. 001	
Rev. No	Doc. Date
0	1/1/2009
Pages	
1 of 1	

Project:		**Notice No:**	
Location:		**Site:**	
Department Issued:		**Date:**	

Section A

Detail of Defaulter

Company:		**Designation:**	
Name of Defaulter:		**Identification No:**	

Brief Description of Default:

Default Signature:

Previous default in past six months:				Name of immediate supervisors:	
Yes		No	X	Supervisor identification No:	
If yes, default Notice No/dates (if applicable)				Supervisor reporting to:	

(If applicable attach photographs of all relevant default performing job activity)

Section B

Recommended Action (Write "Nil" for not applicable)

Correct the default period to restart work		The amount cut's from defaulter	
Causal/verbal warning given to defaulter		Stop the job/task being carried out	
The number days of work from defaulter		Suspend/terminate/other	

Section C

Issuer's Details

Name	Designation	Signature	Date

Safety Default Notice sanctioned by

Name	Designation	Signature	Date
	General/project Manager		
	Safety Manager		

SDTP	14.2 Stop Work Order Notice	Document References	
		SDTP. SWN. 001	
		Rev. No	Doc. Date
		0	1/1/2009
		Pages	
		1 of 1	

Issued to:		Issued By:	
Company:		Designation:	
Name of Defaulter:		Date Issued:	
Position of Defaulter:		Sign of issuer:	
Sign of Acceptance:		Stop work notice No:	

In terms of the Suspension of Work due to:

All work related to........................performed in the (area)...........................
undersupervision of..has been suspended with immediate affect (date)
....due to the following reasons (Circle A, B or C below):
a. Failure to correct identified risk conditions as per violation slip
b. Failing to comply with SHEQ Management System and Operational Controls.
c. Immediate danger placing people's safety, health, and/or the environment at risk.

Identified At Risk Condition:

a.	
b.	
c.	

You must ensure that corrective measures are taken immediately to nullify and control the identified risk (s) to comply with required and agreed standard.

Work may not resume until appropriate corrective measures have been taken and the STOP WORK ORDER/ NOTICE have been cleared/ signed off by the initiator.

Zero Tolerance for Unsafe Conditions and Unsafe Acts

Initiator/ Safety Manager/ Officer follow-up:

Corrective/preventative Action have been completed and work can/may re-start:

Name	Designation	Signature	Date

Information and Study Page		
Project:	ABC Power Project	
Location:	2nd Floor	
Department Issued:	Safety Department	
Notice No.:	SV001	
Site:	FXE1002	
Date:	21/2/2009	
Company:	ABC Company	
Name of Defaulter:	Name of Fabricator	
Designation:	Fabricator	
Identification No.:	PP1112	
Brief description of default:	Found without wearing safety glass, face mask and earplug while doing	
Pipe grinding work by means of a portable electric grinder.		
Supervisors identification No.:	PP 109	
Supervisor reporting to:	Site Mechanical Engineer	
Recommended Action	One day wages cuts	
See the attach document how the Safety violation Default Notice filled in:		

SDTP	14.1 Safety Violation Default Notice	Document References		
		SDTP. SVN. 001		
		Rev. No	Doc. Date	
		0	1/1/2009	
		Pages		
		1 of 1		

Project:	ABC Power Project	Notice No:	SV001
Location:	2nd Floor	Site:	FXE1002
Department Issued:	Safety Department	Date:	21/2/2009

Section A

Detail of Defaulter

Company:	ABC Company	Designation:	Fabricator
Name of Defaulter:	Name of Fabricator	Identification No:	PP 1112

Brief description of default:	Found without wearing safety glass, face mask and earplug

while doing pipe grinding work by means of a portable electric grinder.

Default Signature:	Sign of fabricator		
Previous default in past six months:		Name of immediate supervisor:	Name of Supervisor
Yes		No	X	Supervisor identification No:	pp109
If yes, default Notice No/dates (if applicable)		Supervisor reporting to:	Site Mechanical Engineer

(If applicable attach photographs of all relevant default performing job activity)

Section B

Recommended Action (write "Nil" for not applicable)

Correct the default period to restart work		The amount cut's from defaulter	
Causal/verbal warning given to defaulter		Stop the job/task being carried out	
The number days of work from defaulter	1 days wages	Suspend/terminate/other	

Section C

Issuer's Details

Name	Designation	Signature	Date
Name of Safety Officer	Safety Officer	Sing	Date Signed

Safety Default Notice sanctioned by

Name	Designation	Signature	Date
Name of GM	General/project Manager	Sign	Date Signed
Name of Safety Manager	Safety Manager	Sign	Date Signed

Information and Study Page	
Issued To:	Hot work Supervisor
Issued By :	Name of Safety Officer
Department Issued:	Safety Department
Notice No.:	SW 001
Site:	FXE 1002
Date issued:	23/2/2009
Company:	ABC Company
Name of Defaulter:	Name of Hot work Supervisor
Designation:	Supervisor
Area:	2nd floor, F module

In terms of the Suspension of Work due to:	
All work related to hot work performed in the (area) 2nd floor, F module under supervision of Hot work supervisor has been suspended with immediate affect (date) 21/2/2009 due to the following reasons.	
Reason : C.	Immediate danger placing people's safety, health, and/or the environment at risk.
Identified At Risk Condition:	
a.	Permit-to-work system is not assessed, area not barricaded, signage's not placed.
b.	Painting workers are working in that area with drums full of paint which is highly inflammable.
c.	Extinguishers and fire watchers are not placed.
See the attach document how the Stop Work Notice filled in:	

SDTP	14.2 Stop Work Order Notice	Document References	
		SDTP. SWN. 001	
		Rev. No	Doc. Date
		0	1/1/2009
		Pages	
		1 of 1	

Issued to:	Name of hot work supervisor	**Issued By:**	Name of Safety Officer
Company:	ABC company	**Designation:**	Safety Officer
Name of Defaulter:	Name of hot work supervisor	**Date Issued:**	23/2/2009
Position of Defaulter:	Supervisor	**Sign of issuer:**	Name of Safety Officer
Sign of Acceptance:	Sign of hot work supervisor	**Stop work notice No:**	SW 001

In terms of the Suspension of Work due to:

All work related to **hot work** performed in the (area) **F module** under supervision of **Hot work Supervisor** has been suspended with immediate affect (date) **23/2/2009** due to the following reasons (circle A, B or C below):

a. Failure to correct identified risk conditions as per violation slip..
b. Failing to comply with SHEQ Management System and Operational Controls.
c. Immediate danger placing people's safety, health, and/or the environment at risk.

Identified at Risk Condition:

a.	Permit to work system is not assessed, area not barricaded, signage's not placed.
b.	Painting workers are working in that area with drums full of paint which is highly inflammable.
c.	Extinguishers and fire watchers are not placed.

You must ensure that corrective measures are taken immediately to nullify and control the identified risk (s) to comply with required and agreed standard.

Work may not resume until appropriate corrective measures have been taken and the STOP WORK ORDER/ NOTICE have been cleared/ signed off by the initiator.

Zero Tolerance for Unsafe Conditions and Unsafe Acts

Initiator/ Safety Manager/ Officer follow-up:

Corrective/preventative Action have been completed and work can/may re-start:

Name	Designation	Signature	Date
Name of Safety Officer	Safety Officer	Sign of Safety Officer	Date Signed